AF443072

Skin Stem Cells

Methods and Protocols

Edited by

Kursad Turksen

*Regenerative Medicine Program, Sprott Centre for Stem Cell Research,
Ottawa Hospital Research Institute, Ottawa, ON, Canada*

Humana Press

Editor
Kursad Turksen
Regenerative Medicine Program
Sprott Centre for Stem Cell Research
Ottawa Hospital Research Institute
Ottawa, ON, Canada

ISSN 1064-3745   ISSN 1940-6029 (electronic)
ISBN 978-1-62703-329-9   ISBN 978-1-62703-330-5 (eBook)
DOI 10.1007/978-1-62703-330-5
Springer New York Heidelberg Dordrecht London

Library of Congress Control Number: 2013932713

Humana Press is a brand of Springer
Springer is part of Springer Science+Business Media (www.springer.com)

Preface

During the last decade, an increased interest in somatic stem cells has led to a flurry of research on one of the most accessible tissues of the body: skin. Much effort has focused on such topics as understanding the heterogeneity of stem cell pools within the epidermis and dermis and their comparative utility in regenerative medicine applications. To facilitate studies in this area, I have brought together here some representative but critically important protocols for a variety of skin stem cells. The hope is that by doing so, we will accelerate both the basic understanding of these diverse cell types and the translation of their biological potential to the in vivo setting. I am grateful to all of the contributors for their generosity in describing in great detail their methods for isolation, maintenance, and characterization of stem cell populations from skin.

I would also like to take this opportunity once again to acknowledge Dr. John Walker for his ongoing support of my ideas.

In addition, a special "thank you" goes to Patrick Marton for being a great cheerleader for my projects!

Thank you also to Monica Beaumont for helping me to edit and finalize this volume.

Ottawa, ON, Canada　　　　　　　　　　　　　　　　　　　　　　　　*Kursad Turksen*

Contributors

WADO AKAMATSU • *Department of Physiology, Keio University School of Medicine, Tokyo, Japan*

MEHMET DENIZ AKYUZ • *Department of Dermatology, Cologne Excellence Cluster on Cellular Stress Responses in Aging associated Diseases (CECAD), University of Cologne, Cologne, Germany; Center for Molecular Medicine Cologne, University of Cologne, Cologne, Germany*

BOGI ANDERSEN • *Departments of Medicine and Biological Chemistry, School of Medicine, University of California, Irvine, CA, USA*

JEFF BIERNASKIE • *Faculty of Veterinary Medicine, Department of Comparative Biology and Experimental Medicine, University of Calgary, Calgary, AB, Canada*

EMMANUELLE CADIO • *Laboratory of Genomics and Radiobiology of Keratinopoiesis (LGRK), Institute of Cellular and Molecular Radiobiology (iRCM), Alternative Energies and Atomic Energy Commission (CEA), Evry, France*

MARIANA T. CERQUEIRA • *3B's Research Group—Biomaterials, Biodegradables, and Biomimetics, University of Minho, Headquarters of the European Institute of Excellence on Tissue Engineering and Regenerative Medicine, Guimarães, Portugal; ICVS/3B's - PT Government Associate Laboratory, Braga/Guimarães, Portugal*

LOUBNA CHADLI • *Laboratory of Genomics and Radiobiology of Keratinopoiesis (LGRK), Institute of Cellular and Molecular Radiobiology (iRCM), Alternative Energies and Atomic Energy Commission (CEA), Evry, France*

JIANG CHEN • *Department of Dermatology, University of Colorado Anschutz Medical Campus, Aurora, CO, USA; Charles C. Gates Center for Regenerative Medicine and Stem Cell Biology, University of Colorado Anschutz Medical Campus, Aurora, CO, USA*

DAVID J. CLAYPOOL • *Department of Dermatology, University of Colorado Anschutz Medical Campus, Aurora, CO, USA; Charles C. Gates Center for Regenerative Medicine and Stem Cell Biology, University of Colorado Anschutz Medical Campus, Aurora, CO, USA*

LINA DAGNINO • *Department of Physiology and Pharmacology, Child Health Research Institute University of Western Ontario, London, ON, Canada; Lawson Health Research Institute, University of Western Ontario, London, ON, Canada; Department of Pediatrics, University of Western Ontario, London, ON, Canada*

XING DAI • *Department of Biological Chemistry, School of Medicine, University of California, Irvine, CA, USA*

R. DI PRIMIO • *Facoltà di Medicina e Chirurgia, Dip. Scienze Cliniche e Molecolari, Università Politecnica delle Marche, Ancona, Italy*

STEPHEN E. FEINBERG • *Department of Oral and Maxillofacial Surgery, University of Michigan, Ann Arbor, MI, USA*

DAVID FISHER • *Department of Dermatology, Cutaneous Biology Research Center, Massachusetts General Hospital, Harvard Medical School, Boston, MA, USA*

NICOLAS O. FORTUNEL • *Laboratory of Genomics and Radiobiology of Keratinopoiesis (LGRK), Institute of Cellular and Molecular Radiobiology (iRCM), Alternative Energies and Atomic Energy Commission (CEA), Evry, France*

DANIELA FRANCES • *Center for Molecular Medicine Cologne, University of Cologne, Cologne, Germany*

ANA M. FRIAS • *3B's Research Group—Biomaterials, Biodegradables, and Biomimetics, University of Minho, Headquarters of the European Institute of Excellence on Tissue Engineering and Regenerative Medicine, Guimarães, Portugal; ICVS/3B's - PT Government Associate Laboratory, Braga/Guimarães, Portugal*

MIZUHO FUKUNAGA-KALABIS • *The Wistar Institute, Philadelphia, PA, USA*

S. GALLINAT • *R&D, Skin Research Center, Beiersdorf AG, Hamburg, Germany*

AARON GARDNER • *School of Biological and Biomedical Sciences, Durham University, Durham, UK*

MANON GAUDREAULT • *LOEX/CUO-Recherche Québec, Québec, QC, Canada*

LUCIE GERMAIN • *Génie tissulaire et régénération, LOEX—Centre de recherche FRSQ du Centre hospitalier affilié universitaire de Québec, Québec, QC, Canada; Centre LOEX de l'Université Laval, Québec, QC, Canada; Département d'ophtalmologie, Faculté de médecine, Université Laval, Québec, QC, Canada*

MIKHAIL GEYFMAN • *The Gavin Herbert Eye Institute University of California, Irvine, CA, USA*

SOOSAN GHAZIZADEH • *Department of Oral Biology and Pathology, School of Dental Medicine, Stony Brook University, Stony Brook, NY, USA*

KARL GLEDHILL • *School of Biological and Biomedical Sciences, Durham University, Durham, UK*

SYLVAIN L. GUÉRIN • *LOEX/CUO-Recherche, Génie tissulaire et régénération : LOEX—Centre de recherche FRSQ du Centre hospitalier affilié universitaire de Québec, Quebec, QC, Canada; Département d'ophtalmologie, Faculté de médecine, Université Laval, Québec, QC, Canada*

ERIKA GUEVARA • *Department of Dermatology, University of Colorado Anschutz Medical Campus, Aurora, CO, USA; Charles C. Gates Center for Regenerative Medicine and Stem Cell Biology, University of Colorado Anschutz Medical Campus, Aurora, CO, USA*

ANDREW HAGNER • *Department of Comparative Biology and Experimental Medicine, Faculty of Veterinary Medicine, University of Calgary, Calgary, AB, Canada*

BASIL M. HANTASH • *Escape Therapeutics Inc., San Jose, CA, USA*

MEENHARD HERLYN • *The Wistar Institute, Philadelphia, PA, USA*

YOICHI IMAIZUMI • *Department of Physiology, Keio University School of Medicine, Tokyo, Japan*

KEITA INOUE • *Department of Plastic and Reconstructive Surgery, Shizuoka Cancer Center, Shizuoka, Japan*

TIMOTHY S. IRVINE • *Department of Physiology and Pharmacology, Child Health Research Institute, University of Western Ontario, London, ON, Canada; Lawson Health Research Institute, University of Western Ontario, London, ON, Canada*

KENJI IZUMI • *Division of Oral Anatomy, Graduate School for Medical and Dental Sciences, Niigata University, Niigata, Japan*

COLIN A.B. JAHODA • *School of Biological and Biomedical Sciences, Durham University, Durham, UK*

UFFE B. JENSEN • *Department of Clinical Genetics, Aarhus University Hospital, Aarhus, Denmark*

PRITINDER KAUR • *Epithelial Stem Cell Biology Laboratory, Research Division, Peter MacCallum Cancer Center, St Andrew's Place, Melbourne, Australia*

YUTAKA KAWAKAMI • *Division of Cellular Signaling, Institute for Advanced Medical Research, Keio University School of Medicine, Tokyo, Japan*

JIRO KISHIMOTO • *Shiseido Innovative Science Research and Development Center, Yokohama, Japan*

A. KNOTT • *R&D, Skin Research Center, Beiersdorf AG, Hamburg, Germany*

TAKAHIRO KUNISADA • *Department of Tissue and Organ Development, Regeneration and Advanced Medical Science, Gifu University Graduate School of Medicine, Gifu, Japan*

DANIELLE LAROUCHE • *Génie tissulaire et régénération: LOEX—Centre de recherche FRSQ du Centre hospitalier affilié universitaire de Québec, Québec, QC, Canada; Centre LOEX de l'Université Laval, Quebec, Canada*

LAUREN KIMLIN • *NIH, Bethesda, MD, USA*

BRIANA LEE • *Department of Biological Chemistry, School of Medicine, University of California, Irvine, CA, USA*

LING LI • *The Wistar Institute, Philadelphia, PA, USA*

ALEXANDRA P. MARQUES • *1- 3B's Research Group—Biomaterials, Biodegradables, and Biomimetics, University of Minho, Headquarters of the European Institute of Excellence on Tissue Engineering and Regenerative Medicine, Guimarães, Portugal; ICVS/3B's - PT Government Associate Laboratory, Braga/Guimarães, Portugal*

CYNTHIA L. MARCELO • *Section of Plastic and Reconstructive Surgery, Department of Surgery, University of Michigan, Ann Arbor, MI, USA*

MICHÈLE T. MARTIN • *Laboratory of Genomics and Radiobiology of Keratinopoiesis (LGRK), Institute of Cellular and Molecular Radiobiology (iRCM), Alternative Energies and Atomic Energy Commission (CEA), Evry, France*

TSUTOMU MOTOHASHI • *Department of Tissue and Organ Development, Regeneration and Advanced Medical Science, Gifu University Graduate School of Medicine, Gifu, Japan*

KERRY-ANN NAKRIEKO • *Department of Physiology and Pharmacology, Child Health Research Institute, University of Western Ontario, London, ON, Canada; Lawson Health Research Institute, University of Western Ontario, London, ON, Canada*

CATHERIN NIEMANN • *Center for Molecular Medicine, University of Cologne, Cologne, Germany*

CARIEN M. NIESSEN • *Department of Dermatology, Cologne Excellence Cluster on Cellular Stress Responses in Aging associated Diseases (CECAD), University of Cologne, Cologne, Germany; Center for Molecular Medicine Cologne, University of Cologne, Cologne, Germany*

MICHAELA T. NIESSEN • *Department of Dermatology, Cologne Excellence Cluster on Cellular Stress Responses in Aging associated Diseases (CECAD), University of Cologne, Cologne, Germany; Center for Molecular Medicine Cologne, University of Cologne, Cologne, Germany*

SHIGEKI OHTA • *Division of Cellular Signaling, Institute for Advanced Medical Research, Keio University School of Medicine, Tokyo, Japan*

HIDEYUKI OKANO • *Department of Physiology, Keio University School of Medicine, Tokyo, Japan*

M. ORCIANI • *Dip. Scienze Cliniche e Molecolari, Facoltà di Medicina e Chirurgia, Università Politecnica delle Marche, Ancona, Italy*

DAVID M. OWENS • *Departments of Dermatology and Pathology & Cell Biology, College of Physicians and Surgeons, Columbia University Medical Center, New York, NY, USA*

MONIKA PETERSSON • *Center for Molecular Medicine Cologne, University of Cologne, Cologne, Germany*

JULIA REICHELT • *Institute of Cellular Medicine, Newcastle University, Newcastle upon Tyne, UK; North East England Stem Cell Institute, Newcastle University, Newcastle upon Tyne, UK*

RUI L. REIS • *3B's Research Group—Biomaterials, Biodegradables, and Biomimetics, University of Minho, Headquarters of the European Institute of Excellence on Tissue Engineering and Regenerative Medicine, Guimarães, Portugal; ICVS/3B's - PT Government Associate Laboratory, Braga/Guimarães, Portugal*

DENNIS R. ROOP • *Department of Dermatology, University of Colorado Anschutz Medical Campus, Aurora, CO, USA; Charles C. Gates Center for Regenerative Medicine and Stem Cell Biology, University of Colorado Anschutz Medical Campus, Aurora, CO, USA*

M. RUETZE • *R&D, Skin Research Center, Beiersdorf AG, Hamburg, Germany*

HOLGER SCHLÜTER • *Epithelial Stem Cell Biology Laboratory, Research Division, Peter MacCallum Cancer Center, Melbourne, VIC, Australia*

TSUTOMU SOMA • *Shiseido Innovative Science Research and Development Center, Yokohama, Japan*

PIERRE VAIGOT • *Laboratory of Genomics and Radiobiology of Keratinopoiesis (LGRK), Institute of Cellular and Molecular Radiobiology (iRCM), Alternative Energies and Atomic Energy Commission (CEA), Evry, France*

REHAN M. VILLANI • *Department of Dermatology, Cologne Excellence Cluster on Cellular Stress Responses in Aging associated Diseases (CECAD), University of Cologne, Cologne, Germany; Center for Molecular Medicine Cologne, University of Cologne, Cologne, Germany; Max Planck Institute for Biochemistry, Martinsried, Germany*

VICTORIA VIRADOR • *NIH, Bethesda, MD, USA*

LEE WALLACE • *Institute of Cellular Medicine, Newcastle University, Newcastle upon Tyne, UK; North East England Stem Cell Institute, Newcastle University, Newcastle upon Tyne, UK*

YOULIANG WANG • *State Key Laboratory of Proteomics, Genetic Laboratory of Development and Diseases, Institute of Biotechnology, Beijing, People's Republic of China*

H. WENCK • *R&D, Skin Research Center, Beiersdorf AG, Hamburg, Germany*

XIAO YANG • *State Key Laboratory of Proteomics, Genetic Laboratory of Development and Diseases, Institute of Biotechnology, Beijing, People's Republic of China*

MASAHITO YASUDA • *Department of Dermatology, University of Colorado Anschutz Medical Campus, Aurora, CO, USA; Charles C. Gates Center for Regenerative Medicine and Stem Cell Biology, University of Colorado Anschutz Medical Campus, Aurora, CO, USA*

KOTARO YOSHIMURA • *Department of Plastic Surgery, University of Tokyo, Tokyo, Japan*

LONGMEI ZHAO • *Escape Therapeutics Inc., San Jose, CA, USA*

Chapter 1

Interfollicular Epidermal Stem Cells: Boosting and Rescuing from Adult Skin

Mariana T. Cerqueira, Ana M. Frias, Rui L. Reis, and Alexandra P. Marques

Abstract

Epidermal stem cells isolation struggle remains, mainly due to the yet essential requirement of well-defined approaches and markers. The herein proposed methodology integrates an assemblage of strategies to accomplish the enrichment of the interfollicular epidermal stem cells multipotent fraction and their subsequent separation from the remaining primary human keratinocytes culture. Those include rapid adherence of freshly isolated human keratinocytes to collagen type IV through the β1-integrin ligand and Rho-Associated Protein Kinase Inhibitor Y- 27632 administration to the cultures, followed by an immunomagnetic separation to obtain populations based in the combined CD49f^{bri}/CD71^{dim} expression. Flow cytometry is the supporting method to analyze the effect of the treatments over the expression rate of early epidermal markers keratins19/5/14 and in correlation to CD49f^{bri}/CD71^{dim} subpopulations. The step-by-step methodology herein described indulges the boosting and consecutive purification and separation of interfollicular epidermal stem cells from human keratinocytes cultures.

Key words Epidermal stem cells, Collagen IV, Rock inhibitor, Immunomagnetic separation, Flow cytometry

1 Introduction

Human keratinocytes (hKC) have a limited lifespan in culture that constraints their proliferative capacity and consequently their clinical potential. The long-term function of the skin equivalents generated in a Regenerative Medicine context can be limited by the length of time needed to obtain epithelial sheets in vitro, during which patient is highly susceptible to infection, and also by extensive culture that may lead to terminal differentiation of the hKC to be grafted, thus compromising its success. Therefore, the use of epidermal stem cells (EpSCs) that play an important role in cellular regeneration, wound healing, and neoplasm formation (1) for this purpose enlarges the possibility of providing an alternative and clinically relevant active source of biological material.

Despite a wide effort among stem cells biologists community (2–6), EpSCs isolation difficulty remains, mainly due to the

Kursad Turksen (ed.), *Skin Stem Cells: Methods and Protocols*, Methods in Molecular Biology, vol. 989, DOI 10.1007/978-1-62703-330-5_1, © Springer Science+Business Media New York 2013

insufficiency of molecular markers that distinguish these cells from other proliferative cells within skin basal layer, highlighting the need for defining approaches and a panel of markers to obtain specific and well-characterized cell populations. P63 is abundantly expressed by holoclones and therefore recognized as being also present in EpSCs playing an important role in morphogenesis and in the expression pattern of the cultures (7). It has been also proposed that EpSCs exhibit a characteristic keratin profile that includes the typical K5 and K14 expression of the basal layer but not of K1/10 of the suprabasal layer cells. K19 appears also as an EpSCs-associated marker as it is expressed by cells present in the skin hair follicles bulge and in the deep epidermal rete ridges within thicker epidermis, being also expressed by a subpopulation of hKC in the human basal layer during proliferative lateral skin expansion (8).

Interestingly, the molecules related with cell-substratum adhesion are naturally meaningful as potential EpSCs markers, supported by the hypothesis that EpSCs require strong adherence to the basement membrane to maintain their stem cell characteristics or their position in the stem cell niche. (9). β 1-integrin was firstly identified in highly proliferating KC (holoclones) and was used to distinguish EpSCs and other basal cells (5). However, subsequent studies revealed that the majority of the cells of the basal layer in the human epidermis, EpSCs, and transient amplifying cells exhibit the expression of beta 1 integrin (10) and other putative markers such as the combination of CD49f (α6-integrin) and CD71 (transferrin receptor) (5). Human epidermal cells have been thus divided into three different subsets, $\alpha6^{bri}CD71^{dim}$, $\alpha6^{bri}CD71^{bri}$, and $\alpha6^{dim}$ expressing cells, the first being those with the highest proliferation rate and capability of long-term epidermal renewal (11), even at a limited dilution.

Despite the high importance of EpSCs, they constitute between 1% and 10% of the basal layer cells and, independently of the standardization of a characteristic panel of markers, boosting this population in culture through enrichment methods is a major demand. The involvement of Rho-Associated Protein Kinase (Rock) in tissue homeostasis, namely in the epidermis, is already recognized. Regardless of the unconsciousness of the exact timing of events, the key role that Rock plays in determining hKC fate was clearly demonstrated. By blocking Rock function, an inhibition of hKc terminal differentiation and an increase in cell proliferation was observed (12). It has also been shown that Rock inhibitor (Rocki) leads to an increased number of hKC in primary cultures that can survive and grow forming healthy colonies, thus suggesting its effect in boosting the cells exhibiting stem cell behavior (13) yet retaining the ability to differentiate and to form a stratified epithelium in adequate organotypic models (14). The herein proposed methodology describes an assemblage of strategies to accomplish enrichment and further purification of the EpSCs multipotent fraction present in hKC primary cultures. The procedure combines the rapid adherence of primary hKC to β1-integrin ligand in collagen type IV and the administration of Rho-Associated Protein Kinase (Rock) Inhibitor Y- 27632 to the culture, together with subsequent immunomagnetic separation of subpopulations combining CD49f^{bri}/CD71^{dim} expression.

2   Materials

2.1   Labware
(see Note 1)

Petri dishes (Greiner Bio One, Cat. No. 391-2080).

Forceps (RSG, Cat. No. 311.105).

Surgical scissors (RSG, Cat. No. 101.130).

Cell culture flasks (75 cm^2, 150 cm^2) (BD Falcon, Cat. No. 353136, 353028).

6-well culture plates (BD Falcon, Cat. No. 353224).

15 mL Falcon tubes (BD Falcon, Cat. No. 352097).

50 mL Falcon tubes (BD Falcon, Cat. No. 352070).

Flow cytometry tubes (BD Falcon, Cat. No. 352052).

Pipettes (Corning Science Products, Cat. No. 4489).

Cell strainers of 100 μm pore size (BD Falcon, Cat. No. 352360).

Eppendorf tubes 1.5 mL (Laborspirit, Cat. No. 200400P).

DynaMag™-2 magnet (Invitrogen, Cat. No. 123-21D).

0.22 μm pore membrane filters (Sarsted, Cat. No. 83.1823.101).

2.2   Reagents

Phosphate buffer saline (PBS) (Sigma, Cat. No. P4417).

Distilled water (diH$_2$O).

Antibiotic/antimycotic solution (Gibco, Cat. No. 15240062).

Dispase (BD Biosciences, Cat. No. 354235).

Trypsin–EDTA (Gibco, Cat. No. 25300-062).

Keratinocyte Serum Free Medium (KSFM) Kit with l-Glutamine, EGF, and BPE (Gibco, Cat. No. 17005-075).

Y-27632 dihydrochloride monohydrate (Sigma, Cat. No. Y0503).

Acetic acid (vWR, Cat. No. 20104.334).

Human placenta collagen type IV (Sigma, Cat. No. C5533).

Bovine serum albumin (BSA) (Sigma, Cat. No. A2153).

Dynabeads M-450 Epoxy (Life Technologies, Cat. No. 14011).

Sodium phosphate (Sigma, S0876).

Permeabilization buffer (10×) (eBioScience, Cat No. 00-8333).

CD49f-APC antibody (eBioscience, Cat. No. 17-0495-82).

CD71-PE antibody (BD Biosciences, Cat. No. 555537).

Cytokeratin 19-AF488 antibody (ExBio, Cat. No. A4-120-C100).

Cytokeratin 14-FITC antibody (AbD Serotec, MCA890F).

Keratin 5 antibody (Covance, Cat. No. PRB-160P).

Alexa Fluor 488 Goat anti-Rabbit (Invitrogen, Cat. No. A-11008).

Formaldehyde (vWR, Cat. No. ALFA33314K2).

Sodium azide (Sigma, Cat. No. 13412).

2.3 Reagents Setup

1. *Dispase stock solution (25 U/mL)*: dilute dispase, 1:2 in PBS (see Note 2).

2. *Collagen IV stock solution (1 mg/mL)*: Add 5 mL of 0.25% acetic acid and let to dissolve overnight at 4°C (see Notes 2 and 3).

3. *Rocki stock solution (1 mM)*: Reconstitute 1 mg of Y-27632 dihydrochloride monohydrate in 2.96 mL of diH_2O.

4. *Dynabeads buffer 1*: prepare a buffer of 0.1 M Sodium phosphate in diH_2O and adjust pH to 7.4–8.0 (see Note 4)

5. *Dynabeads buffer 2*: make a 0.1% BSA solution in PBS and adjust the pH to 7.4.

6. Coating of immunomagnetic beads with CD71 and CD49f antibodies (see Note 5)

    6.1 Transfer 10 μL of dynabeads to an eppendorf tube.

    6.2 Place the tube in a magnet for a minute and discard the supernatant. Remove the tube from the magnet.

    6.3 Resuspend the beads in 50 μL of Dynabeads buffer 1 and add 4 μL of CD71 or 2 μL of CD49f antibody.

    6.4 Incubate for 16–24 h at room temperature with gentle tilting and rotation.

    6.5 Repeat step 6.2 and resuspend beads in Dynabeads buffer 2.

7. *Permeabilization buffer*: Dilute permeabilization buffer (10×) in diH_2O to obtain a 1× working solution, store at 4°C.

8. *Labeling buffer*: Prepare a 3% BSA solution in PBS.

9. *Acquisition buffer*: Make a 1% formaldehyde and 0.1% Sodium azide solution in PBS, filter (0.22 μm pore membrane), and store at RT.

10. *Antibiotic/antimycotic solution*: Make a 1% antibiotic solution in PBS

11. *Dispase working solution (2.5 U/mL)*: Dilute 1:10 of stock dispase solution in 1% solution of antibiotic/antimycotic in PBS

12. *Rocki working solution (10 μM)*: Dilute Rocki stock solution (1 mM) 1:100 in KSFM, in order to have KSFM supplemented with 10 μM Rocki.

3   Methods

3.1   Isolation of Human Keratinocytes from Adult Skin

3.1.1   Processing Human Skin

1. Remove the exceeding fat tissue from the dermis with scissors and scalpel.
2. Wash the skin samples with *antibiotic/antimycotic solution* (±20 s).
3. Cut skin into 0.5 cm² pieces.

3.1.2   Epidermal–Dermal Separation by Dispase

Incubate skin pieces in dispase working solution (2.5 U/mL) overnight at 4°C in a 250 mL flask.

After incubation place the skin samples on a Petri dish and peel off epidermis from dermis using two pairs of forceps.

3.1.3   Isolation of Human Keratinocytes: Digestion of Epidermis with Trypsin

1. Place epidermis (dermal side up) in a new Petri dish.
2. Add 0.05% trypsin–EDTA.
3. Incubate the samples at 37°C for 5–7 min.
4. Add an equal amount of KSFM.
5. Scrape of cells carefully with a cell scraper.
6. Pipette rigorous up and down several times.
7. Poor cell suspension trough a 100 μm pore size cell strainer into a 50 mL Falcon tube.
8. Wash with PBS, passing the liquid through the 100 μm pore size cell strainer.
9. Centrifuge for 5 min at $290 \times g$.
10. Wash pellet with 5 mL of PBS.
11. Poor cells trough a 100 μm pore size cell strainer into a 50 mL Falcon tube.
12. Centrifuge for 4 min at $290 \times g$.
13. Resuspend cell pellet (hKC) in KSFM.

3.2   EpSCs Enrichment Strategies (see Note 6)

3.2.1   Rapid Adherence to Collagen IV

1. Coat tissue culture surface with collagen IV, by incubating 5 μg/cm² at 37°C, for at least 1 h (see Note 3).
2. Wash with PBS.
3. Plate 2×10^4 hKC/cm² in KSFM.
4. Change medium every 2–3 days and keep the culture until 80% confluent.

3.2.2   Rock Inhibitor

1. Plate 2×10^4 cells/cm² in Rocki working solution (10 μM).
2. Change medium every 2–3 days; keep them in culture until 80% confluent.

3.2.3   Combined Approach

1. Perform steps 1 and 2 described in section 3.2.1.
2. Proceed as described in section 3.2.2.

3.3 Intermediate Analysis

An intermediate analysis of a fraction of the obtained cells after each treatment should be performed in order to assess the effect in the increased fraction of interest—EpSCs fraction. Therefore in this section a flow cytometry protocol using a combination of CD49f-APC and CD71-PE markers is described.

1. Harvest the adherent cells cultured under the described conditions with trypsin–EDTA.

2. Transfer cells to a 15 mL falcon tube and add labeling buffer up to 10 mL.

3. Centrifuge cell suspension at $200 \times g$ for 5 min.

4. Count cells using a hemocytometer.

5. Discard supernatant and resuspend cell pellet to a concentration of $0.5{-}10^6$ cells/mL in fresh labeling buffer.

6. Add 100 μL of cell suspension to each flow cytometry tube (see Note 7).

7. Add 4 μL of CD71-PE and 2 μL CD49f-APC antibodies; reserve one tube per condition without antibody as control.

8. Incubate 30 min at room temperature.

9. Wash by adding 2 mL of PBS per tube and centrifuge at $250 \times g$ for 3 min.

10. Resuspend cell pellets in 500 μL of acquisition buffer.

11. Acquire data in flow cytometer.

12. Analyze simultaneous expression of CD71 and CD49f (Fig. 1).

3.4 CD71⁻/CD49f⁺ Rescuing: Immunomagnetic Selection

3.4.1 Depletion of CD71⁺ Cells

1. Wash CD71 coated beads by placing the tubes in a magnet for 1 min, discarding the supernatant and adding 1 mL of Dynabeads buffer 2, twice.

2. Incubate the cells harvested on Subheading 3.3, **step 1** and resuspended in Dynabeads buffer 2 with the washed CD71 beads for 30 min at 2–8 °C with gentle tilting and rotation.

3. Place the tubes in a magnet for 2 min.

4. Transfer the supernatant containing the unbound cells to a fresh 15 mL Falcon tube (CD71⁻ cell fraction).

5. Count cells using a hemocytometer.

6. Plate 2×10^4 cells/cm² in new tissue culture vessels/flasks with the correspondent treatments (described in Subheading 3.2).

7. Culture cells until 80% confluent by changing medium 2–3 days.

3.4.2 Positive Selection of CD49f⁺ Cells Among CD71⁻ Population

1. Harvest CD71⁻ cells with trypsin–EDTA and resuspend them in Dynabeads buffer 2.

2. Wash CD49f coated beads by placing the tubes in a magnet for 1 min, discarding the supernatant and adding 1 mL of Dynabeads buffer 2, twice.

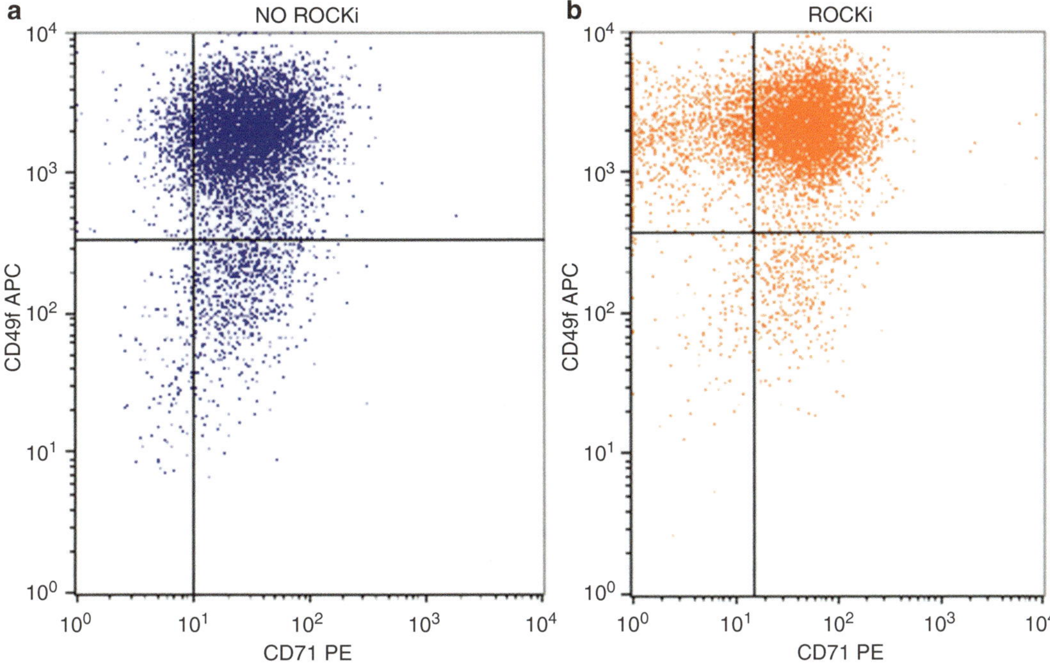

Fig. 1 Dot plots of CD49f/CD71 staining on human keratinocytes isolated from the same human sample and cultured without any treatment (**a**) and after EpSCs enrichment with Rocki (**b**), showing an increase of the population of interest by the differential expression of CD49f^{bri}/CD71^{dim} (6.20% in **a** and 10.37% in **b**)

3. Incubate CD71$^-$ cells for 20 min at 2–8 °C with gentle tilting and rotation.

4. Place the tubes in a magnet for 2 min.

5. Discard the supernatant and gently wash the bead-bounded cells, four times, by adding 1 mL of Dynabeads buffer 2.

6. Place the tubes in the magnet for 1 min and discard the supernatant.

7. Resuspend the cells in fresh KSFM.

8. Plate 2×10^4 cells/cm^2 in new tissue culture vessels/flasks with the correspondent treatments (described in Subheading 3.2) for further cell expansion and analysis.

3.5 Analysis

The analysis of the expression of the early epidermal markers on the obtained cell fraction by flow cytometry is advisable to validate the success of the employed strategies. Thus, this section comprises the protocol for identifying the expression of the intracellular markers using Cytokeratin 19-AF488 and Cytokeratin 14-FITC, and Keratin 5, respectively by direct and indirect staining.

3.5.1 Direct Staining

Follow the protocol from Subheading 3.3, steps 1–11, using 5 µL of Cytokeratin 19-AF488 and 4 µL of Cytokeratin 14-FITC in separate tubes.

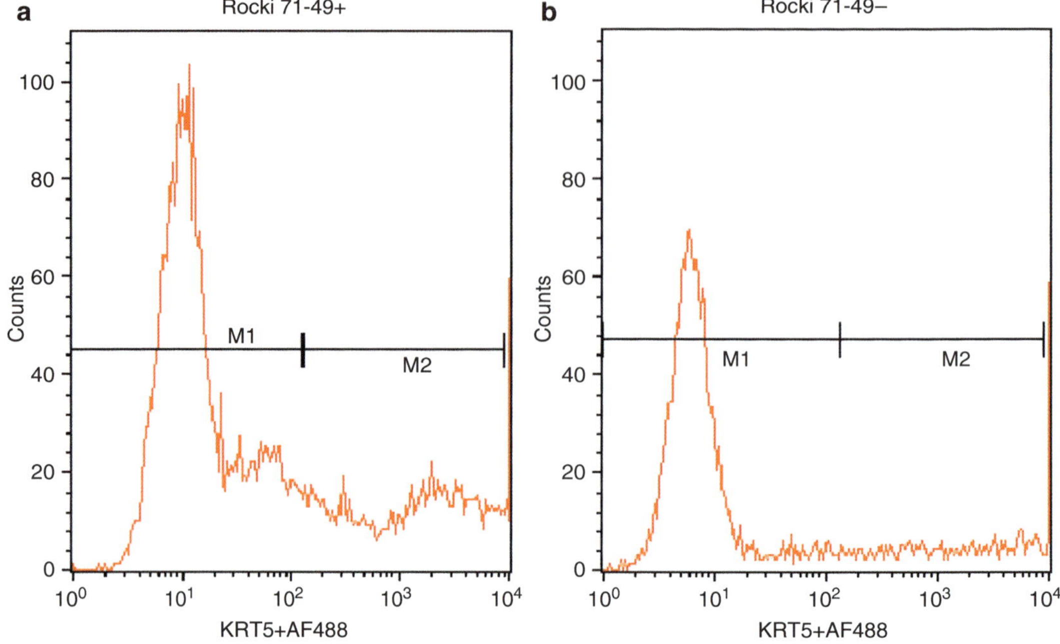

Fig. 2 Expression fluorescence profile of Keratin 5 (early epidermal marker) in both CD71-/α6+ cells (25.45%) (a) and CD71-/α6- (17.42%) (b), showing the higher expression of K5 in the fraction of interest, thus illustrating its early differentiation stage

3.5.2 Indirect Staining

1. Follow the protocol from Subheadings 3.3, steps 1–6.

2. Incubate cells with 200 μL of permeabilization buffer for 10 min at RT.

3. Centrifuge for 5 min at $250 \times g$ and remove supernatant.

4. Resuspend cells in 200 μL of keratin 5 antibody diluted 1:500 in labeling buffer.

5. Incubate for 1 h at room temperature.

6. Wash by adding 2 mL of PBS, centrifuge at $250 \times g$ for 5 min, and remove supernatant.

7. Resuspend cells in 200 μL of Alexa-fluor-labeled secondary antibody diluted 1:500 in labeling buffer.

8. Incubate 45 min at 4°C protected from light.

9. Wash by adding 2 mL of PBS, centrifuge at $250 \times g$ for 5 min, and remove supernatant.

10. Resuspend cell pellets in 500 μL of acquisition buffer.

11. Acquire data in flow cytometer (Fig. 2).

4   Notes

1. All the labware has to be sterilized prior use.

2. It is recommended to make aliquots that should be kept at −20°C, avoiding repeated freeze-thawing.

3. Coating with collagen IV can be previously performed incubating solution overnight at 4°C, without drying.

4. Alternatively, a solution of 0.1 M sodium borate sulfate in diH$_2$O, pH 7.4–8, can be used as buffer 1.

5. This protocol is adapted from the section "coupling of ligands to dynabeads" of the Dynabeads m-450 Epoxy manufacturer's instructions.

6. Cells cultured, in these different treatments, present distinct adherence and proliferation rates. No specific treatment, in which cells are cultured in noncoated plates and in KSFM, should be performed as a control. hKC cultured in Rocki and No treatment take almost 1 week to first adhere.

7. The remaining cells should be used as described in Subheading 3.4 and kept in culture for comparison purposes in the end of the experiment.

References

1. Blanpain C, Lowry WE, Geoghegan A, Polak L, Fuchs E (2004) Self-renewal, multipotency, and the existence of two cell populations within an epithelial stem cell niche. Cell 118:635–648
2. Lau K, Paus R, Tiede S, Day P, Bayat A (2009) Exploring the role of stem cells in cutaneous wound healing. Exp Dermatol 18:921–933
3. Watt FM, Lo Celso C, Silva-Vargas V (2006) Epidermal stem cells: an update. Curr Opin Genet Dev 16:518–524
4. Blanpain C, Fuchs E (2006) Epidermal stem cells of the skin. Annu Rev Cell Dev Biol 22:339–373
5. Li A, Simmons PJ, Kaur P (1998) Identification and isolation of candidate human keratinocyte stem cells based on cell surface phenotype. Proc Natl Acad Sci USA 95:3902–3907
6. Taylor G, Lehrer MS, Jensen PJ, Sun TT, Lavker RM (2000) Involvement of follicular stem cells in forming not only the follicle but also the epidermis. Cell 102:451–461
7. Pellegrini G, Dellambra E, Golisano O, Martinelli E, Fantozzi I, Bondanza S, Ponzin D, McKeon F, De Luca M (2001) p63 identifies keratinocyte stem cells. Proc Natl Acad Sci USA 98:3156–3161
8. Pontiggia L, Biedermann T, Meuli M, Widmer D, Bottcher-Haberzeth S, Schiestl C, Schneider J, Braziulis E, Montano I, Meuli-Simmen C, Reichmann E (2009) Markers to evaluate the quality and self-renewing potential of engineered human skin substitutes in vitro and after transplantation. J Invest Dermatol 129:480–490
9. Alonso L, Fuchs E (2003) Stem cells of the skin epithelium. Proc Natl Acad Sci USA 100(Suppl 1):11830–11835
10. Kaur P, Li A (2000) Adhesive properties of human basal epidermal cells: an analysis of keratinocyte stem cells, transit amplifying cells, and postmitotic differentiating cells. J Invest Dermatol 114:413–420
11. Schluter H, Paquet-Fifield S, Gangatirkar P, Li J, Kaur P (2011) Functional characterization of quiescent keratinocyte stem cells and their progeny reveals a hierarchical organization in human skin epidermis. Stem Cells 29:1256–1268
12. McMullan R, Lax S, Robertson VH, Radford DJ, Broad S, Watt FM, Rowles A, Croft DR, Olson MF, Hotchin NA (2003) Keratinocyte differentiation is regulated by the Rho and ROCK signaling pathway. Curr Biol 13:2185–2189
13. Terunuma A, Limgala RP, Park CJ, Choudhary I, Vogel JC (2010) Efficient procurement of epithelial stem cells from human tissue specimens using a rho-associated protein kinase inhibitor Y-27632. Tissue Eng Part A 16:1363–1368
14. Chapman S, Liu X, Meyers C, Schlegel R, McBride AA (2010) Human keratinocytes are efficiently immortalized by a Rho kinase inhibitor. J Clin Invest 120:2619–2626

Chapter 2

Isolation and Cultivation of Human Scalp Interfollicular Epidermal Stem Cells

Longmei Zhao and Basil M. Hantash

Abstract

Skin regeneration is intricately controlled by epidermal stem cells. In human skin, the long-lived, slow-cycling, and highly proliferative stem cells are located in the basal layer of the interfollicular epidermis (IFE). The ability to isolate and culture human IFE stem cells (IFESCs) offers fascinating therapeutic potential for skin diseases as well as epithelial tissue engineering. Here we describe a straightforward strategy for generation of $\beta1$ integrin$^+$/CD24$^-$ IFESCs from scalp with defined, serum-free, feeder-free medium and collagen I-coated culture plates. The use of defined media throughout isolation and cultivation allows for detailed investigation of the molecular events involved in ESC self-renewal and differentiation as well as provides a safe source for clinical use.

Key words Isolation, Cultivation, Epidermal stem cells, Human scalp, Serum free, Feeder free

1 Introduction

The epidermis forms the outer protective barrier of the body and regenerates throughout life. Skin regeneration is intricately controlled by epidermal stem cells (ESCs) (1), defined here as cells possessing the ability to self-renew—generation of daughter cells capable of undergoing single- or multi-lineage terminal differentiation (2). In human skin, the long-lived, slow-cycling, and highly proliferative stem cell compartment is located in the basal layer of the interfollicular epidermis (IFE) (3–5). Although evidence suggests that IFE, hair follicles, and sebaceous glands are all maintained by their own distinct stem cell compartments (6–10), IFE stem cells (IFESCs) possess the capacity to differentiate into all epidermal lineages. In addition, IFESCs participate in repair of skin after injury (6–9). The ability to isolate and culture human IFESCs offers fascinating therapeutic potential for skin diseases as well as epithelial tissue engineering.

Techniques to isolate and culture keratinocyte progenitor cells have been established, although limitations remain in many of

Kursad Turksen (ed.), *Skin Stem Cells: Methods and Protocols*, Methods in Molecular Biology, vol. 989,
DOI 10.1007/078 1 62703 330 6_2, © Springer Science Business Media New York 2013

these procedures. In particular, many existing lines have been cultured using mouse embryonic fibroblasts (NIH 3T3) as the feeder layer and/or serum-sourced medium components. NIH 3T3 cells support keratinocyte progenitor cell growth from single cells into colonies while suppressing growth of human fibroblasts (11–13). However, continued use of feeders and serum containing culture medium will hinder the development of clinical applications by increasing the risk of transmitting animal viruses and immunogenic antigens as well as pose difficulty with quality control of these undefined components.

Keratinocyte progenitor cells rapidly adhere to collagen-coated dishes (12, 14). Low calcium medium appears to be effective for expanding proliferative keratinocytes while high calcium medium allows these cells to terminally differentiate (15). Cultivation of keratinocyte progenitor cells in serum-free medium led to enhanced expression of $\beta 1$ integrin, an ESC marker, relative to those cultured in serum containing medium (16). The method described here involves cultivation of IFESCs on collagen I-coated plates with defined serum-free medium containing low calcium in order to generate purified IFESCs from human scalp skin. This highly efficient process enables direct generation of keratinocyte progenitors from human skin tissue for potential use in research purposes and clinical applications.

2 Materials

2.1 Tissue

1. Human scalp samples were obtained from facelift procedures with Institutional Review Board approval.

2.2 Reagents and Medium

1. William's E serum-free medium (Life Technologies).

2. Progenitor cell-targeted (PCT) keratinocyte medium (Chemicon, CnT-07).

3. Dulbecco's Modified Eagle Medium (DMEM) (Invitrogen).

4. Phosphate buffered saline (PBS), without Ca^{2+} or Mg^{2+}, pH 7.4 (Invitrogen).

5. 0.25 and 0.05% trypsin-EDTA (Invitrogen).

6. Trypsin neutralizer solution (Invitrogen).

7. 0.4% trypan blue (Invitrogen).

8. Dispase (Invitrogen). Reconstituted in DMEM at 0.125 mg/ml. Aliquot could be stored at −20°C up to 6 months. If stored at 4°C, use within 2 weeks.

9. 10,000 U/ml penicillin and 10,000 mg/ml streptomycin (Invitrogen).

10. 12.5 µg/ml fungazone (amphotericin B) (Invitrogen).

11. Recombinant human collagen I (Fibrogen, RhC1-003). Dissolve 100 mg of recombinant human collagen I in 2 ml of filter-sterilized 10 mM HCl. Aliquot could be stored at 4°C up to 6 months. When coating plates, further dilute to 25 mg/ml with sterile PBS and use 10 mg per 1 cm^2 dish area. Leave freshly coated plates for at least 1 h at room temperature. The coated plates could be stored at 4°C up to 2 weeks.

12. Fixation buffer: To make 8% (w/v) paraformaldehyde (PFA), dissolve 8 mg of PFA (Sigma) in 100 ml of PBS, heat the solution to 60°C and stir for at least 30 min, add a few drop of 10 M NaOH to completely dissolve PFA. Store at –20°C. Dilute to 4% (w/v) in PBS prior to use.

13. Blocking buffer: To make 10% (v/v) goat serum solution, add 1 ml of normal goat serum (Vector Laboratories) in 9 ml of PBS, store at 4°C.

14. Primary antibodies (recommended dilution): mouse anti-b1 integrin monoclonal antibody (1:100, BD Biosciences), mouse anti-CD24 monoclonal antibody (1:100, BD Biosciences).

15. Preparation of ABC complex by adding 50 ml of avidin (Vector Laboratories, PK6102) and 50 ml of avidin-horseradish peroxidase (HRP) (Vector Laboratories, PK6102) in 10 ml of PBS containing 0.1% tween 20, mix immediately and allow to stand for 30 min ahead of application.

16. Prepare 3,3¢-diaminobenzidine (DAB) solution by adding 30 ml of the DAB liquid chromogen solution (Sigma, D3939) to 1 ml of the DAB liquid buffer solution (Sigma, D3939). Use immediately.

17. Mayer's hematoxylin solution (1×, Sigma, MHS1).

18. Crystal mount aqueous mounting medium (Sigma, C0612).

---

3 Methods

The method described here enables production of high-purity ESCs from scalp skin. Initial cultures must be pure as fibroblast-like cells proliferate much faster than keratinocytes thereby dominating the latter in culture. IFESCs immunostain positively for β1 integrin (12) and negatively for CD24 (17) markers. Given use of defined and feeder-free culture media throughout isolation and cultivation, this system provides a unique means of investigating the molecular events involved in ESC self-renewal and differentiation.

3.1 Isolation of Human IFESCs from Scalp Skin

The following procedures are performed in biosafety hoods. Investigators should be trained in the handling of human tissue and human pathogens before initiation of any studies.

1. Human scalp skin is placed in sterile William's E serum-free medium supplemented with penicillin (250 U/ml), streptomycin (250 mg/ml), and fungazone (1.25 mg/ml) immediately after surgery and stored at 4°C until use (see Note 1).

2. Rinse scalp skin several times in sterile PBS with constant agitation to remove clots.

3. Transfer scalp skin to a 100 mm petri dish and trim hairs with a sterile surgical scissor. While turning the dermal side upward, remove as much connective tissue (muscle and fat) as possible with a sterile scalpel and forceps and then cut trimmed scalp skin into pieces (~1 cm^2 in area).

4. Transfer small pieces of scalp skin into a 100 mm petri dish containing 10 ml of 0.125% dispase/DMEM and incubate for 16–18 h at 4°C (see Note 2).

5. Carefully transfer scalp skin to a 100-mm petri dish by turning the epidermal side upward. Gently scrape epidermis away from the dermis using a scalpel and Pasteur pipette (see Note 3).

6. Collect the epidermis into a sterile 50 ml conical tube containing 10 ml of prewarmed 0.25% trypsin-EDTA and trypsinize for 20–30 min at 37°C.

7. Gently shake the tube several times to release the trypsinized cells into solution (see Note 4).

8. Gently pipette cells with 2 ml of pipette and add an equal amount of trypsin neutralizer solution. Centrifuge at 800 rpm for 7 min.

9. Remove the supernatant by vacuum aspiration and resuspend the cells with 1 ml of prewarmed PCT keratinocyte medium.

10. Place 10 ml of cell aliquot into a 500-ml Eppendorf tube and add an equal volume of trypan blue solution, mix and leave at room temperature for 5 min. Determine the number of cells with a hemocytometer and eliminate dead cells from this count.

11. Dilute the cell suspension in an appropriate volume and plate the cells at 1×10^4 cell/cm^2 in collagen I-coated flasks with PCT medium containing 0.07 mM calcium (see Note 5) and incubate at 37°C in a humidified 5% CO$_2$ incubator (see Note 6).

12. Approximately 2 days later (see Note 7), remove the media and nonadherent cells and replace with fresh PCT medium containing 0.07 mM calcium. Change medium every other day.

13. After 10–14 days, tightly packed epithelial colonies should be observed (Fig. 1; see Note 8).

3.2 Subculture Human IFESCs

All solution and medium need be warmed to 37°C before use.

1. Aspirate off the culture medium in a T-75 flask with cells at nearly 80% confluency (see Note 9). Add 5 ml of prewarmed

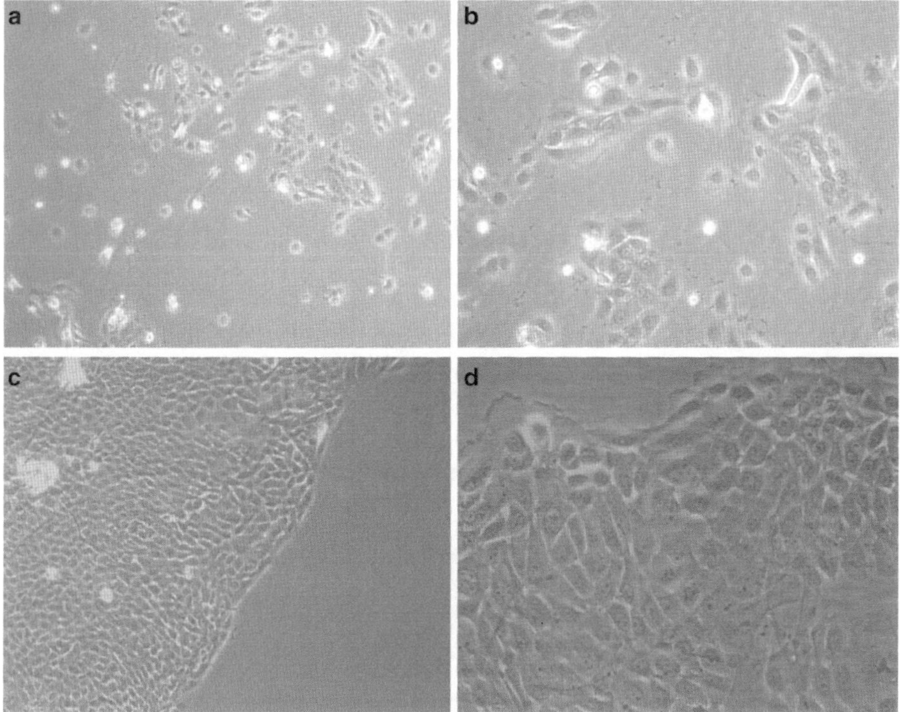

Fig. 1 (**a**, **b**) human IFESCs 3 days after plating. (**c**, **d**) Human IFESCs 10 days after plating. (**a**, **c**) 100× original magnification. (**b**, **d**) 200× original magnification

PBS, gently tilting the flask back and forth several times, then aspirate off PBS.

2. Add 3 ml 0.05% trypsin/EDTA solution per T-75 flask and ensure proper distribution of solution in the flask.

3. Incubate cells for 5–10 min at 37°C in a humidified incubator (see Note 10).

4. Gently tap the flask to dislodge the cells from the surface of the flask, and then add 7 ml of trypsin neutralizer solution per T-75 flask. Wash the flask by pipetting a few times over the cell attachment area (see Note 11).

5. Transfer detached cells to a sterile 15 ml conical tube.

6. Centrifuge cells at $180 \times g$ (or about 1,000 rpm) for 7 min.

7. Aspirate supernatant without dislodging the pellet. Resuspend the cell pellet in an appropriate volume of culture medium and plate the cells at $1 \times 10^4/cm^2$ in collagen I-coated flasks (see Note 12) with PCT medium containing 0.07 mM calcium (Fig. 2).

3.3 Immuno-cytochemistry

1. Plate the cells at 1×10^4 cell/cm^2 in collagen I-coated chamber slides and culture at 37°C in a humidified 5% CO_2 incubator for a few days.

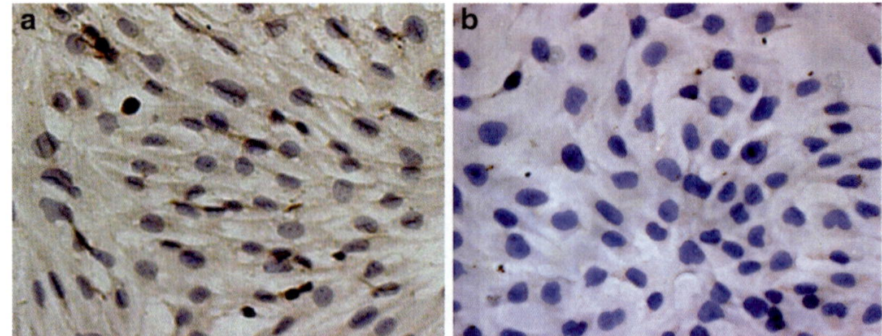

Fig. 2 Immunostaining of primary passage 1 IFESCs cultured for 14 days. (**a**), More than 90% of cells are positive for β1 integrin staining. (**b**) IFESCs are negative for CD24 staining

2. Aspirate the culture medium from each well and gently rinse the cells twice with PBS at room temperature (see Note 13).

3. Fix cells by incubating them in 4% (v/v) PFA in PBS for 20 min at room temperature.

4. Rinse cells three times with PBS (see Note 14).

5. Incubate cells with 1:100 diluted H_2O_2 (30%) for 10 min at room temperature (see Note 15).

6. Rinse cells three times with PBS.

7. Block cells with 10% normal goat serum for 30 min at room temperature.

8. Incubate cells with primary antibodies in fresh blocking buffer at the appropriate dilution (see Note 16) overnight at 4°C or for 2 h at room temperature.

9. Wash cells three times with PBS.

10. Incubate cells with biotinylated secondary antibodies (see Note 17) at a dilution of 1:200 in fresh blocking buffer for 30–60 min at room temperature.

11. Wash cells three times with PBS.

12. Incubate cells with the ABC complex solution for 30 min at room temperature.

13. Drop off the ABC complex solution and cover cells with DAB reagent solution.

14. Monitor carefully under light microcopy during the reaction to prevent overdevelopment and high background (see Note 18).

15. Stop the reaction when the color is satisfactory by dropping off the DAB reagent solution and covering cells with PBS.

16. Rinse cells three times with PBS.

17. Incubate cells for 5 min with Mayer's hematoxylin solution.

18. Rinse cells three times with PBS.

19. Mount slide with mounting medium.

4 Notes

1. For maximum viability, skin tissue should be processed as soon as possible after surgery. However, when necessary tissue can be kept at 4°C for a few days without significant loss of viability. The yield of stem cells per cm^3 skin is affected by the age and health of donor. Younger donors generate more stem cells than older ones.

2. Use enough dispase medium to completely cover tissue. To obtain optimal digestion, tissue should be evenly distributed in dispase medium.

3. The epidermis should peel away quite easily. Take care not to take dermis with the epidermis. This is a critical step to obtain pure populations of epidermal cells. If the epidermis does not peel easily, the tissue should be incubated for up to several hours at 37°C or placed in fresh dispase solution and incubated for an additional 24 h at 4°C. The pieces of peeled epidermis should be kept in calcium- and magnesium-free Hanks' balanced salt solution during this procedure.

4. If the epidermis is not completely digested, keep the tube at room temperature for several minutes to allow the epidermal pieces to float to the surface while cellular debris settles to the bottom of the tube. Using sterile forceps, remove the undigested epidermis into a new 50 ml sterile conical tube. To obtain maximum keratinocyte yield, undigested pieces of epidermis could be incubated in prewarmed trypsin solution. The process may be repeated up to three times. Cells harvested in this way are pooled and counted.

5. PCT medium is defined, serum-free, feeder-free media for superior isolation and growth of progenitor cells. Keratinocyte progenitor cells are selected by using low calcium conditions and avoiding use of fetal calf serum in the culture medium.

6. Ensure proper distribution of cells by quickly moving the flask back and forth several times. Avoid opening the incubator door after plating.

7. There may not be many adherent cells at 24 h. Wait for at least 2 days to allow cells to settle down to the bottom of the dish.

8. At this time point, cells may require daily feeding.

9. Split cells when the culture reaches 60–80% confluency. Avoid splitting cells at more than 80% confluency as cells tend to undergo differentiation when crowded in culture.

10. Incubate the flasks at room temperature until cells have become completely round which usually takes approximately 10–15 min.

11. Sometimes fibroblast-like cells are observed contaminating the culture. To eliminate these contaminants, use two-step enzyme

selection. After washing with PBS, incubate cells with 0.05% trypsin/EDTA at room temperature for 5–10 min, then collect the solution and discard. Incubate cells with fresh 0.05% trypsin/EDTA at room temperature for another 5–10 min and collect cells into a 15-ml sterile conical tube and spin down. If only a few keratinocyte colonies are present in culture, place 10 mm outer diameter borosilicate glass cloning cylinders over regions containing a high percentage of cells with an epithelial morphology and harvest cells from one or more cylinders by trypsinization.

12. A variety of coating matrices may be used to subculture ESCs including collagen I and collagen IV. Precoated culture plates/flasks are also available commercially.

13. Do not let cells dry out during immunostaining.

14. Cells can be stored in 0.02% (w/v) sodium azide in PBS at 4°C for several days after this step.

15. Blocking of endogenous peroxidase can be performed after primary antibody incubation.

16. The researcher should test serial concentration of primary antibody to determine the optimal concentration.

17. Be sure that the correct isotype specific secondary antibody for each primary antibody is used.

18. This process should not take longer than 2 min.

Acknowledgments

The protocols described here were developed under the support of a generous gift from the Minnesota Jewish Foundation (2006) and an ASDS Cutting Edge Grant (BMH). The authors thank Shu Jiang, M.D., Ph.D. for her technical assistance.

References

1. Quesenberry P, Levitt L (1979) Hematopoietic stem cells (second of three parts). N Engl J Med 301:819–823

2. Lajtha LG (1979) Stem cell concepts. Differentiation 14:23–34

3. Ghazizadeh S, Taichman LB (2005) Organization of stem cells and their progeny in human epidermis. J Invest Dermatol 124:367–372

4. Lavker RM, Sun TT (1982) Heterogeneity in epidermal basal keratinocytes: morphological and functional correlations. Science 215:1239–1241

5. Lavker RM, Sun TT (2000) Epidermal stem cells: properties, markers, and location. Proc Natl Acad Sci USA 97:13473–13475

6. Levy V, Lindon C, Zheng Y et al (2007) Epidermal stem cells arise from the hair follicle after wounding. FASEB J 21:1358–1366

7. Ghazizadeh S, Taichman LB (2001) Multiple classes of stem cells in cutaneous epithelium: a lineage analysis of adult mouse skin. EMBO J 20:1215–1222

8. Ito M, Liu Y, Yang Z et al (2005) Stem cells in the hair follicle bulge contribute to wound repair but not to homeostasis of the epidermis. Nat Med 11:1351–1354

9. Levy V, Lindon C, Harfe BD et al (2005) Distinct stem cell populations regenerate the follicle and interfollicular epidermis. Dev Cell 9:855–861

10. Owens DM, Watt FM (2003) Contribution of stem cells and differentiated cells to epidermal tumours. Nat Rev Cancer 3:444–451

11. Rheinwald JG, Green H (1975) Serial cultivation of strains of human epidermal keratinocytes: the formation of keratinizing colonies from single cells. Cell 6:331–343

12. Jones PH, Watt FM (1993) Separation of human epidermal stem cells from transit amplifying cells on the basis of differences in integrin function and expression. Cell 3: 713–724

13. Roh C, Roche M, Guo Z et al (2008) Multipotentiality of a new immortalized epithelial stem cell line derived from human hair follicles. In Vitro Cell Dev Biol Anim 44:236–244

14. Jones PH, Harper S, Watt FM (1995) Stem cell patterning and fate in human epidermis. Cell 80:83–93

15. Ji L, Allen-Hoffmann BL, de Pablo JJ et al (2006) Generation and differentiation of human embryonic stem cell-derived keratinocyte precursors. Tissue Eng 12:665–679

16. Lorenz K, Rupf T, Salvetter J et al (2009) Enrichment of human beta 1 bri/alpha 6 bri/ CD71 dim keratinocytes after culture in defined media. Cells Tissues Organs 189:382–390

17. Jiang S, Zhao L, Purandare B et al (2010) Differential expression of stem cell markers in human follicular bulge and interfollicular epidermal compartments. Histochem Cell Biol 133:455–465

Chapter 3

Isolation of Hair Follicle Bulge Stem Cells from YFP-Expressing Reporter Mice

Kerry-Ann Nakrieko, Timothy S. Irvine, and Lina Dagnino

Abstract

In this article we provide a method to isolate hair follicle stem cells that have undergone targeted gene inactivation. The mice from which these cells are isolated are bred into a *Rosa26*-yellow fluorescent protein (YFP) reporter background, which results in YFP expression in the targeted stem cell population. These cells are isolated and purified by fluorescence-activated cell sorting, using epidermal stem cell-specific markers in conjunction with YFP fluorescence. The purified cells can be used for gene expression studies, clonogenic experiments, and biological assays, such as viability and capacity for directional migration.

Key words Stem cells, Keratinocyte, Hair follicle, YFP, Cre recombinase

1 Introduction

The epidermis is the outermost layer of the skin and provides an essential protective barrier against infectious pathogens, chemicals, UV radiation, and mechanical insults (1). The epidermis is a stratified squamous epithelium composed of keratinocyte stem cells and their committed or differentiated progeny. Stem cells provide this tissue with its enormous capacity for continuous self-renewal and regeneration following injury (2–4). Multipotent keratinocyte stem cells reside at various sites, including the bulge region of the hair follicle and in the interfollicular epidermis (5–7). In the hair follicle bulge, specific markers of keratinocyte stem cells have been identified, including the surface proteinCD34 (8, 9) and keratin 15 (10). Hair follicle stem cells have been isolated by fluorescence-activated or magnetic cell sorting, taking advantage of their high expression of both CD34 and α6 integrin (11–13). In this protocol, we describe the isolation from reporter mice of hair follicle stem cells specifically targeted to express yellow fluorescent protein (YFP). We originally used this method to identify and purify hair follicle stem cells in which the *Ilk* gene had been

Kursad Turksen (ed.), *Skin Stem Cells: Methods and Protocols*, Methods in Molecular Biology, vol. 989,
DOI 10.1007/978-1-62703-330-5_3, © Springer Science+Business Media New York 2013

inactivated (14). In these mice, expression and RU486-induced nuclear translocation of Cre recombinase fused to a modified progesterone receptor (Cre/PR) specifically in hair follicle bulge stem cells was achieved. These animals had *Ilk* alleles flanked by loxP sites and were bred into a *Rosa26-YFP* reporter background. As a result, targeted, ILK-deficient keratinocyte stem cells were identified by their expression of YFP. This protocol can also be used to purify and characterize keratinocyte stem cell populations in which a gene of interest is flanked by loxP sites, following inactivation by Cre/PR (15).

2   Materials

2.1 *Keratinocyte Stem Cell Isolation and Culture*

1. Depilatory cream (e.g., Neet, Premier Consumer Products Inc.).

2. Electric shaver.

3. 1% (w/w) RU486: Mix thoroughly 1 g RU486 (Mifepristone, TCI America) with 99 g Neutrogena hand cream.

4. Dulbecco's modified Eagle's medium (DMEM, Gibco Invitrogen).

5. Ham's F12 nutrient medium (Gibco Invitrogen).

6. Ca^{2+}-free DMEM/F12 medium: 3:1 (v/v) mixture of calcium-free Dulbecco's modified Eagle's medium and Ham's F12 nutrient medium; special order from Gibco Invitrogen, Cat. # 90-5010EA.

7. Sterile Milli-Q, 18.2 MΩ H_2O.

8. Sterile Ca^{2+}-free phosphate-buffered saline (PBS).

9. 10,000 U/ml Penicillin and 10 mg/ml streptomycin (Gibco Invitrogen).

10. Fungizone® (100×, Gibco Invitrogen).

11. l-glutamine (Invitrogen).

12. Transferrin (SIGMA Aldrich): 5 mg/ml dissolved in sterile PBS and stored in single-use aliquots at –20°C.

13. Hydrocortisone (SIGMA Aldrich): 4 mg/ml dissolved in 95% ethanol and stored at –20°C.

14. Insulin (SIGMA Aldrich): 5 mg/ml dissolved in 18.2 MΩ H_2O; pH adjusted to 3.0 with HCl, stored at –20°C in single-use 500-µl aliquots.

15. Cholera toxin (List Biological Laboratories): 1 mg/ml dissolved in 18.2 MΩ H_2O, stored at –20°C.

16. 3, 3′, 5-Triiodo-l-thyronine (T3, SIGMA Aldrich): 340 µg/ml dissolved in 0.1 N NaOH, stored at –20°C.

17. Epidermal growth factor (EGF, PreproTech): 100 µg/ml dissolved in sterile 18.2 MΩ H_2O, stored at –20°C in single-use 100-µl aliquots.

18. Sterile 0.06 M Calcium chloride solution.

19. Chelex-treated fetal bovine serum (FBS, Qualified grade, Gibco Invitrogen, see Notes 1 and 2): Mix 40 g of Chelex 100 chelating ion exchange resin (200–400 mesh, sodium salt, Biorad) with 400 ml 18.2 MΩ H_2O, and stir at room temperature for 1 h. Completely decant the H_2O and add 500 ml FBS. Stir at room temperature for 1 h, filter through filter paper to remove resin particles, and follow by filtration through 0.45-µm bottle top filters under sterile conditions.

20. Ca^{2+}-free keratinocyte stem cell (KSC) growth medium: Ca^{2+}-free DMEM/F12 medium containing 15% chelex-treated FBS, penicillin/streptomycin (100 U penicillin and 100 µg/ml streptomycin), hydrocortisone (0.5 µg/ml), cholera toxin (10^{-10} M), EGF (10 ng/ml), insulin (5 µg/ml), T3 (7.0 ng/ml), and transferrin (50 µg/ml).

21. Ca^{2+}-containing KSC medium: Ca^{2+}-free KSC medium supplemented with sterile 0.06 M $CaCl_2$ to a final concentration of 0.3 mM.

22. FACS buffer: Sterile Ca^{2+}-free PBS containing 1% chelex-treated FBS.

23. Trypsin-EDTA (0.25%, 1×, Gibco Invitrogen).

24. Trypsin (2.5%, Gibco Invitrogen): Dilute to 0.25% with PBS.

25. Mitomycin C (Roche): 0.4 mg/ml dissolved in PBS, filter-sterilized, and stored protected from the light at 4°C.

26. 95 and 70% ethanol.

27. 0.4% Trypan blue stain (ICN BioMed), dissolved in 0.85% NaCl.

28. Bacterial grade Petri dishes.

29. Dissecting scissors, scalpel, two forceps (Fine Science Instruments).

30. 50-ml conical tubes (BD Falcon BD Biosciences).

31. 5-ml round-bottom snap cap tubes (BD Falcon BD Biosciences).

32. 5-ml cell strainer round-bottom tubes (BD Falcon BD Biosciences).

33. 40-µm nylon cell strainers (BD Falcon BD Biosciences).

34. 70-µm nylon cell strainers (BD Falcon BD Biosciences).

35. 20 gauge syringe needles (BD).

36. Improved Neubauer hemocytometer (VWR).

37. 100-mm Primaria tissue culture dishes (BD Falcon BD Biosciences).

38. 24-well, flat bottom Primaria tissue culture plates (BD Falcon BD Biosciences).

39. Sterile Acrodisc® 0.2-μm HT Tuffryn® syringe filters (Pall Life Sciences).

40. Nalgene bottle-top sterile filter units (VWR).

41. Collagen type I from rat tail (BD Biosciences): Diluted to 40 μg/ml in sterile PBS.

2.2   3T3 Feeder Cell Culture

1. DMEM/F12 medium: 3:1 (v/v) mixture of calcium-containing Dulbecco's modified Eagle's medium and Ham's F12 nutrient medium (Gibco Invitrogen).

2. Bovine calf serum (Gibco Invitrogen).

3. 3T3 growth medium: DMEM/F12 medium supplemented with 10% bovine calf serum.

4. Swiss 3T3 mouse fibroblasts (ATCC).

2.3   Fluorescence-Activated Cell Sorting (FACS)

1. Alexa Fluor® 649-conjugated anti-Cd34 antibody (Biolegend).

2. Phycoerythrin (PE)-conjugated rat anti-human integrin α6 subunit antibody (clone GoH3, BD Pharmingen).

3. 7-Amino-actinomycin D (7-AAD) staining solution (25 μg/ml, Biolegend).

4. Becton Dickinson FACSAria cell sorter (BD Biosciences) or equivalent.

2.4   Mice

The experiments we describe are conducted with reporter mice generated by breeding B6:SJL-Tg(Krt1-15-cre/PGR)22Cot/J mice (16) with B6.129X1-*Gt(ROSA)26Sor*$^{tm1(EYFP)Cos}$/J, both available from the Jackson Laboratory (see Note 3). These reporter mice can be bred into other backgrounds in which a gene of interest is flanked by loxP sites and can therefore be used to select for bulge stem cells in which a gene of interest has been inactivated (14, 15).

3   Methods

3.1   Treatment of Mice with RU486

1. Anesthetize female mice during the second telogen (see Note 4).

2. Using an electric shaver, clip the dorsal hair as close to the skin as possible, taking care of not causing any wounds.

3. Completely remove the remaining hair by spreading depilatory cream and waiting 5 min before carefully removing it completely with a damp or wet towel.

4. Spread a thin layer of 1% RU486 onto the area where hair has been removed (see Note 5).

5. Monitor the mice until fully recovered from anesthesia, and house singly.

6. Repeat topical treatments with RU486 daily for 4 more days (see Note 6).

3.2 Maintenance of 3T3 Fibroblasts

1. Seed 1.5×10^5 3T3 cells in a T75 culture dish, using 3T3 growth medium.

2. Change the medium every second day.

3. When cultures reach 80% confluence, trypsinize using 0.25% trypsin/EDTA solution and passage, seeding as in Subheading 3.2, step 1 (see Note 7).

3.3 Preparation of 3T3 Feeder Cells

1. Seed 3T3 fibroblasts in 100-mm culture dishes and culture to confluence.

2. One day before isolation of keratinocyte stem cells, remove the growth medium from confluent 3T3 cultures and replace with 5 ml of fresh 3T3 growth medium containing 8 µg/ml mitomycin C.

3. Culture the feeder cells for 2 h to render them mitotically inactive.

4. Remove the mitomycin C-containing medium and rinse each culture dish twice with 5–10 ml PBS.

5. Add fresh 3T3 growth medium and culture the feeder cells overnight.

6. The day of isolation of epidermal cells, and at least 2 h prior to plating the keratinocyte stem cells, coat the surface of the culture dishes desired for the stem cells with sufficient Collagen I solution to completely cover the surface (e.g., 1 ml/well for 6-well culture dishes) and incubate at 37°C. After 2 h, remove the collagen solution and rinse the culture dishes with PBS.

7. At least 1 h prior to plating keratinocyte stem cells, trypsinize the mitotically inactive 3T3 cells.

8. Transfer the cells to a 50-ml conical tube and add 5 ml of 3T3 growth medium.

9. Pellet the cells by centrifugation ($500 \times g$, 5 min), resuspend them in 3T3 medium, and seed them in the desired collagen-coated keratinocyte culture dish (see Note 8).

10. Culture the feeder cells, allowing them to attach (at least 1 h).

3.4 Isolation of Adult Epidermal Keratinocytes as Single-Cell Suspensions

1. Euthanize mice previously treated with RU486 by CO_2 inhalation or following the appropriate institutionally approved protocol.

2. Wash the skin with 70% ethanol to disinfect, and blot dry.

3. In a laminar flow hood, to ensure sterility, use a sterile scalpel to make an incision on the dorsal skin from the base of the tail to the head, just inside the boundary between the hair-free and the hair-containing regions of the skin. All subsequent steps that involve handling the tissue or cell suspensions should be carried out in a laminar flow hood, to maintain sterility.

4. Make two additional incisions perpendicular to the first within the boundary between the hair-free skin and the hair-containing skin. The first incision is behind the head and the other just above the tail end.

5. Remove the hair-free dorsal skin with sterile forceps and place on a sterile Petri dish, epidermis-side down.

6. Remove subcutaneous fat patches and blood vessels by gently scraping with a scalpel (see Note 9).

7. Spread the skin, dermis side down, in a clean 100-mm Petri dish, ensuring there are no folds (see Note 10).

8. Gently add 10 ml of 0.25% trypsin to the Petri dish, avoiding the skin. Ensure that the skin floats on the solution and that the epidermis never contacts the trypsin.

9. Digest the skins at 4°C overnight.

10. The next morning, place the Petri dishes at 37°C for 10 min.

11. Transfer each skin to a clean Petri dish, dermis side down.

12. Hold the skin in one corner using a sterile forceps and scrape the epidermis using a scalpel. The epidermal tissue will often separate from the dermis as an intact sheet (see Note 11).

13. Mince the epidermis obtained from a single skin to produce fragments of approximately 2×2 mm and transfer the tissue into sterile 50-ml conical tubes containing 20 ml ice-cold Ca^{2+}-free KSC medium. Gently rock at 4°C for 10 min.

14. Pour the suspension through a sterile 70-μm strainer placed on a sterile 50-ml conical tube. Wash the strainer with 5 ml cold Ca^{2+}-free KSC medium.

15. Pour the suspension through a sterile 40-μm strainer placed on a sterile 50-ml conical tube. Keep the cell suspension on ice.

16. Repeat steps 14 and 15 of Subheading 3.4 for each skin. Keep the tubes on ice until all skins are processed.

17. Centrifuge the tubes at $250 \times g$ for 5 min at 4°C. Aspirate the supernatants, ensuring that the cell pellets are not disturbed.

18. Resuspend each cell pellet in 5 ml of freshly made, ice-cold FACS buffer, keeping the tubes on ice (see Note 12).

19. Label each of four FACS tubes as follows (see Table 1): α6, CD34, 7-AAD, Unstained.

    Add 275 µl of ice-cold FACS buffer to each tube.

20. Transfer a 25-µl sample of cells to each of the FACS tubes prepared in Subheading 3.4, step 19, keeping them on ice. These samples will be used as controls to set up the sorting parameters (see Note 13).

21. Transfer a second 20-µl aliquot of the cell suspension to a microfuge tube containing 20 µl of trypan blue stain. Incubate for 5 min at room temperature and determine the number of viable cells using a hemocytometer.

22. Centrifuge the remainder of the cell suspension of Subheading 3.4, step 18 at $250 \times g$ for 5 min at 4°C. Aspirate the supernatants, ensuring that the cell pellets are not disturbed. Resuspend the cell pellets in sufficient ice-cold FACS buffer to obtain a cell suspension containing 1×10^7 cells/ml, and transfer to a FACS tube. Label this tube as "Sort" (see Note 14).

3.5 Antibody Labeling of Adult Epidermal Keratinocytes for FACS

1. To label the cells for α6 integrin, add anti-α6 antibody to the control sample labeled "α6" (from Subheading 3.4, step 19) and "Sort" to a final dilution of 1:25, as summarized in Table 1 (see Note 15).

2. Incubate cells on ice for 30 min, gently mixing the cell suspensions every 10 min. Keep the remaining, unstained samples on ice (see Note 16)

3. Collect the cells by centrifugation ($250 \times g$, 5 min, 4°C).

4. Gently remove the supernatant and resuspend the cells in the same volume of ice-cold FACS buffer used in Subheading 3.4, step 22 for the "Sort" sample, or in 300 µl FACS buffer for the "α6" control sample.

5. To label the cells for CD34, add anti-CD34 antibody to the control sample labeled "CD34" (from Subheading 3.4, step 19) and "Sort" to a final dilution of 1:50, as summarized in Table 1 (see Note 15).

6. Incubate cells on ice for 30 min, gently mixing the cell suspensions every 10 min. Keep the remaining, unstained samples on ice.

7. Collect the cells by centrifugation ($250 \times g$, 5 min, 4°C).

8. Gently remove the supernatant, and resuspend the cells in the same volume of ice-cold FACS buffer used in Subheading 3.4, step 22 for the "Sort" sample, or in 300 µl FACS buffer for the "CD34" control sample.

Table 1
Summary of samples and antibodies for FACS purification of hair follicle stem cells

Tube	Sample	Vol. FACS buffer	Vol. cells	Antibody	Antibody dilution	7-AAD dilution
1	Unstained control[a]	275	25	None	n/a[b]	None
2	α6[a]	275	25	Integrin α6	1:25	1:100
3	CD34[a]	275	25	CD34	1:50	1:100
4	7-AAD[a]	275	25	None	n/a[b]	1:100
5	Sort	[c]	[c]	Integrin α6 CD34	1:25 1:50	1:100

Samples 1–4 are controls to set up gating parameters in the flow cytometer. Sample 5 is the experimental sample to be sorted
[a]Cells from Subheading 3.4, step 20
[b]Not applicable
[c]Cells from Subheading 3.4, step 22 are suspended in FACS buffer to a final density of 1×10^7 cells/ml

9. Prepare sterile tubes for collection of sorted cells, by filling them with 50% chelex-treated FBS diluted in PBS. Incubate at 22°C for at least 30 min. Discard the solution and add Ca^{2+}-free KSC medium.

10. To stain apoptotic and necrotic cells, add 7-AAD (1:100 dilution) to all samples, except the "unstained" control, 15 min prior to cell sorting, and mix gently.

11. Filter the cell suspension through 40 μm strainers and place in clean, sterile FACS tubes. Keep cells on ice and proceed to sorting.

3.6 Flow Cytometric Isolation of Targeted Bulge Epidermal Stem Cells

1. Set up sortin g parameters for keratinocytes using a 100-μm nozzle at low pressure (20 psi), which allows sorting at about 10–12,000 events per second, and preserves cell viability.

2. Set up the flow cytometer so that excitation is available from the 488 nm to the 633 nm lasers. Recording for all fluorescence channels should also be set up.

3. Run the unstained control first. Basal epidermal keratinocytes have low forward (FSC) and side (SSC) scatter, and this is the cell population that should be selected for further analysis, excluding cell debris and larger, differentiated cells. The FSC and SSC detectors are set in a linear scale, whereas the fluorescence channels are set to logarithmic scale. Use the appropriate voltage on each detector to generate a cell population with a wide distribution, but ensuring

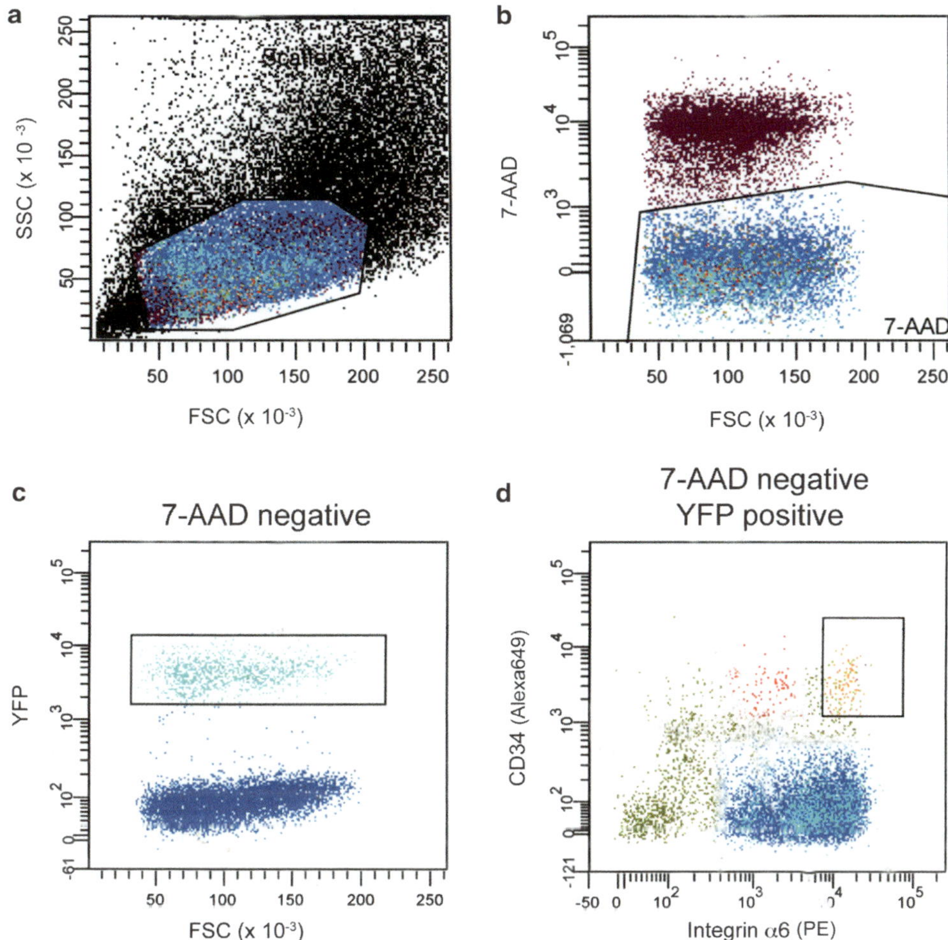

Fig. 1 Purification by FACS of targeted hair follicle stem cells from single cell suspensions of epidermal keratinocytes. Keratinocytes isolated from the epidermis of RU486-treated mice in the second telogen were labeled with anti-CD34 and anti-integrin α6 antibodies conjugated, respectively, with Alexa Fluor ™ 649 and PE. (**a**) Forward scatter (FSC) vs. SSC plot of unstained cells. (**b**) Cells with low 7-AAD staining are selected to exclude dead cells. (**c**) Cells that show YFP fluorescence are selected to identify the population targeted by RU486 treatment. (**d**) Cells gated for joint high CD34 and integrin α6 fluorescence are selected as targeted bulge stem cells

that all events remain visible in the FSC vs. SSC plots for each of the fluorescence channels, as exemplified in Fig. 1a (see Note 17).

4. Use the 7-AAD vs. FSC plot to select a region with low 7-AAD fluorescence, which corresponds to viable cells, as shown in Fig. 1b.

5. Run the unstained control again, selecting for YFP-positive and negative populations. YFP-negative unstained cells should low fluorescence units, as illustrated in Fig. 1c.

6. Run each of the "α6" and "CD34" single-label controls, ensuring there is no overlap in the fluorescence signals.

7. Create a region of YFP-positive, 7-AAD-negative cells and display on a plot of PE (for integrin α6) versus Alexa Fluor ™ 649 (for CD34) as Fig. 1d. These are the viable bulge keratinocytes that have been appropriately targeted (see Note 18).

8. Collect the sorted cell populations into sterile tubes containing Ca^{2+}-free KSC medium, to avoid formation of cell clumps. Keep the cells on ice, until placed in culture.

9. Check the purity of the YFP^{HIGH}/CD34^{HIGH}/α6^{HIGH} sorted population. Calculate the cell density in this suspension, based on the number of sorted cells recovered.

3.7 Culture of Sorted Epidermal Stem Cells

1. Replace the culture medium used for the feeder cells in Subheading 3.3, step 10 with prewarmed Ca^{2+}-containing KSC medium.

2. Using a hemocytometer, determine the number of sorted viable cells from a small aliquot of trypan-blue stained keratinocytes.

3. Plate the bulge stem cells onto the feeder cultures, at a cell density of 10^4 trypan blue-excluding keratinocytes/2 cm^2.

4. Replace the growth medium every other day. Under these conditions, colonies should become apparent after 7–10 days of culture.

---

4   Notes

1. The serum used to culture epidermal keratinocytes must be batch tested, as not all lots support keratinocyte proliferation and keratinocyte stem cell expansion. Once an appropriate lot is found, it is best to have a sufficient supply for all experiments planned.

2. Primary keratinocytes respond to extracellular Ca^{2+} concentrations ≥0.1 mM by activating their differentiation program. To maintain low extracellular Ca^{2+} concentrations, ions are chelated and removed from FBS using Chelex resin. The chelated FBS is then stored at –20°C in single-use aliquots.

3. Optimal isolation of bulge keratinocyte stem cells is achieved using 50–55-day-old female mice, during the second telogen. Depending on the experiment to be conducted, three to nine mice generally suffice. The ability to identify and isolate YFP-expressing cells during FACS experiments increases substantially if the mice are homozygous for the (*ROSA*)*26Sortm1*$^{(EYFP)Cos}$ locus.

4. Ketamine–xylazine combinations are effective anesthetic agents for mice. Suggested starting doses are 100 mg/kg ketamine and 10 mg/kg xylazine, jointly administered intraperitoneally. Generally, this combination will produce a surgical level of

anesthesia for about 15–30 min. Sedation will last 1–2 h. Follow Standard Operating Procedures from the appropriate Animal Care organization in your institution.

5. About 0.1 g of the 1% (w/w) RU486 mixed with the Neutrogena cream are sufficient to cover the hair-free dorsal skin of a mouse (16). RU486 induces nuclear translocation of Cre/PR fusion proteins, allowing inactivation of loxP-flanked genes. The transgenic line used expresses Cre/PR under the control of the keratin 15 promoter, which is specific to hair follicle bulge keratinocytes (10).

6. The length of time necessary following the last RU486 treatment for targeted cells to express YFP may vary, depending on the strain used, but can be as short as one day.

7. It is critical to maintain an exponentially proliferating cell population. To this end, cultures should never be allowed to reach confluence.

8. In general, the mitomycin C-treated 3T3 cells obtained from one 100-mm culture dish are sufficient to prepare a 48-cm^2 surface (e.g. one 24-well plate) for keratinocyte stem cells.

9. Before removal of fat patches, the tissue will appear pink and glossy, whereas after removal of the adipose tissue, the dermis will appear white and rather dull. Removal of the fat and blood vessels is very important to allow proper penetration into and digestion of the tissue by the trypsin. Proper tissue digestion is essential to obtain a final cell preparation of high quality.

10. The skin can be cut in strips approximately 1 cm wide, which will improve trypsin digestion and cell yield.

11. Scrape the epidermis holding the scalpel perpendicular to the tissue and avoiding excessive force. If the epidermis does not separate from the dermis, rescrape the latter to remove any fat tissue that may still remain and trypsinize for 30–60 min at 37°C.

12. The presence of 1% chelex-treated FBS improves cell viability. Cell suspensions from similar samples can be pooled at this time.

13. Samples containing each about 50,000 cells are sufficient as controls and to set up gating for cell sorting.

14. To isolate sufficient cells for experiments, we suggest labeling at least 1×10^7 cells.

15. The concentration of antibodies should be determined by dilution experiments. We have provided suggested initial concentrations.

16. Samples containing fluorescently labeled antibodies should be kept protected from the light at all times.

17. We have used a Becton Dickinson FACSAria cell sorter, using FACSDiVa software. The sorter is equipped with a 13-mW

Coherent Sapphire solid state 488 nm blue laser and 11 mW JSD Uniphase HeNe 633 nm red laser. The blue laser octagon was used to detect YFP (detector E, 530/30 bandpass filter), phycoerythrin (detector D, 575/40 bandpass filter), and 7-AAD to discard apoptotic/necrotic cells (detector B, 695/20 bandpass filter). The red laser trigon was used to detect Alexa Fluor ™ 649 (detector B or C, 660/20 bandpass filter).

18. When bulge stem cells are isolated from mice in the second telogen, two populations of CD34-high cells are apparent and can be distinguished by their relative levels of α6 integrin. The bulge stem cells will show a high fluorescence signal for both CD34 and α6 integrin.

Acknowledgments

We thank K. Chadwick for expert help with FACS experiments. This work was supported with funding from the Canadian Institutes of Health Research to L.D.

References

1. Rutter N (2000) Clinical consequences of an immature barrier. Semin Neonatol 5:281–287

2. Fuchs E, Raghavan S (2002) Getting under the skin of epidermal morphogenesis. Nat Rev Genet 3:199–209

3. Byrne C, Hardman MJ, Nield K (2003) Covering the limb - formation of the integument. J Anat 202:113–124

4. Alonso L, Fuchs E (2003) Stem cells of the skin epithelium. Proc Natl Acad Sci USA 100:11830–11835

5. Cotsarelis G, Sun TT, Lavker RM (1990) Label-retaining cells reside in the bulge area of pilosebaceous unit: implications for follicular stem cells, hair cycle, and skin carcinogenesis. Cell 61:1329–1337

6. Taylor G, Lehrer MS, Jensen PJ, Sun TT, Lavker RM (2000) Involvement of follicular stem cells in forming not only the follicle, but also the epidermis. Cell 102:451–461

7. Morris RJ, Potten CS (1999) Highly persistent label-retaining cells in the hair follicles of mice and their fate following induction of anagen. J Invest Dermatol 112:470–475

8. Trempus CS, Morris RJ, Bortner CD, Cotsarelis G, Faircloth RS, Reece JM, Tennant RW (2003) Enrichment for living murine keratinocytes from the hair follicle bulge with the cell surface marker CD34. J Invest Dermatol 120:501–511

9. Blanpain C, Lowry WE, Geoghegan A, Polak L, Fuchs E (2004) Self-renewal, multipotency, and the existence of two cell populations within an epithelial stem cell niche. Cell 118:635–648

10. Liu Y, Lyle S, Yang Z, Cotsarelis G (2003) Keratin 15 promoter targets putative epithelial stem cells in the hair follicle bulge. J Invest Dermatol 121:963–968

11. Jensen KB, Driskell RR, Watt FM (2010) Assaying proliferation and differentiation capacity of stem cells using disaggregated adult mouse epidermis. Nat Protoc 5:898–911

12. Nowak JA, Fuchs E (2009) Isolation and culture of epithelial stem cells. Methods Mol Biol 482:215–232

13. Lorz C, Segrelles C, Garin M, Paramio JM (2010) Isolation of adult mouse stem keratinocytes using magnetic cell sorting (MACS). Methods Mol Biol 585:1–11

14. Nakrieko KA, Rudkouskaya A, Irvine TS, D'Souza SJ, Dagnino L (2011) Targeted inactivation of integrin-linked kinase in hair follicle stem cells reveals an important modulatory role in skin repair after injury. Mol Biol Cell 22:2532–2540

15. Kim DJ, Kataoka K, Rao D, Kiguchi K, Cotsarelis G, Digiovanni J (2009) Targeted disruption of stat3 reveals a major role for follicular stem cells in skin tumor initiation. Cancer Res 69:7587–7594

16. Morris RJ, Liu Y, Marles L, Yang Z, Trempus C, Li S, Lin JS, Sawicki JA, Cotsarelis G (2004) Capturing and profiling adult hair follicle stem cells. Nat Biotechnol 22:411–417

Chapter 4

Analysis of Bulge Stem Cells from the Epidermis Using Flow Cytometry

Rehan M. Villani, Mehmet Deniz Akyuz, Michaela T. Niessen, and Carien M. Niessen

Abstract

The epidermis is a multilayered epithelium consisting of multiple different progenitor cell populations, all of which are important to epidermal function. In order to study these populations, several techniques have been developed that enable specific purification of the different progenitor cell populations. The best characterized stem cell population in the epidermis, and likely the most pluripotent, are the quiescent stem cells in the hair follicle bulge. In this chapter, we provide a method for isolating bulge stem cells from skin of adult mice using fluorescence-activated cell sorting of immunofluorescently labeled keratinocytes. We use the cell surface markers CD34 and α6-integrin for the enrichment of bulge stem cells. This method also contains notes on how to adjust the cytometer settings for a reproducible analysis.

Key words Epithelial stem cells, Bulge stem cells, CD34, α6-Integrin, Fluorescence activated cell sorting

---

1   Introduction

Flow cytometry (FC) was originally developed as a method for counting hematopoietic progenitor cells in the blood. Using FC and the closely related Fluorescent-activated cell sorting (FACS), cells can now be sorted into live, pure cell populations which can be cultured or used for further experimental techniques. Using this technique, in combination with methods for isolation of live keratinocytes from skin, the purification of epidermal stem cell populations can be achieved and is routine methodology in a number of laboratories (1–3).

The general technique of FC relies on the detection of single cells traveling past an excitation laser beam. Light deflection occurs when the cell passes through the laser beam enabling the detection of "forward scatter," a measure of cell size, "side scatter," a measure

---

Rehan M. Villani, Mehmet Deniz Akyuz, and Michaela T. Niessen have contributed equally to this work.

Kursad Turksen (ed.), *Skin Stem Cells: Methods and Protocols*, Methods in Molecular Biology, vol. 989,
DOI 10.1007/970-1-62703-330 5_4, © Springer Science+Business Media New York 2013

of the granularity, or one or more fluorescence signals, by detection of specific wavelengths. Cells are labeled with antibodies specific to cell-surface markers each with unique fluorophore conjugates, and the number of positive cells, their size, and characteristics can be measured. Using a combination of nonoverlapping fluorescent tags, the measurement and isolation of single and double labeled populations can be achieved.

In the epidermis, alpha-6 integrin and CD34 are common markers used to identify the epidermal bulge stem cell population (4). However, the techniques we describe here are by no means limited to this population and with the correct controls the protocol can be used to analyze many different epidermal progenitor cell populations. In the following protocol, we describe the basic protocol for measurement and isolation of epidermal bulge stem cells using flow cytometry, using alpha-6 integrin and CD34 markers. The protocol requires the isolation of keratinocytes, followed by labeling with alpha-6 and CD34 specific antibodies and finally the analysis of flow-sorted cell populations. We will include notes regarding the important points for a successful isolation, an accurate and reliable data analysis, and some downstream applications.

2 Materials

2.1 Solutions

All solutions must be prepared in a sterile environment, preferably under a laminar flow hood freshly before starting the experiment.

1. 0.25% Trypsin, without EDTA (GIBCO, Life Technologies, Darmstadt, Germany), cool down to 4°C (Solution A).

2. Dulbecco's Modified Eagle Medium (DMEM) (GIBCO, Life Technologies, Darmstadt, Germany) + 10% Fetal calf serum (FCS) (PAA, Pasching, Austria) chelated with Chelex 100 (Bio-Rad, Munich, Germany) (see Note 1) + Penicillin–Streptomycin (P/S) (Biochrom, Berlin, Germany), cool to 4°C (Solution B).

3. Phosphate buffered saline (PBS) containing 2% FCS (chelated), cool to 4°C (Solution C).

2.2 Instruments and Supplies

1. Sterile 4 in. dissecting forceps.

2. Sterile 4 in. Iris scissors.

3. #20 blade steel scalpel or equivalent.

4. Electric shaver or equivalent.

5. 50-mL conical tubes or equivalent (Greiner Bio One, Frickenhausen, Germany).

6. 40-µM nylon cell strainers (BD Falcon, Franklin Lakes, NJ, USA).

7. 70-µM nylon cell strainers (BD Falcon, Franklin Lakes, NJ, USA).

8. 5-mL round-bottom cap tubes (BD Falcon, Franklin Lakes, NJ, USA).

9. 5-mL cell strainer round-bottom cap tubes (BD Falcon, Franklin Lakes, NJ, USA).

10. Tissue culture dish, 100×20 mm or equivalent (TPP, Trasadingen, Switzerland).

11. 4-color flow cytometer; BD FACSalibur or equivalent (BD Biosciences, Franklin Lakes, NJ, USA).

12. Laminar flow hood or equivalent (Kojair Biowizard Golden Line GL-130, Vilppula, Finland).

13. Dissection pad.

14. 20-g syringe needles.

15. Sterile and disposable 5- and 25-mL tissue culture pipettes (TPP, Trasadingen, Switzerland).

16. Pipetus Pipetboy (Hirschmann Laborgeraete, Ebertstadt, Germany).

17. Centrifuge or equivalent (HERMLE Z 383K, Wehingen, Germany).

18. Vacuum pump.

19. 15 mL conical tube (Greiner Bio One, Frickenhausen, Germany).

20. Shaker or equivalent (Heidolph Duomax 1030, Schwabach, Germany).

21. 1.5-mL tubes.

2.3 Antibodies (see Note 2) and Controls (see Note 3)

1. CD34-ALEXA 647, 510341 (eBiosciences, San Diego, CA, USA).

2. α6-Integrin–PE, 555736 (BD pharminogen, San Diego, CA, USA).

3. 7-AAD (BD pharminogen, San Diego, CA, USA).

4. CD34-ALEXA 647 isotype control, 514321 (eBiosciences, San Diego, CA, USA).

5. α6-Integrin–PE isotype control, 555844 (BD pharminogen, San Diego, CA, USA).

2.4 Mice

1. 21 days or 52–59-day-old mice from any laboratory strain (see Note 4).

3 Methods

3.1 Preparation of Mice

The whole experiment must be carried out in a sterile environment, preferably under a laminar flow hood.

Day 1

1. Sacrifice mice according standard laboratory protocols.

2. Shave the whole corpus of the mice with an electric shaver. Shave the hair against the direction of growth to cut it as short as possible. Do not harm the skin during this procedure.

3. Wash the skin with 70% ethanol to remove any residual hair.

4. Make two dorsolateral incisions with the scissors, one just behind the forepaws and the other one just behind the hindpaws.

5. Peel off the skin carefully using the forceps. Try to remove the skin in one piece.
   In the next step you have the option to choose either 5.1 or 5.2.

   5.1. Place the skin (epidermal side down) on a dissection pad, stretch and fix with needles.

   5.2. Place the skin (epidermal side down) on a 100-mm tissue culture dish, add 5-mL PBS.

6. Gently scratch the dermal side of the skin with a scalpel. Try to get rid of the fat and muscle tissue without damaging the skin. Scratch until you can see the shiny dermis (see Note 5).

7. Cut the scratched skin into two pieces.

8. Fill a 100-mm tissue culture dish with 10-mL 0.25% trypsin, 2 mM EDTA.

9. Float the skin (dermal shiny side down) on the trypsin making sure that the epidermis does not come into contact with trypsin, stays dry and floats unobstructed.

10. Incubate skin overnight at 4°C.

3.2 Isolation of Epidermal Cells

Day 2

Important: The experiment must be continued at 4°C, in a cold room or all solutions and equipment must be kept on ice for the duration of the procedure.

1. Prepare 50-mL conical tubes and put on ice. Scalpels, forceps, 40 and 70-µM nylon cell strainers, 25-mL tissue culture pipettes must also be ready before starting the cell isolation.

2. Take out solution B from the fridge and put on ice.

3. Take out the 100-mm tissue culture dish with the skin floating inside.

4. Put the tissue culture dish on ice and aspirate gently the trypsin with 10-mL pipette.

5. Add 7-mL of Solution B with 2-mM EDTA (see Note 6) to the 100-mm tissue culture dish

6. Use the forceps to hold one edge of the skin and start scratching off the epidermis with a scalpel into the ice cold Solution B. Do not scratch the same area more than three times. Stop scratching when you can see the shiny dermis from the whole epidermal part.

7. Discard dermis.

8. Transfer everything to a 15-mL conical tube.

9. Wash the tissue culture dish with 2-mL Solution B and add this to the rest.

10. Pipette the solution up and down ten times with a 25-mL tissue culture pipette.

11. Rotate slowly at 4°C for 15 min.

12. Pipette solution up and down ten times this time with a 5-mL tissue culture pipette.

13. Filter the solution through a 70-μM nylon cell strainer (see Note 7) into a 50-mL conical tube with a 5-mL tissue culture pipette.

14. Add 2-mL of Solution B to the empty cell strainer to wash off remaining epidermal cells.

15. Filter the solution through a 40-μM nylon cell strainer into a 50-mL conical tube.

16. Transfer the cell suspension in a 15-mL conical tube.

17. Spin at 4°C at 2,000×g for 10 min.

3.3 Staining Epidermal Cells for FC Analysis or Sorting

1. Aspirate supernatant carefully with a vacuum pump from the 15-mL conical tube.

2. For blocking, add 4-mL of Solution C and resuspend cells carefully via shaking, tapping, flipping, and slowly pipetting.

3. Meanwhile, count cells. You should have primarily round-shaped cells, total cell number per mouse should be around 20×10^6 cells.

4. Filter cells again through a 40-μM nylon cell strainer.

5. Label the 5-mL round-bottom cap tubes (FC tubes), transfer the cells (see Table 1). Each mouse should have a separate FC tube. Cells from wildtype mice can be used for technical controls.

6. Spin at 4°C at 2,000×g for 7 min.

7. Meanwhile prepare antibody dilutions for controls and experimental samples (see Table 2). Antibodies need to be diluted in

Table 1
Suggested sample set for set up and experiment

Sample	Antibody	Minimum number of cells (see Note 8)
1	Unstained negative control	2.5×10^4
2	7-AAD	2.5×10^4
3	CD34-ALEXA 647 isotype control	2.5×10^4
4	α6-Integrin–PE isotype control	2.5×10^4
5	CD34-ALEXA 647	2.5×10^4
6	α6-Integrin–PE	2.5×10^4
7	α6-Integrin–PE/7-AAD	2.5×10^4
Experimental samples	CD34-ALEXA 647/α6-integrin–PE/7-AAD	1×10^5

Table 2
Required antibody dilutions for cell sample labeling

Antibody	Dilution
CD34-ALEXA 647	1:20
α6-Integrin–PE	1:25
CD34-ALEXA 647 isotype control	1:20
α6-Integrin–PE isotype control	1:25

50-µL (see Note 9) of solution C per sample, made ready to add to cells.

8. Put the antibody dilutions onto the cells.

9. Put the FC tube on an appropriate rack holder and incubate for 30 min in the dark in either the cold room or on ice. The FC tubes must be carefully rotated every 15 min.

10. Meanwhile prepare a bag with items needed during FC sorting. The necessary items are liquid nitrogen, 5-mL cell strainer round-bottom cap tubes, TRIZOL, 1,000 µL pipettes and tips, Rubber pipette bulb filler, labeled 1.5-mL Eppendorf tubes, ice, solution C.

11. Add 1-mL of solution C and 15-µL 7-AAD to the staining samples.

12. Add 200-µL of solution C and 5-µL 7-AAD to the isotope control samples.

13. Go to the FACS machine.

3.4 FC Sorting or Analysis

Note regarding the calibration of flow cytometer (see Note 10).

1. Set up software ready for cell counting and so that settings can be optimized. You will need to set which settings to record/filter based on your fluorophore emission spectra. Create dot plots of FSC (x axis) vs. SSC (y axis), FSC (x) vs. PI (7-AAD) (y), two plots with ALEXA 647 and PE as the (x) and signal counts (y) and finally a plot with ALEXA 647 (vs. PE).

2. Sort cells with low pressure to increase the viability, filtering cells every half hour.

3. Setup of experiment (see Note 11). First look at unstained control in the FSC/SSC, adjust the detector voltage in order to see a well separated cell population (Fig. 1a) (see Note 12). Gate these cells as P1 (see Note 13). Then using the FSC-1 (x) vs. PI (7-AAD) (y) plot (see Note 14), adjust the PI control so that all the counts are within the plot and gate the negative cells as P2 (Fig. 1b). Then, for each fluorophore conjugate, run the single labeled sample (Samples 5 and 6, Table 1) adjust laser intensity parameters. Gate P3 as positive signals (Fig. 1c), look for distinct peaks and minimize bleed through using the other single labeled sample (Fig. 1d). It is important to validate that the isotype control (Sample 3 and 4, Table 1) does not detect significant signals. Repeat the process with the second fluorophore, gate P4 indicates the positive signals for the second channel (Fig. 1e). All parameters should be less than one hundred. Change voltage of parameters to ensure all cell populations are within the detection limits and load compensation controls (see Note 15).

4. Analysis. On the dot plot SSC-1 (y) vs. FSC-1 (x) designate gate1 P1: your cells of interest (similar to Fig. 1a).

5. Out of Gate 1 make up new dot-plot FSC-1 (x) vs. PI (7-AAD) (y) and gate the negative cells (Gate P2) as indicated in Fig. 1b.

6. Out of Gate P2 make up new dot plot PE-A (x) vs. ALEXA 647-A (y). Create a gate around the PE high/ALEXA 647 high (α6-integrin/CD34 high stem cell) population (Fig. 1f, gate P5). The PE low/ALEXA 647 low population represents the more differentiated keratinocyte populations (Fig. 1f, arrow) and PE low/ALEXA 647 low populations are indicated by gate P6. Record a minimum of 1×10^4 counts (from gate P2) for each experimental sample.

3.5 Notes for Further Techniques Post Counting/Sorting

3.5.1 RT-PCR Analysis of RNA

This can be performed easily using the separated cell populations. To extract RNA, take sorted populations and spin down at 800 rpm for 5 min. Remove supernatant and isolate RNA with trizol and glycogen, purifying RNA with RNAeasy kit (Qiagen, Hilden, Germany). cDNA can then be prepared using superscript III enzyme (Invitrogen, Carlsbad, CA, USA) or quantiTect

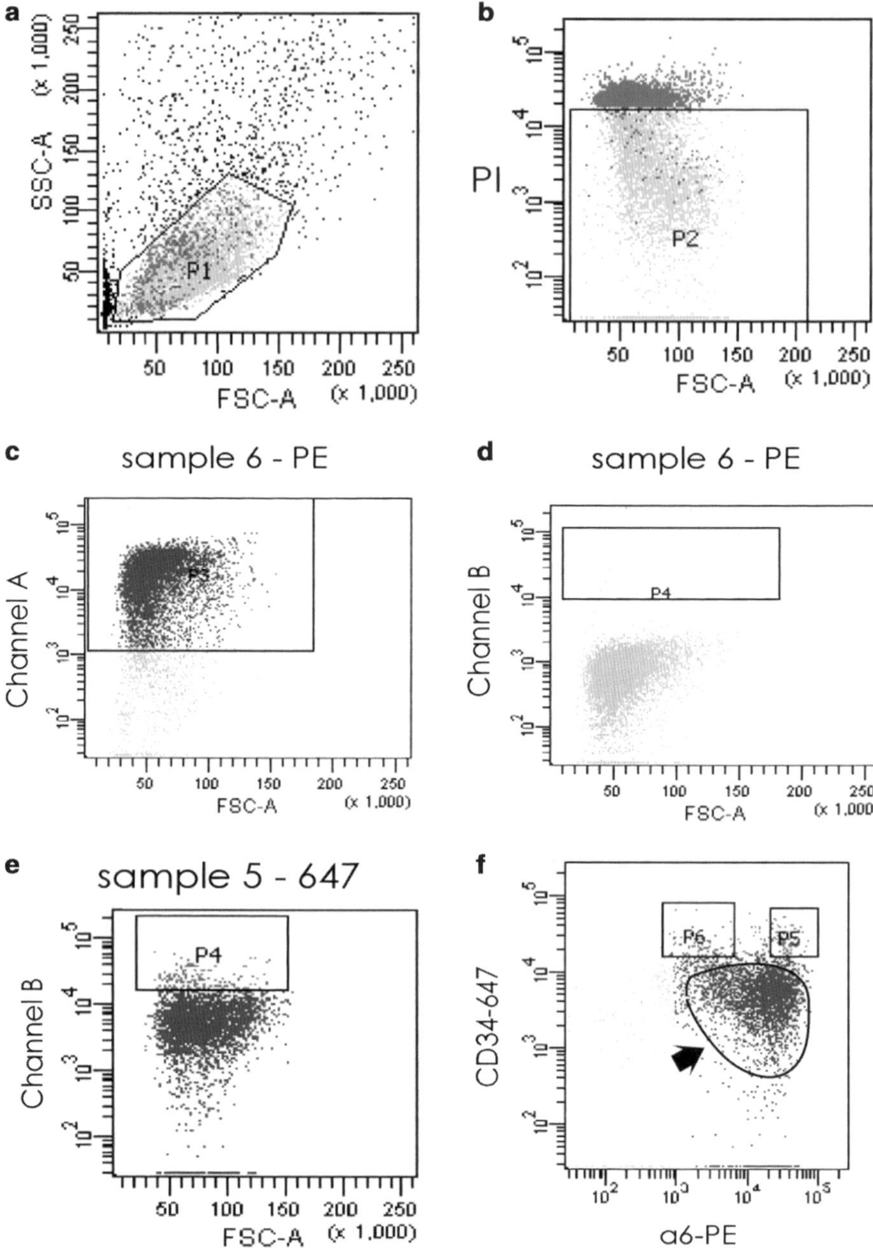

Fig. 1 Example FACS plots for counting keratinocyte populations. (**a**) Dot plot SSC-1 (*y*) vs. FSC-1 (*x*), gate P1 indicates counted events in following analysis. (**b**) FSC-1 (*x*) vs. PI (7-AAD) (*y*) where gate P2 indicates the negative cells. (**c**) FSC-1 vs. Channel A, when PE is Channel A, gate P3 indicates counted PE positive cells. (**d**) FSC-1 vs. Channel B, when PE is Channel A, gate P4 indicates counted PE negative cells. (**e**) FSC-1 vs. Channel B, when 647 is Channel B gate P4 indicates counted 647 positive cells. (**f**) PE-A (*x*) vs. ALEXA 647-A (*y*). Gate P5 indicates PE high/ALEXA 647 high (α6-integrin high/CD34 high stem cell) population and gate P6 indicates the PE low/ALEXA 647 high (α6-integrin low/CD34 high) stem cell population and the *circled* (*arrow*) region indicates the more differentiated keratinocyte populations (α6-integrin low/CD34 high)

cDNA kit (Qiagen, Hilden, Germany) following the manufactures specifications and RT-PCR performed using the Taqman system (Applied biosciences, Carlsbad, CA, USA).

3.5.2 Colony Forming Assay

The proliferative potential of a cell population can be analyzed by plating 1,000 cells in a 60-mm dish, according to standard keratinocyte plating procedures (5). These cells maintained for 1-month post plating will form colonies. Quantification of colony number and the proportion of small, medium, and large size colonies indicate the proliferative potential of the original population (6).

3.5.3 In Vitro Culture

FACS cell populations can also be returned to in vitro culture. Keratinocytes should be plated onto coated plates, as per previously published methods for keratinocyte culture (6).

4 Notes

1. Chelexing the Fetal calf serum is very important to remove calcium, as levels beyond 0.7-mM in media have been shown to induce keratinocyte differentiation and therefore, to maintain cell pluripotentiality calcium must be tightly controlled.

2. Many different antibodies or antibody combinations are possible; however, fluorophores with nonoverlapping emission spectra must be used. The anti-α6-integrin antibody used is conjugated to Phycoerythrin (PE) and the anti-CD34 antibody is conjugated with ALEXA 647. These two conjugates have emission spectra that can be separated by FACS detectors, enabling the separation of the signal and population separation.

3. Isotype controls are very important to validate the base level of fluorescence from nonspecific antibody binding. Isotype controls should be exactly matched, and the controls used here are the same antibody isotype and carry the same fluorescent conjugate as their respective experimental antibody.

4. The age of the skin collected is critical. At postnatal day 21 and 52–59 murine skin should be in telogen phase of the hair cycle, at this stage the hair follicles are much shorter and therefore the epidermis can be isolated from the dermis easier.

5. During scraping of the dermis it is important to remove the fatty layer completely, so that the trypsin can perfuse properly through to the basement membrane.

6. The addition of EDTA to solution B helps to prevent cells clumping.

7. It is important that the cells are filtered regularly during the process to remove cell clumps.

8. The minimum cell numbers provided are a suggestion only. After validating that the required number for each experimental sample has been achieved, the remaining cells/samples can be divided amongst the remaining control labeled samples. It is important that for each labeling experiment at least one sample of each control is included.

9. For the single label and isotype controls, ensure that the corresponding isotype control is made to the same dilution as the experimental antibody. Furthermore, it is best to prepare a master mix for the experimental samples, as this ensures minimal variation between samples (50-µL per sample or mouse).

10. The flow cytometer should be regularly calibrated to check laser alignment, follow the manufacturer's instructions.

11. Compensation set up refers to the procedure that accounts for bleed-through fluorescence signals from one fluorophore channel to another. You can enter compensation settings that will apply to all further readouts, which will automatically deduct the false positive population, as calculated by your single stained compared to your double stained samples.

12. Adjust the voltage settings so that the majority of cells are within the dot plot. You should be able to distinguish the very small signals, in the lower left hand, these are the cell debris. The majority of signals should be concentrated in the center of the dot plot. If there is a group of signals corresponding to a very large SSC/FSC and suggesting large, granular cells, these are likely to be cell doublets or cell clumps resulting from inefficient keratinocyte isolation. These cells should not be included in the gating for analysis.

13. Gating cells refers to marking the population boundary of your cell population on the plot and designating the gate number, referred to as P1, P2, P3, etc. Subsequent plots can then be set to show only the defined gated population or to represent only the cell counts that fall within these gate settings.

14. 7-AAD enables the gating of live cells for analysis. The stain is selectively taken up by dead cells and therefore the 7-AAD negative population represents the live cells.

15. Compensation controls account for bleed-through between fluorophores. By loading the compensation controls, which is easiest performed using the software settings, the signal for each fluorophore will automatically be modified to account for nonspecific signal from the other channel. This is performed by detecting the signal from one fluorophore (such as PE) in the other fluorophore's channel (such as the ALEXA 647 signal in the PE single labeled sample).

Acknowledgments

Work in the laboratory is funded by the DFG SFB829 and SFB832, the German Cancer Aid, and Koeln Fortune.

References

1. Redvers RP, Li A, Kaur P (2006) Side population in adult murine epidermis exhibits phenotypic and functional characteristics of keratinocyte stem cells. Proc Natl Acad Sci U S A 103(35):13168–13173

2. Jensen KB, Driskell RR, Watt FM (2010) Assaying proliferation and differentiation capacity of stem cells using disaggregated adult mouse epidermis. Nat Protoc 5(5):898–911

3. Blanpain C, Lowry WE, Geoghegan A et al (2004) Self-renewal, multipotency, and the existence of two cell populations within an epithelial stem cell niche. Cell 118(5):635–648

4. Trempus CS, Morris RJ, Bortner CD et al (2003) Enrichment for living murine keratinocytes from the hair follicle bulge with the cell surface marker CD34. J Invest Dermatol 120(4):501–511

5. Barrandon Y, Green H (1985) Cell size as a determinant of the clone-forming ability of human keratinocytes. Proc Natl Acad Sci U S A 82(16):5390–5394

6. Morris RJ, Potten CS (1994) Slowly cycling (label-retaining) epidermal cells behave like clonogenic stem cells in vitro. Cell Prolif 27(5):279–289

Chapter 5

Lineage Tracing of Hair Follicle Stem Cells in Epidermal Whole Mounts

Monika Petersson, Daniela Frances, and Catherin Niemann

Abstract

Lineage tracing of tissue stem cells represents a powerful tool to address fundamental questions of development, differentiation and cellular renewal in a natural tissue environment.

The Cre/lox site-specific recombination system is increasingly used to genetically label specific cell populations to perform cell lineage tracing or fate mapping experiments in sophisticated mouse models. Here we describe a method of labeling and subsequent tracking stem cells of the hair follicle bulge region in mouse skin. Hair follicle stem cells are specifically labeled by expressing the Cre recombinase under control of keratin15 (K15) regulatory sequences and by crossing the Cre-containing animals with Cre-sensitive Rosa26R (R26R) reporter mice. To achieve a temporal control of recombinase activity in stem cells, Cre is fused to a modified estrogen receptor (CreER(G)T2). In the K15CreER(G)T2/R26R mouse model, hair follicle stem cells (HFSCs) are specifically labeled after Cre activation upon treatment of mice with tamoxifen. By analyzing the skin tissue at different time points following genetic labeling, important information on stem cell behavior and contribution of labeled stem cells to epidermal structures during tissue homeostasis and hair follicle regeneration are obtained. Combining the lineage tracing approach with the whole mount technique allows examining large areas of the epidermis containing many hair follicles and sebaceous glands and reveals the complex three-dimensional relationship of labeled stem cell clones within the tissue.

Key words Cre recombinase, Epidermis, GFP, Hair follicle, LacZ, Lineage tracing, Skin, Stem cell, Tissue regeneration, Whole mount

1 Introduction

Studying stem cell behavior over time provides important insights into fundamental processes like tissue morphogenesis, renewal, differentiation, cell migration, and aging. A major methodical advantage of recent years was the generation of elegant mouse models using a genetic site-specific recombination technique that is temporally controlled and specifically activated in tissue stem cells (1, 2).

Kursad Turksen (ed.), *Skin Stem Cells: Methods and Protocols*, Methods in Molecular Biology, vol. 989, DOI 10.1007/978-1-62703-330-5_5, © Springer Science+Business Media New York 2013

Initially, individual stem cells are labeled by this genetic approach and the label is faithfully transmitted to all daughters of the initial cell. By analyzing number, location, and phenotypes of the clonally derived cells at increasingly greater times following genetic labeling, the types of cells, their movements, and destiny can be deduced. This approach has proved its value in various recent studies of stem cell populations of different mammalian tissues, including the skin epithelium (3–8). Importantly, the generation and tracing of labeled stem cell clones can be combined with genetic manipulation to facilitate functional analysis of a specific genetic modification in tissue stem cells.

In general, the protocol involves the generation of a transgenic mouse line expressing the Cre recombinase under control of a stem cell-specific promoter or regulatory sequence. To gain the temporal control, Cre is fused to a modified estrogen receptor sequence (ER(G)T2). This modified ligand-binding domain of the human estrogen receptor carries a triple mutation in G400V/M543A/L544A. The CreER(G)T2 fusion protein requires the estrogen receptor ligand tamoxifen for efficient nuclear recombinase activity (9, 10) (for more details, see also ref. 11). To label tissue stem cells, CreER(G)T2 transgenic mice are crossed to a Cre-sensitive R26R reporter strain such as Rosa26RLacZ (or Rosa26EYFP), which ubiquitously expresses the LacZ (or EYFP) gene interrupted by an LoxP site-flanked ("floxed") stop sequence (12, 13) (Fig. 1). Upon treatment of these mice with tamoxifen, Cre recombinase eliminates the stop sequence in a pool of stem cells possessing the activated form of the enzyme. This constitutive recombination results in LacZ (or EYFP) reporter expression specifically in stem cells and subsequently in all their progeny (Fig. 1).

Clearly, the success of this experimental approach relies on the stem cell-specific expression and the tight temporal control of Cre activity under physiological conditions and requires extensive testing for the recombination properties of the particular Cre mouse line in combination with the Cre reporter mouse.

Previously, different stem cell populations have been identified in the epithelium of mammalian skin, including stem cells of the interfollicular epidermis and the bulge, hair germ, isthmus, and junctional zone of the hair follicle (14). The stem cell compartments of the hair follicle have been isolated and further characterized due to the expression of distinct marker molecules. For instance, the best characterized stem cell compartment of the hair follicle bulge is distinguished by expression of keratin15 (K15), Sox9, CD34, Tcf3, Lgr5, Lhx2, and other molecules (15–20). In contrast, stem cells of the isthmus express Lgr6 and the MTS24 antigen Plet1 (21, 22). Several groups have performed sophisticated studies to target CreER(G)T2 to different stem cell populations of the skin by generating Lgr5–IRESGFP–CreER(G)T2 (23), Lgr6–IRESGFP–CreER(G)T2 (21), K14CreER(G)T2 (10), and K15CreER(G)T2 (24) (Fig. 1).

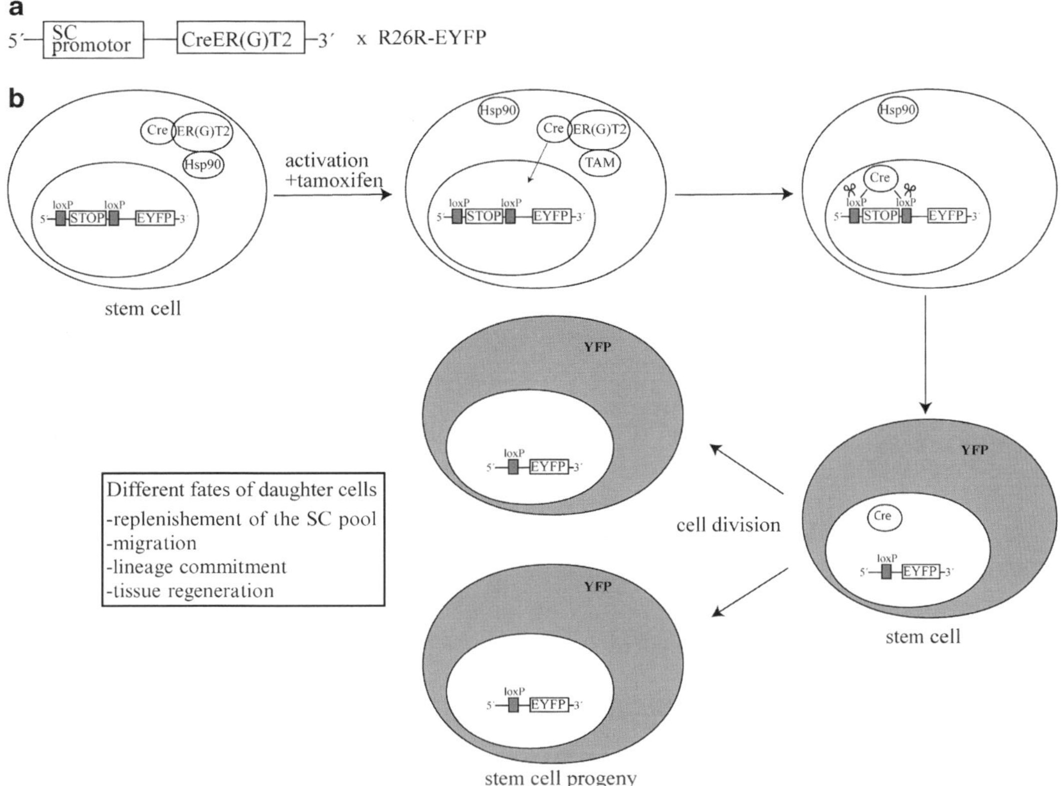

Fig. 1 Tamoxifen-induced Cre-mediated reporter gene activation used for lineage tracing analyses. (**a**) Schema of CreER(G)T2 driven by a stem cell (SC)-specific promotor. In this construct, Cre recombinase is fused to a modified tamoxifen-responsive estrogen-receptor (ER(G)T2). To visualize recombination events and to label stem cells, SC-CreER(G)T2 mice are crossed to Cre-reporter, e.g., R26REYFP mice. (**b**) Cre(G)T2 fusion protein is in complex with heat shock protein 90 (Hsp90) within the cytoplasm of stem cells in SC-CreER(G)T2/R26REYFP mice. SC-CreER(G)T2/R26REYFP mice carry a EYFP sequence downstream of a terminal sequence (STOP) being flanked by two loxP sited (*gray boxes*). Upon treatment, tamoxifen (TAM) binds to ER and subsequently, Cre enzyme translocates to the nucleus. Cre recombinase recognizes loxP sites and excises the STOP cassette which leads to constitutive expression of EYFP. Following division of an YFP positive stem cell, genetic modification is transduced to the daughter cells and YFP is also expressed by stem cell progeny. YFP positive daughter cells can migrate and adopt different fates

Here, we present the lineage tracing approach for stem cells of the hair follicle bulge using the low expressing K15CreER(G)T2 transgenic mouse line crossed with R26R Cre reporter mice (12, 13, 24). We describe methods to detect LacZ and EYFP reporter gene expression following activation of the Cre enzyme by tamoxifen application. In addition, labeled stem cells and their descendants can be analyzed for their proliferative potential and the expression of marker molecules depicting stem cell characteristics or distinct differentiation programs.

In principle, this method can be applied for analyzing other epidermal stem cell compartments, e.g., located to the isthmus or junctional zone of the hair follicle. We provide a robust method to analyze tissue stem cells and their progeny in epidermal whole mounts which allows investigating stem cell behavior in a complex three-dimensional relationship under natural conditions. The protocols give important details for the isolation of epidermal whole mounts from tail and back skin of adult animals as well as neonatal mice (24–26).

2 Materials

2.1 Generation of Reporter Mice for Tracing of Hair Follicle Stem Cells

1. The following mouse lines are required: stem cell-specific Cre line (e.g., K15CreER(G)T2-mice (24) expressing inducible Cre recombinase under the keratin15 (K15) promoter) and R26RLacZ-reporter mice (Jackson strain name B6.129S4-*Gt(ROSA)26Sor*^{tm1Sor}/J) (12) or R26REYFP-reporter mice (Jackson strain name B6.129X1-*Gt(ROSA)26Sor*$^{tm1(EYFP)Cos}$/J) (13). All mouse strains are maintained on the same genetic background, e.g., C57/Bl6 (see Notes 1 and 2).

2.2 Activation of Cre Recombinase In Vivo

1. Syringes (1 ml syringes with 100 µl graduation).

2. Needle (not too fine because of viscosity of sunflower oil, 0.6×25 mm, BD Microlance™3).

3. Sunflower oil (Sigma).

4. Tamoxifen (Sigma, catalogue number T5648).

2.3 Isolation of Epidermal Whole Mounts from Mouse Skin

1. Electric clipper and dissecting tools: forceps, scissor, and scalpel.

2. Petri dishes for tissue preparation, 15 ml falcon tubes (for incubation of skin samples), and 24-well plates for fixation, cardboard.

3. Phosphate buffered saline (PBS, Dulbecco without Ca^{2+} and Mg^{2+}).

4. For the separation of the epidermis from the underlying dermal tissue, the following solutions are required: 5 mM EDTA pH 8.0 in PBS for preparation of tail skin and 20 mM EDTA pH 8.0 in PBS for back skin samples. The thermolysin (Sigma) solution is prepared freshly and is diluted in PBS to give a final concentration of 0.25 mg/ml.

5. Water bath for incubation steps.

6. Fixatives:

    (a) For immunofluorescence staining: 3.4% formaldehyde in PBS (v/v) (formalin).

    (b) For the detection of the reporter gene: 0.2% glutaraldehyde/2% formaldehyde in PBS (v/v), prepare freshly and use within 1 week, store at 4°C.

2.4 Detection of Reporter Gene Activation in Epidermal Whole Mounts

1. 96-well plates are used for the staining procedure.

2. See materials for immunofluorescence staining in Subheading 2.6 (see Tables 2 and 3).

2.4.1 Detection of YFP Positive Cells in Epidermal Whole Mounts

2.4.2 Detection of LacZ in Epidermal Sheaths

1. Potassium phosphate buffer (KPP), 0.5 M K_2HPO_4/KH_2PO_4 pH 7.4.

2. Washing buffer, 0.05 M ethylene glycol tetraacetic acid (EGTA), 0.02% NP-40 in 0.1 M KPP pH 7.4.

3. 5-Bromo-4-chloro-indolyl-β-D-galactopyranoside (X-Gal, Sigma).

4. X-Gal solution, 0.5 mg/ml X-Gal, 10 mM $K_3(Fe(27)_6)$, 10 mM $K_4(Fe(27)_6)$, 0.02% NP-40 in 0.1 M KPP, pH 7.4. Needs to be prepared freshly.

5. Mayer's haematoxylin.

2.4.3 Detection of the Cre Recombinase Enzyme in Epidermal Sheaths

1. 96-well plates for the staining procedure.

2. See materials for immunofluorescence staining in Subheading 2.6 (see Tables 2 and 3).

2.5 Cell Fate Mapping of Labeled Stem Cells in Epidermal Whole Mounts

1. For collecting tissue samples, see Subheading 2.3.

2. For LacZ detection use a light microscope.

3. For YFP detection, use confocal microscope and software for generating and analyzing Z-projections.

2.6 Analysis of Epidermal Whole Mounts by Immunofluorescent Stainings

1. 96-well plates are used for the staining procedure.

2. Whole mount buffer (WMB), 20 mM Hepes pH 7.2, 0.9% NaCl. Store the buffer at 4°C.

3. Blocking solution, 0.5% milk powder, 0.25% fish skin gelatin, 0.5% Triton X-100 in WMB. Prepare this solution freshly and use no longer than 2 days. Store at 4°C.

4. Washing buffer for whole mounts, 0.2% Tween20 in PBS.

5. Distilled water.

6. To embed the epidermal whole mounts, use glass slides (Histobond), cover slips, and Mowiol/Dabco, 4.8 g Mowiol, 12 g Glycerin, 12 ml H_2O stir 3–4 h, add 24 ml 0.2 M Tris–HCl pH 8.0, heat to 50°C and stir overnight, add 1.2 g Dabco (2.5%), dissolve, and centrifuge. This solution can be stored at −20°C for several months.

7. Primary and secondary antibodies (see Tables 2 and 3).

2.7 Analysis of Cell Proliferation and EdU-Pulse Chase Experiments

1. For EdU or BrdU injections: syringes (1 ml syringes with 100 µl graduation).

2. 96-well plates are used for the staining procedure.

3. EdU Click-it Kit (Invitrogen).

4. 1% Triton X-100 in PBS for the permeabilization of the tissue.

3 Methods

3.1 Generation of Reporter Mice for the Tracing of Hair Follicle Stem Cells

The invention of inducible Cre/loxP systems has paved the way to label tissue cells and to follow their fate over a certain period of time. A combination of lineage tracing experiments and genetic manipulation, such as cell type-specific deletion of a gene is feasible. However, recombination events represented by reporter gene expression do not necessarily correspond to those of the floxed target gene.

Irrespective of the lineage tracing or genetic manipulation, different doses and durations of tamoxifen treatment have to be assayed to find the optimal protocol to activate Cre recombinase specifically. Moreover, it is essential to treat genotype matched littermate controls with vehicle (oil) to validate specificity of stem cell labeling.

3.2 Activation of Cre Recombinase In Vivo

1. Dissolve tamoxifen (4 mg/200 µl for one dose) in sunflower oil by vigorous shaking for 1 h.

   Prepare this solution freshly and do not heat to facilitate dissolution. If crystalline residues are visible, it is recommended to continue vortexing rigidly.

2. In order to label individual bulge HFSCs, apply a single dose of tamoxifen intraperitoneally (i.p.) (see Note 3).

   Depending on the aim of the study, K15CreER(G)T2/R26R mice can be treated at a distinct stage of the hair regeneration cycle (Fig. 2a). For instance, to label K15 positive bulge cells during the resting phase of the hair cycle, mice are treated at P19-P21, when the hair cycle is synchronized (see Note 4). As a control, treat mice of same age and sex with vehicle only (oil).

3. Cre-mediated recombination can be monitored 2–3 days following tamoxifen i.p. injection.

3.3 Isolation of Epidermal Whole Mounts from Mouse Skin

3.3.1 Epidermal Whole Mounts Derived from Tail Skin

1. Sacrifice mice and cut off the tail (see Notes 5 and 6).

2. Fix tail with forceps and slit tail on ventral side lengthwise by using a scalpel (see Fig. 3a and Note 7).

3. Separate skin from the bone at the anterior part of the tail. Grab bone with forceps and peel off the skin along the whole of the tail (Fig. 3b, c).

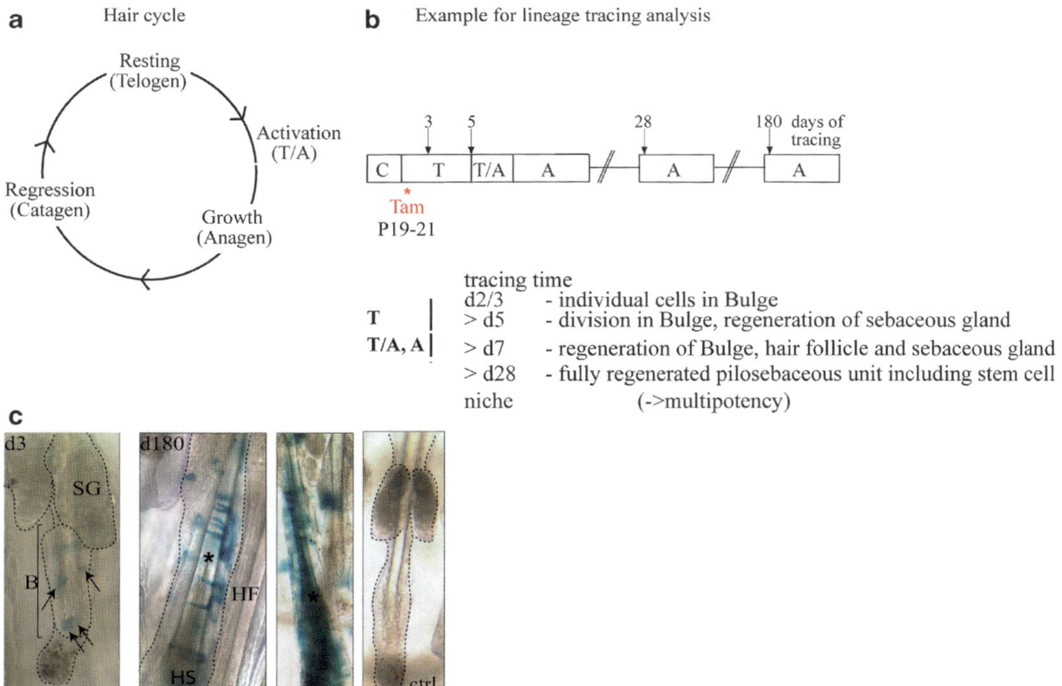

Fig. 2 Lineage tracing of epidermal stem cell progeny during hair regeneration. (**a**) For renewal, hair follicles undergo cycles of resting, growth, and regression. Note that stem cells are more quiescent and become activated and mobilized during distinct phases of the cycle. (**b**) Example of a lineage tracing analysis throughout the hair cycle and the expected compartments to be replenished by stem cell progeny following a certain tracing time. (**c**) LacZ-stained epidermal whole mounts derived from tamoxifen treated K15CreER(G)T2/R26RLacZ mice. Individual stem cells can be labeled in the hair follicle bulge by 3 days of chase. Fully regenerated LacZ positive HF can be detected, e.g., 180 days following Cre activation. Oil-treated littermate control mice remain negative for LacZ. B = bulge, SG = sebaceous gland, HF = hair follicle

4. Stretch skin tissue on a piece of cardboard with dermal site down. Cut the tail skin in pieces of about 5 mm in length using a scalpel (see Fig. 3d and Note 8).

5. Transfer pieces of skin tissue into falcon tubes and incubate with 10 ml 5 mM EDTA/PBS at 37°C in a water bath (see Note 9 and Table 1).

6. Retrieve piece of skin from the EDTA solution and put the dermal site down onto a plastic dish (see Fig. 3e and Note 10).

7. Gently peel off the epidermis as sheet from the underlying dermal tissue with the help of two fine bended forceps (Fig. 3e, f).

8. Incubate epidermal sheaths with formaline in 24-well plate (see Note 11) to fix epidermal tissue for 1.5 h at room temperature (RT). For detection of YFP or LacZ expressing cells, proceed to step 1 of Subheadings 3.4.1 or 3.4.2, accordingly.

9. Transfer epidermal whole mounts into 24-well plate filled with PBS and store the tissue at 4°C (see Note 12).

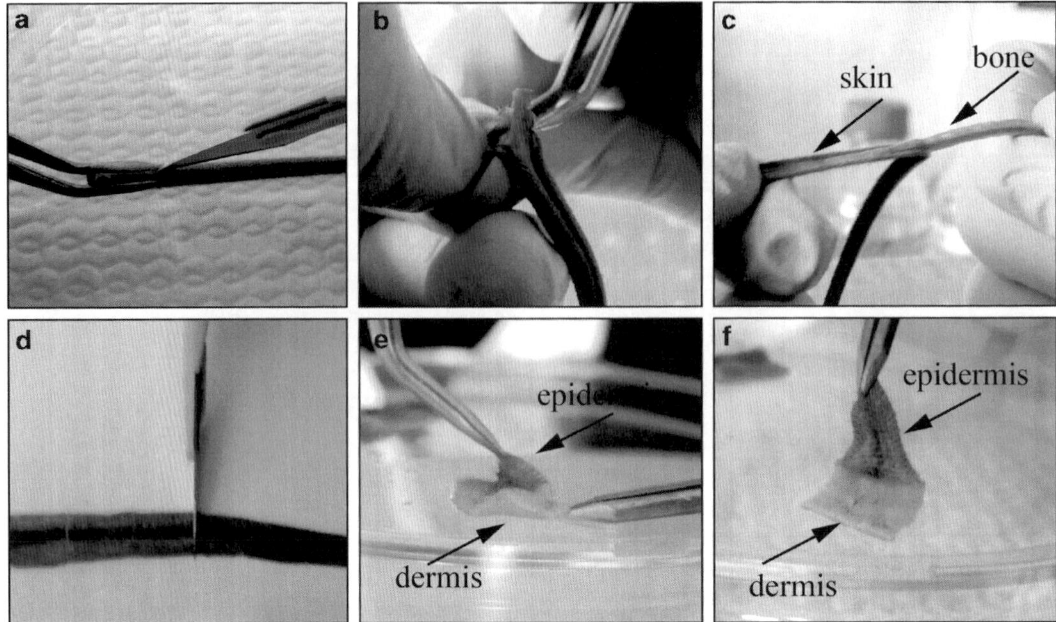

Fig. 3 Isolation steps of epidermal whole mounts from mouse tail skin. (**a**) The tail is fixed with forceps and slit open on the ventral side with a scalpel. (**b**) The bone is separated from the skin at the anterior part of the tail. (**c**) Skin is peeled of the bone. (**d**) Skin tissue is stretched on cardboard and cut in pieces using a scalpel. (**e, f**) For isolation of epidermal sheaths, dermis is fixed with one forceps and epidermis grabbed with another forceps. Epidermis is gently peeled off the dermis

Table 1
Isolation of epidermal whole mounts at different time points of mouse development

Age of mice	Skin tissue	Tissue separation	Incubation time (at 37°C) (h)
E17-P2	Tail	5 mM EDTA	1
	Back	20 mM EDTA	1.5
P3-P5	Tail	5 mM EDTA	1–2
	Back	20 mM EDTA	2
P6-P16	Tail	5 mM EDTA	2
	Back	20 mM EDTA	2–3
>P16	Tail	5 mM EDTA	3–4
	Back	Thermolysin 0.25 mg/ml	1

3.3.2 Whole Mounts from Back Skin	1. Shave the back of the mouse with electric clipper before harvesting skin tissue. Depending on the age of the mice (>P3), fat tissue has to be removed prior EDTA digest. Therefore, put skin tissue with epidermal side down into a Petri dish containing PBS and remove fat tissue rigidly (skin becomes translucent) (see Note 13).

2. Stretch skin on a cardboard with dermal site down.

3. Cut the skin into several small pieces rectangularly (approx. 0.5–1 cm^2).

   (a) Incubate skin in 20 mM EDTA/PBS at 37°C (see Table 1).

   (b) Float whole adult skin in thermolysin (0.25 mg/ml in PBS) and digest the tissue for 1 h at 37°C.

4. Proceed as described in step 7 Subheading 3.3.1.

3.4 Detection of Reporter Gene Activation in Epidermal Whole Mounts

3.4.1 Detection of YFP Positive Cells in Epidermal Whole Mounts

1. Fix epidermal whole mounts derived from AK15CreER(G)T2/R26REYFP mice in 0.2% glutaraldehyde/2% formaldehyde for 1.5 h at RT.

2. Store in PBS at 4°C for minimum of 2 h before staining (see Note 12).

3. In order to enhance the YFP signal within the cells, stain epidermal sheaths with an antibody against GFP (see Table 2). Apply the protocol as described in Subheading 3.6 (see Notes 14 and 15).

3.4.2 Detection of LacZ in Epidermal Sheaths

1. Fix epidermal whole mounts derived from K15CreER(G)T2/R26RLacZ mice in 0.2% glutaraldehyde/2% formaldehyde for 1.5 h at RT.

2. Store in PBS at 4°C for minimum of 2 h before staining (see Note 12).

3. Incubate epidermal whole mounts for 30 min in washing buffer.

4. Stain tissue samples with X-Gal solution for 16 h at RT in a 24-well plate under gentle agitation (see Note 16).

5. Wash o/n at RT in washing buffer under gentle agitation.

6. Perform the counterstaining of the stained whole mounts with haematoxylin. Stain whole mounts 1–5 min depending on the dilution of the haematoxylin. Wash whole mounts with H$_2$O for 5 min and finally, wash with dH$_2$O for 5 min (see Note 17).

7. Mount whole mounts on glass slides with Mowiol/Dabco and glass coverslip (see Note 18). Store at 4°C.

8. Immediate evaluation of the stainings is recommended due to diffusion of the dye.

Table 2
Primary antibodies for marker detection in epidermal whole mounts

Antigen	Host	Dilution	Reference/company
Cre	Rabbit	1:750	Covance
GFP	Chicken	1:4,000	Abcam
GFP-488	Goat	1:500	Rockland
GFP	Rabbit	1:2,500	Molecular Probes
Lrig1	Goat	1:100	R&D Systems
MTS24	Rat	1:50	(28)
Plet1	Rat	1:3	(29)
K15	Mouse	1:1,500	Neomarkers
CD34	Rat	1:25	E-Biosciences
Sox9	Rabbit	1:3,000	(30)

3.4.3 Detection of the
Cre Recombinase Enzyme
in Epidermal Sheaths

1. Fix epidermal whole mounts derived from K15CreER(G)T2 mice in 0.2% glutaraldehyde/2% formaldehyde for 1.5 h at RT.

2. Store in PBS at 4°C for minimum 2 h before staining (see Note 12).

3. Apply the staining protocol and dilute Cre antibody according to the method provided in Subheading 3.6 (see Table 2).

4. Active Cre recombinase following tamoxifen treatment is detected in the nucleus of Cre expressing cells. In the vehicle controls, the Cre enzyme is seen diffuse cytoplasmic (see Fig. 1 and Note 19).

3.5   Cell Fate Mapping of Labeled Stem Cells in Epidermal Whole Mounts

1. Collect tissue samples (for the isolation of epidermal whole mounts, see Subheading 3.3) at different time points following tamoxifen treatment. Exemplary, to test the contribution of HFSCs for different hair follicle lineages, activation of the Cre recombinase is required before anagen growth phase of the hair cycle and the tissue should be analyzed at the end of anagen phase at the earliest (Fig. 2b, c).

2. To analyze labeled HFSCs and contribution of all their descendants to different cell lineages within the epidermal tissue, labeled cells can be visualized by staining for LacZ, EYFP, or nuclear Cre as described in Subheading 3.3.

3. To evaluate localization and differentiation of stem cell-derived progeny, co-immunofluorescent studies with different marker molecules can be performed (see Subheading 3.6, Tables 2 and 3 and see Note 20).

Table 3
Secondary antibodies for whole mount immunostaining

Species	Conjugate	Dilution	Reference/company
Anti-chicken	Alexa488	1:500	Molecular probes
Anti-goat	Alexa488, Alexa594, Cy5	1:500	Molecular probes
Anti-mouse (IgG2A)	Alexa488	1:500	Molecular probes
Anti-rat	Alexa488, 594, Cy5	1:400	Molecular probes
Anti-rabbit	Alexa488, Alexa594, Cy5	1:500	Molecular probes

3.6 Analysis of Epidermal Whole Mounts by Immunofluorescent Staining

1. Incubate epidermal whole mounts with blocking solution for 1 h. All steps are carried out at RT under gentle agitation.

2. Apply primary antibody at the appropriate dilution o/n (see Note 21 and Table 2).

3. Wash epidermal whole mounts with 0.2% PBS-Tween20 for 45 min. Repeat washing step 3–4 times.

4. Add secondary antibody in blocking solution o/n and protect the whole mounts (plates) from light (see Note 22 and Table 3).

5. Wash epidermal whole mounts with 0.2% PBS-Tween20 for 45 min. Repeat washing step 3–4 times (see Note 23).

6. For the staining of nuclei, apply DAPI (1 μg/ml) for 30 min at RT.

7. Wash epidermal whole mounts with 0.2% PBS-Tween20 for 15 min. Repeat the washing 2–3 times.

8. Rinse with distilled water.

9. Mount with Mowiol/Dabco (see Note 18). Store the samples at 4°C.

3.7 Analysis of Cell Proliferation and EdU-Pulse Chase Experiments

The in vivo labeling with the thymidine analogue EdU allows detection of cells in the S-phase of the cell cycle under physiological conditions. Furthermore, pulse chase experiments following EdU application provide an insight into the proliferation rate of labeled cells and their progeny. In addition, the cell fate of an EdU positive proliferative cell can be monitored until EdU is diluted from the tissue due to high proliferative potential of the cells (see Note 24).

1. Administrate EdU 40 mg/kg body weight by i.p. injection.

2. Isolate epidermal whole-mounts at time point of interest (see Subheadings 3.3.1 or 3.3.2).

For analysis of proliferation of epidermal cells, sacrifice mice 1–3 h following EdU administration. For EdU-pulse chase experiments and tracing of cell fate of EdU labeled cells, epidermal whole mounts are isolated at a later time point depending on the aim of the study (see Note 25).

3. Use EdU Click-it Kit according to the manufacturer's instructions with slight modifications (see Note 26).

4 Notes

1. It is recommended to keep the Cre reporter mice homozygous to assure maximal visualization of positive recombinants.

2. Mouse experiments should be carried out following institutional and ethical guidelines and according to the animal license given by the State Office.

3. Alternatively, a maximal dose of 3.5 mg can be administered for two or more consecutive days to increase the number of labeled HFSCs. Topical and oral applications (as food supplement) of tamoxifen are also possible. However, i.p. injections seem to be more efficient than topical application of the same dose.

4. Hair cycling may be altered in males and females and differ between different genetic backgrounds and mouse strains.

5. Take a piece of tissue (for instance of the limb or ear) for genotyping if required.

6. The earliest time point for successful isolation of epidermal whole mounts from mouse tail skin is E17.5.

7. For preparation of whole mounts from neonatal mice, it is recommended to fix the whole tail between thumb and index finger and carefully slit the skin open on the ventral side of the tail.

8. For isolation of whole mounts from mice up to P8, it is not imperatively necessary to stretch the tissue on a cardboard and to cut tail skin in pieces before incubation in EDTA solution.

9. Note that incubation time varies dependent on the age and strain of the mice. For reference time points, see Table 1. The incubation of tail skin from neonatal mice up to P8 can be done in 2 ml Eppendorf reaction cups on a 37°C rotating shaker.

10. To recognize dermal site: skin tends to fold towards dermal site.

11. When transferring epidermal sheaths into fixation solution make sure that the samples are plane and unfolded.

12. Epidermal whole-mounts can be stored for at least 2 months at 4°C. PBS needs to be refilled regularly. For long-term storage, epidermal sheaths can be placed into 0.2% sodium azide in PBS. However, be aware that sodium azide is not compatible with some antibody staining procedures.

13. Take into account that during the growth phase of hair regeneration cycle the isolation of epidermal sheaths from back skin is not feasible due to high density and length of hair follicles. The earliest time point for successful isolation of whole mounts from back skin is E17.5.

14. Depending on the batch of the antibody and isolation procedure, a background autofluorescence can be observed occasionally.

15. YFP expression can additionally be examined by conducting quantitative PCR analyses using RT^2 RT SYBR Green qPCR Master Mix (SuperArray Bioscience Cooperation, USA) and a StepOnePlus real time PCR system (Applied Biosystems). Use YFPfor 5′-gtggtgcccatcctggtcga-3′ and YFPrev 5′-taggc-cgaaggtggtcacgagg-3′ oligonucleotides and apply the following PCR conditions: 10 min 95°C initial denaturation, cyclic denaturation at 95°C for 15 s followed by an annealing step at 60°C for 1 min.

16. Incubation time may vary between different mouse strains. Check regularly for LacZ positive cells under the light microscope. Depending on the incubation time, crystalline residues become visible in the hair shaft. Always stain a vehicle-treated control in parallel to validate specificity of the LacZ signal.

17. Optimize haematoxylin staining procedure to obtain a light counterstain. Test different dilutions and incubation times.

18. For mounting, place epidermal surface onto the glass slide (the hair follicles should point to the cover slip for further analysis by microscopy). For epidermal tissue from newborn or young animals, it can be difficult to distinguish the epidermal and dermal site. Therefore, check orientation through the light microscope before mounting.

19. In addition, DNA recombination in AK15CreER/R26REYFP mice can be detected by conducting conventional semi-quantitative PCR. Employ 20 ng DNA template for PCR-reaction using HotStarTaqPlus DNA Polymerase (Qiagen, Germany) and following oligonucleotides: Recfor 5′-ggttgaggacaaactct-tcgcggt-3′ and Recrev 5′-taggccgaaggtggtcacgagg-3′. To validate for equal amounts of template DNA, run GapDH PCR (mGapDHfor 5′-acctttggcattgtggaagg-3′, mGapDHrev 5′-aca-cattggggggtaggaaca-3′). Apply following PCR conditions: 5 min, 94°C initial denaturation, 30 s, 94°C cyclic denaturation, 60 s cyclic annealing at the 61°C; 45 s, 72°C, cyclic elongation for a

total of 35 cycles followed by a 72°C elongation step for 5 min. All PCR reactions are carried out either in T-Gradient Thermocycler or Personal Thermocycler (Biometra). Analyze PCR products by agarose gel electrophoresis. If recombination occurred, a fragment of approximately 410 bp will be amplified.

Detailed analysis for recombination events or YFP expression in a particular (stem) cell population can be performed, if keratinocytes were sorted for specific markers using FACS prior to DNA isolation and PCR reaction.

20. The statistical analysis of number and distribution of LacZ positive stem cell clones in epidermal whole-mounts of tamoxifen-treated K15CreER(G)T2/R26RLacZ mice and control animals is highly recommended. Evaluate hair follicles of biological triplicates for each tracing time point and classify LacZ positive cells concerning their location within the pilosebaceous unit. Calculate standard deviation and test for significance. Employ student's t-test to determine the p-value.

21. All incubation steps of the immunofluorescent staining protocol are performed in 96-well plates using a minimal volume of 100 µl per well.

22. For some immunofluorescent studies, it might be sufficient to apply secondary antibodies for 3–4 h.

23. For staining with more than one primary antibody, apply next primary antibody after extensively washing the whole mounts and proceed with step 2. If the double staining does not work, it might be worthwhile to change the order of the antibodies. For nuclear antibody staining, treat epidermal sheath with 2N HCl for 10 min at 37°C. Prewarm HCl and be sure to rinse carefully with 0.2% Tween20 in PBS.

24. Other thymidine analogues, e.g., BrdU can be used. However, EdU can be applied in a lower concentration and is easier to detect due to Click-it reaction avoiding HCl treatment.

25. Number of dividing cells and proliferation rate within the tissue are dependent on stage of the hair cycle and age of the mice.

26. Permeabilize whole mounts with 1% Triton X-100 in PBS for 1 h. The duration of the Click-it reaction is 30 min. Wash accurately with 0.2% Tween20 in PBS o/n. If required, co-immunofluorescent staining with antibodies can be performed (see Subheading 3.6) but should be done prior EdU Click-it reaction.

References

1. Fox DT, Morris LX, Nystul T, Spradling AC (2008) Lineage analysis of stem cells in stem book. Cambridge: Harvard Stem Cell Institute; 2008–2009

2. Feil R, Wagner J, Metzger D, Chambon P (1997) Regulation of Cre recombinase activity by mutated estrogen receptor ligand-binding domains. Biochem Biophys Res Commun 237(3):752–757

3. Barker N, Huch M, Kujala P, van de Wetering M, Snippert HJ, van Es JH, Sato T, Stange DE, Begthel H, van den Born M, Danenberg E, van den Brink S, Korving J, Abo A, Peters PJ, Wright N, Poulsom R, Clevers H (2010) Lgr5(+ve) stem cells drive self-renewal in the stomach and build long-lived gastric units in vitro. Cell Stem Cell 6(1):25–36

4. Clayton E, Doupe DP, Klein AM, Winton DJ, Simons BD, Jones PH (2007) A single type of progenitor cell maintains normal epidermis. Nature 446(7132):185–189

5. Morris RJ, Liu Y, Marles L, Yang Z, Trempus C, Li S, Lin JS, Sawicki JA, Cotsarelis G (2004) Capturing and profiling adult hair follicle stem cells. Nat Biotechnol 22(4):411–417

6. Livet J, Weissman TA, Kang H, Draft RW, Lu J, Bennis RA, Sanes JR, Lichtman JW (2007) Transgenic strategies for combinatorial expression of fluorescent proteins in the nervous system. Nature 450(7166):56–62

7. Chen J, Kwon CH, Lin L, Li Y, Parada LF (2009) Inducible site-specific recombination in neural stem/progenitor cells. Genesis 47(2):122–131

8. Brack AS, Conboy MJ, Roy S, Lee M, Kuo CJ, Keller C, Rando TA (2007) Increased Wnt signaling during aging alters muscle stem cell fate and increases fibrosis. Science 317(5839):807–810

9. Forde A, Constien R, Grone HJ, Hammerling G, Arnold B (2002) Temporal Cre-mediated recombination exclusively in endothelial cells using Tie2 regulatory elements. Genesis 33(4):191–197

10. Indra AK, Warot X, Brocard J, Bornert JM, Xiao JH, Chambon P, Metzger D (1999) Temporally-controlled site-specific mutagenesis in the basal layer of the epidermis: comparison of the recombinase activity of the tamoxifen-inducible Cre-ER(T) and Cre-ER(T2) recombinases. Nucleic Acids Res 27(22):4324–4327

11. Feil S, Valtcheva N, Feil R (2009) Inducible Cre mice. Methods Mol Biol 530:343–363

12. Soriano P (1999) Generalized lacZ expression with the ROSA26 Cre reporter strain. Nat Genet 21(1):70–71

13. Srinivas S, Watanabe T, Lin CS, William CM, Tanabe Y, Jessell TM, Costantini F (2001) Cre reporter strains produced by targeted insertion of EYFP and ECFP into the ROSA26 locus. BMC Dev Biol 1:4

14. Jaks V, Kasper M, Toftgard R (2010) The hair follicle-a stem cell zoo. Exp Cell Res 316(8):1422–1428

15. Liu Y, Lyle S, Yang Z, Cotsarelis G (2003) Keratin 15 promoter targets putative epithelial stem cells in the hair follicle bulge. J Invest Dermatol 121(5):963–968

16. Merrill BJ, Gat U, DasGupta R, Fuchs E (2001) Tcf3 and Lef1 regulate lineage differentiation of multipotent stem cells in skin. Genes Dev 15(13):1688–1705

17. Nguyen H, Rendl M, Fuchs E (2006) Tcf3 governs stem cell features and represses cell fate determination in skin. Cell 127(1):171–183

18. Nowak JA, Polak L, Pasolli HA, Fuchs E (2008) Hair follicle stem cells are specified and function in early skin morphogenesis. Cell Stem Cell 3(1):33–43

19. Rhee H, Polak L, Fuchs E (2006) Lhx2 maintains stem cell character in hair follicles. Science 312(5782):1946–1949

20. Trempus CS, Morris RJ, Bortner CD, Cotsarelis G, Faircloth RS, Reece JM, Tennant RW (2003) Enrichment for living murine keratinocytes from the hair follicle bulge with the cell surface marker CD34. J Invest Dermatol 120(4):501–511

21. Snippert HJ, Haegebarth A, Kasper M, Jaks V, van Es JH, Barker N, van de Wetering M, van den Born M, Begthel H, Vries RG, Stange DE, Toftgard R, Clevers H (2010) Lgr6 marks stem cells in the hair follicle that generate all cell lineages of the skin. Science 327(5971):1385–1389

22. Nijhof JG, Braun KM, Giangreco A, van Pelt C, Kawamoto H, Boyd RL, Willemze R, Mullenders LH, Watt FM, de Gruijl FR, van Ewijk W (2006) The cell-surface marker MTS24 identifies a novel population of follicular keratinocytes with characteristics of progenitor cells. Development 133(15):3027–3037

23. Jaks V, Barker N, Kasper M, van Es JH, Snippert HJ, Clevers H, Toftgard R (2008) Lgr5 marks cycling, yet long-lived, hair follicle stem cells. Nat Genet 40(11):1291–1299

24. Petersson M, Brylka H, Kraus A, John S, Rappl G, Schettina P, Niemann C (2011) TCF/Lef1 activity controls establishment of diverse stem and progenitor cell compartments in mouse epidermis. EMBO J 30(15):3004–3018

25. Braun KM, Niemann C, Jensen UB, Sundberg JP, Silva-Vargas V, Watt FM (2003) Manipulation of stem cell proliferation and lineage commitment: visualisation of label-retaining cells in wholemounts of mouse epidermis. Development 130(21):5241–5255

26. Frances D, Niemann C (2012) Stem cell dynamics in sebaceous gland morphogenesis in mouse skin. Dev Biol 363(1):138–146

27. Chan EF, Gat U, McNiff JM, Fuchs E (1999) A common human skin tumour is caused by activating mutations in beta-catenin. Nat Genet 21(4):410–413

28. Gill J, Malin M, Hollander GA, Boyd R (2002) Generation of a complete thymic microenvironment by MTS24(+) thymic epithelial cells. Nat Immunol 3(7):635–642

29. Raymond K, Richter A, Kreft M, Frijns E, Janssen H, Slijper M, Praetzel-Wunder S, Langbein L, Sonnenberg A (2010) Expression of the orphan protein Plet-1 during trichilemmal differentiation of anagen hair follicles. J Invest Dermatol 130(6):1500–1513

30. Stolt CC, Lommes P, Sock E, Chaboissier MC, Schedl A, Wegner M (2003) The Sox9 transcription factor determines glial fate choice in the developing spinal cord. Genes Dev 17(13):1677–1689

Chapter 6

Isolation and Characterization of Cutaneous Epithelial Stem Cells

Uffe B. Jensen, Soosan Ghazizadeh, and David M. Owens

Abstract

During homeostasis, adult mammalian skin turnover is maintained by a number of multipotent and unipotent epithelial progenitors located either in the epidermis, hair follicle, or sebaceous gland. Recent work has illustrated that these various progenitor populations reside in regionalized niches and are phenotypically distinct from one another. This degree of heterogeneity within the progenitor cell landscape in the cutaneous epithelium complicates our ability to target, purify, and manipulate cutaneous epithelial stem cell subpopulations in adult skin. The techniques outlined in this chapter describe basic procedures for the isolation and purification of murine epithelial progenitors and assessing their capacity for ex vivo propagation.

Key words Skin, Epidermis, Hair follicle, Keratinocyte, Tissue regeneration

1   Introduction

The perpetual renewal of mammalian skin is known to be maintained by permanently residing stem cells that are able to sustain three principal differentiated lineages: the interfollicular epidermis (IFE), sebaceous gland (SG), and hair follicle (HF) (1, 2). In addition, recent studies identified Merkel cell mechanoreceptors residing in specialized epithelial structures termed touch domes in the hairy skin as a fourth lineage maintained by keratinocyte progenitors (3). While it has long been accepted that skin homeostasis is dependent on the ability of stem cells to replenish the turnover of these mature epithelial lineages, it is the work over the last decade that has significantly enhanced our understanding the location and function of multiple stem or progenitor niches in the skin. These findings have dramatically changed our view of the cutaneous epithelial stem cell landscape rendering a highly compartmentalized epithelium maintained by multiple classes of phenotypically distinct regional niches (2).

Kursad Turksen (ed.), *Skin Stem Cells: Methods and Protocols*, Methods in Molecular Biology, vol. 989,
DOI 10.1007/978-1-62703-330-5_0, © Springer Science+Business Media New York 2013

In some cases progenitor niches have been labeled using mouse genetics approaches and characterized under normal conditions to be long lived and able to sustain the cellular input to certain epithelial structures including the interfollicular epidermis (4, 5), sebaceous gland (6, 7), and hair follicle (8–11). In other cases, antibodies against cell surface proteins have been utilized to mark and isolate epithelial progenitors located in the IFE (3, 5, 12) and HF (13–16). These efforts have facilitated our understanding of the relative proliferative capacity of progenitor pools as well as their capacity to regenerate IFE, HF, SG, or Merkel lineages in surrogate assays. Collectively, these studies have illustrated the role of epithelial progenitors during skin homeostasis as well as their ability to respond to skin insult.

As new biomarkers have been implemented to better define the profile of progenitor cell subsets in the IFE and HFs, the individual cell of interest becomes less frequent. This can be a major technical challenge to functional studies such as skin and hair reconstitution and clonogenic studies where a significant number of cells may be required. In this chapter, we will outline some basic methods for isolation and functional assessment of keratinocyte clonogenicity, multipotency, and self-renewal capabilities from freshly isolated single-cell suspensions of murine epidermal keratinocytes that have been subjected to FACS sorting. In particular, we will focus on clonogenic and skin and hair reconstitution assays. Methodologies to establish cultures of epidermal keratinocytes at clonal densities have been established for more than three decades and were developed by Rheinwald and Green (17). While there have been many modifications added in this method over the years (18), we observe the highest success rates when maintaining Rheinwald and Green's principle of using a feeder layer of mitotically arrested mouse 3T3 fibroblasts. The development of the hair reconstitution assay (19, 20) revealed the shortcomings of in vitro assays, which typically do not account for stem cell potency. Importantly, we feel the ability to conduct skin and hair reconstitution assays from freshly isolated FACS-sorted keratinocyte subsets provides a robust platform to identify and distinguish unipotent, bipotent, and multipotent epithelial progenitors.

2 Materials

2.1 Skin Cell Isolation Solutions

1. 0.25% trypsin/1 mM EDTA stock solution (Invitrogen).

2. 1× PBS, pH = 7.6 (Invitrogen), sterilized.

3. Fibroblast growth medium: DMEM (Invitrogen) supplemented with 10% Donor Bovine Serum (Invitrogen) and 2% penicillin–streptomycin (Invitrogen).

4. Collagenase Type I (Worthington Biochemical), 10 mg/ml stock solution in PBS.

5. DNAse I (Worthington Biochemical), 20,000 U/ml stock solution in PBS.

6. 70 μm cell strainer (Fisher Scientific).

7. Hank's Balanced Salt Solution (HBSS) (Invitrogen).

8. Betadine 1% solution in water.

9. 70% EtOH solution.

2.2 Antibodies

1. α6 Integrin (CD49f, BD Biosciences) (see Note 1).

2. Sca-1 (Ly6G, BD Biosciences).

3. CD34 (RAM 34, BD Biosciences).

4. CD200 (OX-2, BD Biosciences).

2.3 Clonogenic Assay

1. Complete FAD growth medium: Three parts DMEM (Invitrogen), one part Ham's F12 Supplement (Invitrogen), 10% Defined Fetal Bovine Serum (HyClone), 10 ng/ml EGF (Peprotech), 0.5 mg/ml hydrocortisone (Sigma-Aldrich), 10^{-10} M cholera enterotoxin (Sigma-Aldrich), 5 mg/ml insulin (Sigma-Aldrich), 1.8×10^{-4} M adenine (Sigma-Aldrich), 100 U/ml penicillin (Invitrogen), and 100 mg/ml streptomycin (Invitrogen).

2. Cnt-57 serum-free medium (CELLnTEC).

3. 3T3 fibroblasts (ATCC) mitotically arrested with either mitomycin c (Sigma) or γ-radiation.

4. 0.25% trypsin/1 mM EDTA stock solution (Invitrogen).

5. Nunclon 6-well dishes (Fisher Scientific).

6. 0.5 mM EDTA (Fisher Scientific), a.k.a. versene.

7. Rhodamine B, 1% solution in H_2O (Sigma).

2.4 Skin Reconstitution Assay

1. Silicon culture chambers—Upper F2U #30-268; Lower F2L #30-269 (Renner Gmbh).

2. Surgical instruments including forceps, curved scissor, stapler, and staple remover (all from Temin); sterile drapes; alcohol swabs; and anesthetics.

3. Immunodeficient mice Nude mice (NCR nude), male, 7–9 weeks old, supplied by Taconic or we prefer NSWNU-M (homozygote females) (see Note 2).

4. Small heating pad.

3 Methods

3.1 Epidermal Keratinocyte Isolation

1. Under a biological cabinet, submerge euthanized 8-week-old mice in 1% betadine for 2 min and wash in sterile H_2O. Submerge mice in 70% EtOH for 1 min and wash in sterile H_2O.

2. Under a biological cabinet, surgically excise the dorsal skin from using sterile forceps and scissors and float skin dermis side up in sterile PBS in a sterile Petri dish. Scrape away the subcutaneous fat and muscle using a sterile scalpel and forceps using multiple dishes with new PBS as necessary (see Note 3).

3. Float skins epidermis side up in 0.25% trypsin/1 mM EDTA in a 10-cm Petri dish for 1.5–2 h (see Note 4).

4. Aspirate trypsin and recover skins in 10 ml fibroblast growth medium. Gently detach epidermis from dermis using a scalpel and mince epidermal scrapings into small pieces.

5. Transfer scrapings into a sterile 100 ml bottle and add 30 ml fibroblast growth medium and a stir bar. Recover epidermal cells with gentle stirring for 30 min and filter cells through a 70-μm cell strainer into a 50-ml Falcon tube. Spin cells for 10 min at 1,000 rpm. Wash cells in 10 ml PBS, spin and resuspend cells in 10 ml fibroblast growth medium and spin for 10 min at 1,000 rpm.

6. For antibody labeling (see below), resuspend cell pellets at 5–10×10^6 cells/ml in fibroblast growth medium. Typically 10–12×10^6 viable basal keratinocytes are harvested from a single dorsal skin.

3.2 Preparation of Highly Inductive Dermal Fibroblasts

1. Surgically excise the dorsal skin from euthanized postnatal day 1–2 mice using sterile forceps and scissors (see Note 5).

2. In a dry Petri dish, lay skins flat dermis side down with no folded edges. Slowly pour in ice-cold 0.25% trypsin/1 mM EDTA and avoid getting the tops of the skins wet. Incubate skins overnight at 4°C.

3. Remove skins, one at a time, from trypsin and place on dry p150 plate, dermis side up. Flatten it again and use fine forceps to separate epidermis from dermis, starting at one edge of skin and flipping the dermis up and off the epidermis, which should stay on the plate.

4. Transfer dermis one at a time to a plate containing 10 ml of media on ice.

5. For eight dermises, use 0.5 ml collagenase stock solution (10 mg/ml in H_2O) plus 12 ml HBSS in a 50–100 ml sterile beaker. Transfer dermises into the beaker and mince into small pieces using sharp scissors.

6. Transfer to a 250-ml flask with a magnetic stir bar. Stir at 37°C for 30 min (Optional: For the last 5 min add 20 μl of DNAse I (stock at 20,000 U/ml in PBS)).

7. Dilute 3–4-fold with media and filter through sterile gauze or 70 μm filter. Rinse the flask with 5 ml media and pass through the filter.

8. Spin the cells at $450 \times g$ for 5 min at 4°C.

Table 1
Markers of epithelial progenitor cells in adult mouse skin

Progenitor location	Markers	References
IFE	Lrig1$^+$	(5)
IFE	$\alpha6^{bright}CD71^{dim}$	(12)
IFE and HF infundibulum	$\alpha6^+CD34^-Sca-1^+$	(16)
IFE touch dome	$\alpha6^+CD34^-Sca-1^+CD200^+$	(3)
Junctional zone	Lrig1$^+$	(5)
HF isthmus and infundibulum	$\alpha6^+MTS24(Plet-1)^+$	(15)
Sebaceous duct	Blimp1$^+$	(7)
HF isthmus	$\alpha6^{low}CD34^-Sca-1^-$ Lgr6$^+$	(11, 16)
HF bulge	$\alpha6^{dim+bright}CD34^+$ Krt15$^+$	(13, 14, 21)
Lower HF bulge and hair germ	Lgr5$^+$	(10)
Hair germ	P-cadherin$^+$	(22)

9. Resuspend and wash the pellet once in HBSS.

10. Count cells, use at 2×10^6 fresh dermal cells per graft, cryopreserve or plate at 1×10^6 cells per 10 cm dish for later use.

3.3 Antibody Labeling and FACS Analysis

1. Select an appropriate panel of antibodies for the target cells of interest (Please see Table 1 for examples of published progenitor marker profiles). When possible, select antibodies directly conjugated to fluorescent dyes.

2. Count cells and aliquot equal amounts into experimental and control (single-stained and unstained tubes) in complete growth medium.

3. Incubate antibodies at concentrations according to manufacturer guidelines for 30 min to 1 h in complete growth medium on ice (see Note 9). Spin and wash cells, resuspend in growth medium supplemented with DAPI or an alternative nuclear stain.

4. Conduct sorting (see Note 10). Typically, $10–20 \times 10^6$ viable $\alpha6^+$ basal keratinocytes can be sorted from a single dorsal skin.

3.4 Clonogenic Analysis

Clonogenic assays typically require fewer cells compared to skin and hair reconstitution assays and harvested cells from a single animal will usually suffice. Cells can be grown with a confluent layer of mitotically arrested 3T3 fibroblasts in complete FAD growth medium or defined serum-free medium in the absence of fibroblasts, although the clonogenic efficiency is generally lower. Serum-free medium can also be used in control experiments to facilitate assays that quantify the number of adherent cells shortly after plating (12–16 h).

1. Preparing the 6-well plates one day in advance will allow the fibroblasts to fully attach, spread, and condition the growth medium. Plate 1×10^6 mitotically arrested 3T3 fibroblasts in 3 ml complete FAD growth medium per well.

2. The next day, harvest mouse keratinocytes from a single mouse dorsal skin and FACS sort desired keratinocyte subpopulations as described above. The FACS instrument can be optimized towards purity and accuracy in counting since cell numbers will be in excess. propidium iodide or DAPI should be used to exclude dead cells. Many of the dead cells are post-mitotic suprabasal cells that are sensitive to the 70% ethanol washes during cell harvesting. Use either the FACSAria Automated Cell Deposition Unit (ACDU) function or manually seed 1×10^3 (a range of 0.5 to 2×10^3 cells can be used) sorted keratinocytes per well.

3. Incubate cells for 2 weeks at 32 °C in a humidified incubator with 5% CO_2 and change the medium every 48 h.

4. After 2 weeks, aspirate off the culture medium and replace with 3 ml versene per well. After 1–2 min at room temperature, detach the feeder layer by repeated pipetting of versene over the plate (keratinocytes will not detach). Gently wash plates two times with PBS (take care not to detach colonies).

5. Stain wells with rhodamine B for 1 h (see Note 6) at room temperature (just enough rhodamine solution to cover the cells is sufficient). Aspirate off rhodamine solution and gently wash wells three times with PBS (take care not to detach colonies).

6. Aspirate off the final wash and allow wells to completely dry by turning plates upside down. Afterwards, plates can be imaged and colonies may be manually counted. Typically the total number of colonies as well as the number of colonies greater than 4 mm in diameter are counted and compared between keratinocyte subpopulations.

3.5 Skin and Hair Reconstitution Assay

1. Clip hair with electric clippers, if necessary, and clean the dorsal skin with 1% betadine and place anesthetized mice on heating pad.

2. Use scissors to make a small incision on the back of the mouse (approximately 1 cm in diameter). Better areas for chamber placement are interscapular or suprapelvic. Do not make incisions directly on the spinal protrusion.

3. Assemble the upper and lower grafting chambers together and insert through the incision so that the rims of the chamber are under skin (see Note 7).

4. Secure the chamber to the skin with surgical stapler clips (two staples is usually enough).

5. Allow the chamber to adhere to the dorsal surface overnight prior to implanting cells.

6. Mix the desired number of epidermal cells and 2×10^6 dermal cells together as a slurry in HBSS. We have successfully implanted 1×10^5 to 6×10^6 epidermal cells per graft. Spin cells at 1,000 rpm for 5 min, resuspend the pellet in 100 µl HBSS, and store on ice until use.

7. Gently mix cell suspensions before pipetting entire aliquot into chamber of hat, through the hole on top.

8. Replace each mouse in individual cages (on belly and away from the spout of the water bottle).

9. After one week (see Note 8), anesthetize mice and remove staples and gently tug on chamber to release it from mouse's back. Use tweezers to loosen skin around edge of chamber. Grafted area may be moist and oozy, leave it be and replace mouse in cage, as before.

10. Chambers are retained, cleaned (soak overnight in soapy water), and autoclaved for reuse.

11. Grafts are usually biopsied at 5–10 weeks post-grafting. Hair usually appears after approximately 2–3 weeks.

4 Notes

1. The use of directly conjugated antibodies is recommended whenever possible.

2. In our hands, hairless immunodeficient mouse strains such as Nude are more amenable to skin grafting procedures.

3. We typically use three Petri dishes with clean PBS per skin.

4. After 1.5 h of trypsin digestion, check the skins for detachment by gently scraping the epidermis with a scalpel. If the epidermis is easily removed, then no further digestion is required. If the epidermis does not detach, check again every 15 min.

5. Euthanized pups are washed in sterile water once, followed by two washes in 70% EtOH. EtOH is removed completely and clean pups are placed in sterile Petri dish in the hood. When processing multiple pups, place detached skins in PBS until all skins are harvested.

6. The cells can be stained from 1 h to 1 week in Rhodamine B.

7. Prior to implanting the chambers, make sure there is a hole in top half of hat. A small hole punch can be used.

8. If necessary, the grafting caps can be left on the skin for 2 weeks.

9. Depending on affinity and purification, antibody labeling concentrations typically range from 0.25 to 1.0 μg per 1×10^6 cells.

10. A high-speed sorting device is required, i.e., BD FACSAria; however, many factors contribute to a successful cell sort that require consideration. The length of the trypsin digest required to separate the epidermal and dermal compartments renders the isolated keratinocytes fragile to further mechanical stress. Keratinocytes also have a tendency to aggregate and, although it is possible to pre-filter the sample and gate out most aggregates electronically, they threaten the number of events that can be analyzed and sorted due to clogging of the sample line. To account for these issues, we prefer at larger size nozzle (100 μm) in order to obtain an uninterrupted sort. This on the other hand decreases the sheet pressure and limits the drop drive to around 24,000 drops/s on the BD FACSAria system. The resulting events that can be analyzed per second therefore are around 8,000. If the population of interest is 3% of the total sample, 240 target events can be identified of which between 5 and 15% will be aborted electronically as they present a conflict to the purity of the sample. Thus, acquiring 1×10^6 target cells would require more than 1.5–3 h of efficient sorting time. If more cells are required, it may be necessary to sacrifice the animals and harvest the cells at multiple time points in order to maintain viable cells.

Acknowledgment

This work was supported by NIH R21ES020060 research grant (D.M.O.).

References

1. Fuchs E, Tumbar T, Guasch G (2004) Socializing with the neighbors: stem cells and their niche. Cell 116:769–778

2. Yan X, Owens DM (2008) The skin: a home to multiple classes of epithelial progenitor cells. Stem Cell Rev 4:113–118

3. Woo S-H, Stumpfova M, Jensen UB, Lumpkin EA, Owens DM (2010) Identification of epidermal progenitors for the Merkel cell lineage. Development 137:3965–3971

4. Clayton E, Doupe DP, Klein AM, Winton DJ, Simons BD, Jones PH (2007) A single type of progenitor cell maintains normal epidermis. Nature 446:185–189

5. Jensen KB, Collins CA, Nascimento E, Tan DW, Frye M, Itami S, Watt FM (2009) Lrig1 expression defines a distinct multipotent stem cell population in mammalian epidermis. Cell Stem Cell 4:427–439

6. Ghazizadeh S, Taichman LB (2001) Multiple classes of stem cells in cutaneous epithelium: a lineage analysis of adult mouse skin. EMBO J 20:1215–1222

7. Horsley V, O'Carroll D, Tooze R, Ohinata Y, Saitou M, Obukhanych T, Nussenzweig M,

Tarakhovsky A, Fuchs E (2006) Blimp1 defines a progenitor population that governs cellular input to the sebaceous gland. Cell 126:597–609

8. Levy V, Lindon C, Harfe BD, Morgan BA (2005) Distinct stem cell populations regenerate the follicle and interfollicular epidermis. Dev Cell 9:855–861

9. Ito M, Liu Y, Yang Z, Nguyen J, Liang F, Morris RJ, Cotsarelis G (2005) Stem cells in the hair follicle bulge contribute to wound repair but not to homeostasis of the epidermis. Nat Med 11:1351–1354

10. Jaks V, Barker N, Kasper M, van Es JH, Snippert HJ, Clevers H, Toftgard R (2008) Lgr5 marks cycling, yet long-lived, hair follicle stem cells. Nat Genet 40:1291–1299

11. Snippert HJ, Haegebarth A, Kasper M, Jaks V, van Es JH, Barker N, van de Wetering M, van den Born M, Begthel H, Vries RG, Stange DE, Toftgård R, Clevers H (2010) Lgr6 marks stem cells in the hair follicle that generate all cell lineages of the skin. Science 327: 1385–1389

12. Tani H, Morris RJ, Kaur P (2000) Enrichment for murine keratinocyte stem cells based on cell surface phenotype. Proc Natl Acad Sci U S A 97:10960–10965

13. Trempus CS, Morris RJ, Bortner CD, Cotsarelis G, Faircloth RS, Reece JM, Tennant RW (2003) Enrichment for living murine keratinocytes from the hair follicle bulge with the cell surface marker CD34. J Invest Dermatol 120:501–511

14 Blanpain C, Lowry WE, Geoghegan A, Polak L, Fuchs E (2004) Self-renewal, multipotency, and the existence of two cell populations within an epithelial stem cell niche. Cell 118: 635–648

15. Nijhof JG, Braun KM, Giangreco A, van Pelt C, Kawamoto H, Boyd RL, Willemze R, Mullenders LH, Watt FM, De Gruijl FR, van Ewijk W (2006) The cell-surface marker MTS24 identifies a novel population of follicular keratinocytes with characteristics of progenitor cells. Development 133:3027–3037

16. Jensen UB, Yan X, Triel C, Woo SH, Christensen R, Owens DM (2008) A distinct population of clonogenic and multipotent murine follicular keratinocytes residing in the upper isthmus. J Cell Sci 121:609–617

17. Rheinwald JG, Green H (1975) Serial cultivation of strains of human epidermal keratinocytes: the formation of keratinizing colonies from single cells. Cell 6:317–330

18. Morris RJ (1994) Procedure for harvesting epidermal cells from the dorsal epidermis of adult mice for primary cell culture in "high calcium" defined medium. In: Leigh IM, Watt FM (eds) Keratinocyte methods. Cambridge University Press, Cambridge, pp 25–31

19. Weinberg WC, Goodman LV, George C, Morgan DL, Ledbetter S, Yuspa SH, Lichti U (1993) Reconstitution of hair follicle development in vivo: determination of follicle formation, hair growth, and hair quality by dermal cells. J Invest Dermatol 100:229–236

20. Kamimura J, Lee D, Baden HP, Brissette J, Dotto GP (1997) Primary mouse keratinocyte cultures contain hair follicle progenitor cells with multiple differentiation potential. J Invest Dermatol 109:534–540

21. Lyle S, Christofidou-Solomidou M, Liu Y, Elder DE, Albelda S, Cotsarelis G (1998) The C8/144B monoclonal antibody recognizes cytokeratin 15 and defines the location of human hair follicle stem cells. J Cell Sci 111:3179–3188

22. Greco V, Chen T, Rendl M, Schober M, Pasolli HA, Stokes N, Dela Cruz-Racelis J, Fuchs E (2009) A two-step mechanism for stem cell activation during hair regeneration. Cell Stem Cell 4:144–169

Identification and Analysis of Epidermal Stem Cells from Primary Mouse Keratinocytes

Youliang Wang and Xiao Yang

Abstract

The skin, one of the largest organs of the body, is a dynamic tissue in which terminally differentiated keratinocytes are replaced by the proliferation and differentiation of epidermal stem cells. Epidermal stem cells are relatively undifferentiated, retain a high capacity for self-renewal throughout their lifetime, and normally have a slow cell division cycle in vivo. Furthermore, they have a high proliferation potential in vitro, and it is often desirable to isolate and culture them from adult mice to use in conjunction with in vivo studies. However, the isolation of these cells has been problematic. Here, we describe reliable methods for identifying a population of isolated bulge stem cells by flow cytometry and for measuring the growth and differentiation potential of primary mouse keratinocytes by clonal analysis.

Key words Epidermal stem cells, Clonogenicity, Flow cytometry, Immunofluorescence

1 Introduction

The mouse epidermis is comprised primarily of keratinocytes, of which stem cells form a subpopulation. Epidermal stem cells within the hair follicle are known as hair follicle stem cells (HFSCs) and reside in a structure within the outer root sheath (ORS) of the hair follicle called the "bulge." Cells in this compartment exist in a relatively undifferentiated state and their normal role is to regenerate the hair follicle and regrow the hair shaft during every hair cycle. In adult skin, bulge cells divide relatively infrequently and preferentially retain nucleotide label over time (1). They are multipotential, being able to differentiate into interfollicular epidermis and sebocytes, and contribute to all the differentiated cell types involved in forming the hair follicle, including the outer root sheath, inner root sheath, and hair shaft (2, 3).

The bulge is marked by embryonic expression of the transcription factors NFATc1, Lhx2, Sox9, and TCF3 (2, 4–6). In the adult, these markers are accompanied by CD34 and keratin 15(K15) promoter activity, which begin to occur shortly before

Kursad Turksen (ed.), *Skin Stem Cells: Methods and Protocols*, Methods in Molecular Biology, vol. 989,
DOI 10.1007/978-1-62703-330-5_7, © Springer Science+Business Media New York 2013

the first postnatal hair cycle. In contrast, neonatal bulge cells do not express CD34, nor do they express a lacZ reporter gene under the control of the K15 promoter, a marker for the adult bulge. HFSCs also express high levels of β1- and α6-integrins and low levels of CD71 (7). In fact, some of the best-characterized stem cell markers are integrins: β1- and α6-integrin expression increases in cultured keratinocytes with a high proliferative capacity and appears to be enhanced in the bulge area relative to the lower and upper segments of the anagen phase follicle in vivo. In the adult, α6-integrin and CD34 are effective HFSC markers for anagen follicles in mice (8, 9).

Undifferentiated, self-renewing epidermal stem cells also produce daughter transient amplifying (TA) cells, which make up the majority of the proliferative cell population. TA cells undergo a finite number of cell divisions in vivo before leaving the proliferative compartment and becoming terminally differentiated. The proliferative potential of individual multiplying epidermal stem cells is best evaluated by clonal analysis, which can be performed either with freshly isolated skin cells or with cultured primary cells (6, 10–12). In clonal analysis experiments, three types of colonies can be easily distinguished: (1) "holoclones," which are large and round with a smooth perimeter and contain mostly small cells, are formed by keratinocytes with a significant growth potential; (2) "meroclones," which are small, have an irregular shape, and contain large, flattened differentiated squame-like cells, are formed by cells with restricted growth potential; and (3) "paraclones," which have a jagged perimeter and contain intermingled areas of small and large differentiated cells, drive from cells with limited growth potential (13, 14).

We present methods for isolating primary adult mouse keratinocytes from the dorsal skin of mice and for identifying a subpopulation of bulge stem cells using flow cytometry. In addition, we provide protocols for measuring the growth potential of individual cells within a population of primary mouse keratinocytes using a clonogenic cell assay and for determining their differentiation potential by immunofluorescent labeling (15, 16).

2 Materials

Prepare all solutions using endotoxin-free Milli-Q dH_2O (prepared by purifying deionized water to attain a sensitivity of 18 mΩ-cm resistances at 25°C).

1. C57BL/6J mice. All animal studies were approved by the Review Board of Institute of Biotechnology.

2. Phosphate-buffered saline (PBS): To make 1 L of PBS, dilute 0.2 g of KCl, 0.2 g KH_2PO_4, 8 g NaCl, and 2.16 g

$Na_2HPO_4\cdot7H_2O$ in 600 mL ddH_2O; adjust to pH 7.4; and make up to 1 L final volume. Filter sterilize (using a 0.22 µm pore size) and store in a sterile bottle at 4°C.

3. 0.25% Trypsin (Fisher Scientific/HyClone, South Logan, UT, USA).

4. Mouse embryonic fibroblast (MEF) media: DMEM (high glucose, HyClone, South Logan, UT, USA), 10% (v/v) FCS (HyClone, South Logan, UT, USA ,store aliquots at –20°C), 1/100 (v/v) L-glutamine (HyClone, South Logan, UT, USA, store aliquots at –20°C), and 1/100 (v/v) pen/strep (HyClone, South Logan, UT, USA, store aliquots at –20°C).

5. MEF freezing medium: 10% DMSO (Sigma, Saint Louis, MO, USA) in MEF medium.

6. Tissue culture flasks, multi-well plates, and petri dishes (Corning, Shanghai, China).

7. Mitomycin C (Cat. No. 107409, Roche): dissolve to 0.5 mg/mL in PBS, filter-sterilize, and store at 4°C.

8. Dispase (Cat. No. 17105, Gibco).

9. CnT-57 medium (CELLnTEC, Bern, Switzerland, www.cellntec.com): add supplements prior to use and store in the dark.

10. TypLE Select (Cat.No.12563-011, Invitrogen).

11. Antibodies: anti-α6-integrin antibody conjugated to phycoerythrin and anti-CD34 conjugated to FITC (eBioscience, San Diego, CA, USA); Ki67 (Abcam, Cambridge, MA, USA); K14 and involucrin antibodies (Covance Laboratories, Madison, WI, USA); Alexa Fluor488™ goat anti-rabbit IgG (Cat. No. A-11008 Invitrogen).

12. Keratinocyte-SFM medium (Cat. No. 17005–075, Invitrogen): add EGF and BPE prior to use.

3 Methods

3.1 *Isolation and Handling of Primary Mouse Embryonic Fibroblasts*

3.1.1 Isolation of Primary Mouse Embryonic Fibroblasts

1. Sacrifice a pregnant female mouse at day 14.5 post coitum by cervical dislocation. Dissect out the uterine horns, rinse briefly in 70% (v/v) ethanol, and place into a 10-cm petri dish containing PBS (see Note 1).

2. Separate each embryo from its placenta and surrounding membranes. Remove the head and limbs, scoop out the internal organs, and wash with fresh PBS to remove as much blood as possible.

3. Place each embryo into a single well of a 6-well plate containing PBS and transfer the plate into a tissue culture hood. Transfer the embryo to new sterile 6-well plate containing 1 mL trypsin per well.

4. Mince each embryo with a sterile razor blade to the consistency of sludge and then further homogenize by pipetting using a 1-mL pipette. Incubate at 37°C for 10 min.

5. Remove the plate from the incubator and pipette again to break up the homogenate further. There should be some viscous material on the bottom of the wells. Immediately add 2 mL MEF medium to each well and mix (see Note 2).

6. Divide the cell mixture between three 10-cm dishes and incubate at 37°C. Cells should adhere to the wells within 2 days. Change the medium once the cells have attached. These are passage number 0 or P0 cells.

7. Aspirate to remove any remaining cartilaginous chunks (these will be very clearly visible when you change medium).

8. Grow to confluency and then freeze the cells from two 10-cm dishes to make P0 stocks. Subculture the cells in the third dish (see Note 3).

3.1.2 Splitting MEFs

1. Wash MEFs once with PBS and then trypsinize (using 1–2 mL 0.25% trypsin per 10 cm dish). After cells become detached, inactivate trypsin by adding 4 mL MEF medium. Pipette up and down several times to dissociate into single cells and then pellet at low speed for 5 min.

2. Remove the supernatant, resuspend cells in fresh MEF medium, and plate out at a 1:5 dilution. Return cells to the incubator and grow until confluent.

3.1.3 Freezing MEFs

1. Wash confluent P0 cells once with PBS and trypsinize as described in Subheading 3.1.2, step 1.

2. Remove supernatant and resuspend cells in cold freezing medium. Transfer 1 mL cell suspension into each cryo-vial (usually one embryo equivalent per vial) and chill on ice before transferring the vials to a −80°C freezer in a precooled styrofoam box.

3. After 24 h, transfer the cells to liquid nitrogen for long-term storage.

3.1.4 Thawing and Replating MEFs

1. Quickly thaw a vial of frozen MEFs in a 37°C water bath. Clean the outside of the vials with 70% (v/v) ethanol, transfer the cells into 5 mL of prewarmed MEF medium, and then pellet by centrifugation (5 min, low speed).

2. Remove supernatant, resuspend cells in MEF medium, and plate out (=P1, if thawing P0 cells).

3.1.5 Making Feeder Cells

1. Remove medium from an almost confluent plate of MEFs, replace with fresh medium containing 10 μg/mL mitomycin C, and incubate for 2 h at 37°C.

2. Remove the medium and wash twice with PBS, then trypsinize, centrifuge, resuspend, and replate cells as above (see Note 4).

3.2 Isolation of Mouse Epidermal Keratinocytes

3.2.1 Extraction of Mouse Epidermal Keratinocytes

1. Euthanize four or five 6- to 8-week-old mice by CO_2 inhalation for 1.5 min followed by cervical dislocation. Clip off approximately 15–18 cm^2 of the dorsal fur with electric hair clipper and place the mice in a jar containing sufficient povidone-iodine solution to cover them. Shake the jar well to distribute of the solution evenly over the mice. Pour off the solution and rinse with distilled water until clear. Repeat the iodine wash and rinse again with water. Wash twice with 70% ethanol accompanied with shaking. After the final rinse, add sufficient 70% ethanol to cover the mice, and soak for 5–10 min (see Note 5).

2. Inside a tissue culture hood, remove the dorsal skin using thumb forceps and scissors, and place the pieces of skin in a baker containing PBS with 2× gentamycin.

3. Remove a piece of the skin with forceps and scalpel, and place on a petri dish with the hairy side facing downwards. Quickly scrape all subcutaneous tissue from the skin using a scalpel blade held perpendicular to the skin until the skin is translucent. Do not allow the skin to dry out. Return the piece of skin to the PBS until all pieces are processed.

4. Place a piece of scraped skin onto a petri dish hairy side facing upwards and spread it out. Bisect the skin along its length and cut into 1.0–1.5 cm strips.

5. Put the strips into a 50 mL centrifuge tube containing 20–25 mL CnT-57 medium, 5 mg/mL dispase, and 2× antibiotics/antimycotics and incubate overnight (~15 h) at 4°C.

6. Transfer the strips of skin to a new petri dish containing fresh CnT-57 medium to wash away excess dispase.

7. While the skin is submerged in the medium, gently separate the dermis from the epidermis using two pairs of curved forceps (see Note 6).

8. Place 5 mL TrypLE Select into a new petri dish. Lift the epidermis out of the CnT-57 medium using forceps and carefully transfer it onto the surface of TrypLE Select with the basal layer facing downward (see Note 7).

9. Incubate for 20–30 min at room temperature until the medium becomes turbid (cover with a lid to prevent evaporation).

10. Tilt the petri dish at 30° angle (by resting one side on the lid) and add 5 mL of CnT-57 medium to the epidermis (to dilute the remaining TrypLE). Keeping the dish tilted to minimize the surface area of the liquid and make the collection of cells more efficient, gently rub the epidermis on a small area of the base of the petri dish to gently separate single cells from the cell sheet. During this process, the medium will become more turbid.

11. Use 1–2 mL fresh medium to wash disaggregated cells to the bottom of the dish and transfer the single-cell solution to a 50-mL

centrifugation tube. Repeat the rubbing process after adding another 2 mL of CnT-57 medium and add the second cell suspension to the same tube.

12. Spin the cells at $200 \times g$ at room temperature for 5 min and discard the supernatant.

13. Resuspend cells in 2 mL CnT-57 medium.

14. Dilute 20 μL cell suspensions with equal volume of 0.4% Trypan blue and count the number of viable and dead cells using a hemocytometer. Typically 2×10^7 cells can be obtained per mouse, with >90% viability.

3.2.2 Primary Mouse Keratinocyte Culture (see Note 8)

1. Seed cells into culture flasks at a density of 4–5×10^4 cells/cm^2 (see Note 9).

2. Culture the cells at 35°C and 5% CO_2, changing the medium each second to third day.

3. Passage the cells just before they reach confluency which is around 1 week after seeding. If only a single colony grows, discard the flask to avoid clonal selection.

3.2.3 Subculturing Primary Mouse Keratinocytes

1. Aspirate the medium.

2. Wash cells with PBS (optional).

3. Add 1 mL of TrypLE Select per 12.5 cm^2 flask surface area to the cells and incubate for about 10–15 min at 35°C. Check for cell detachment under the microscope.

4. Dilute the TrypLE Select with 2.5 times more CnT-57 medium and pipette 2–3 times up to break up any remaining cell clumps (this prevents cells from remaining trapped in the DNA of dead cells).

5. Transfer the cell suspension into a 15-mL centrifuge tube.

6. Spin the cells at $200 \times g$ for 5 min.

7. Aspirate the supernatant and resuspend the pellet in 5 mL CnT-57 medium.

8. Count cells and seed at 7.5×10^5 cells in 4 mL CnT-57 medium into a 25-cm^2 flask.

9. Return cells to the incubator and change the medium every 3 days.

10. Passage (or subculture) the cells when they reach confluency or before older cells at the center of the colonies begin to differentiate (see Note 10).

3.3 Flow Cytometry

1. Spin the cells from step 12 of Subheading 3.2.1 at $400 \times g$ at room temperature for 5 min and discard the supernatant.

2. Wash the cell pellet once with PBS containing 0.1% BSA.

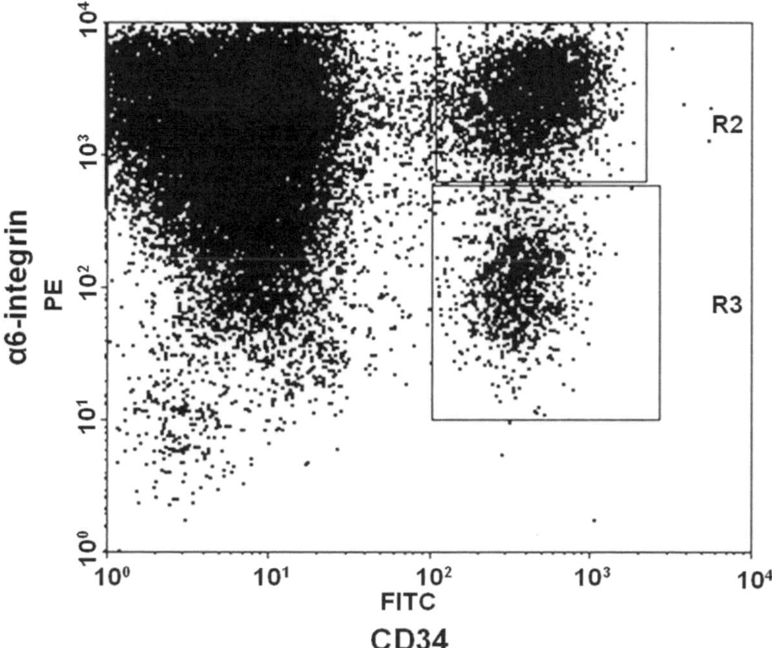

Fig. 1 Fluorescence profiles of mice back skin keratinocytes stained with α6-integrin and CD34. R2 represents the basal bulge stem cells population and R3 represents the suprabasal bulge stem cells

3. Aspirate the supernatant and resuspend the cells by flicking the tube. Add sufficient PBS containing 0.1% BSA to get a concentration of 1×10^7 cells/mL.

4. Aliquot 1×10^6 cells(in 100 μL) into an Eppendorf tube, add 1 μg anti-α6-integrin antibody conjugated to phycoerithrin and 0.125 μg anti-CD34 conjugated to FITC, incubate for 45 min at room temperature with rotation.

5. Wash cells twice with PBS containing 0.1% BSA.

6. Sort the cells using the flow cytometer according to local protocols (see Fig. 1).

3.4 Clonogenicity Assays

1. Calculate the number of fibroblasts obtained from step 2 of Subheading 3.1.5 and add 5×10^5 cells to each 35-mm dishes. Allow cells to attach for 24 h prior to seeding the primary keratinocytes.

2. Spin the keratinocytes from step 12 of Subheading 3.2.1 at $400 \times g$ at room temperature for 5 min and discard the supernatant Resuspend cells in 5 mL of keratinocyte-SFM medium and seed 5×10^4 or 1×10^5 viable cells into each 35-mm dish. Place 2 mL of keratinocyte-SFM medium.

3. Incubated in a 35°C, 100% humidified incubator containing 5% CO_2.

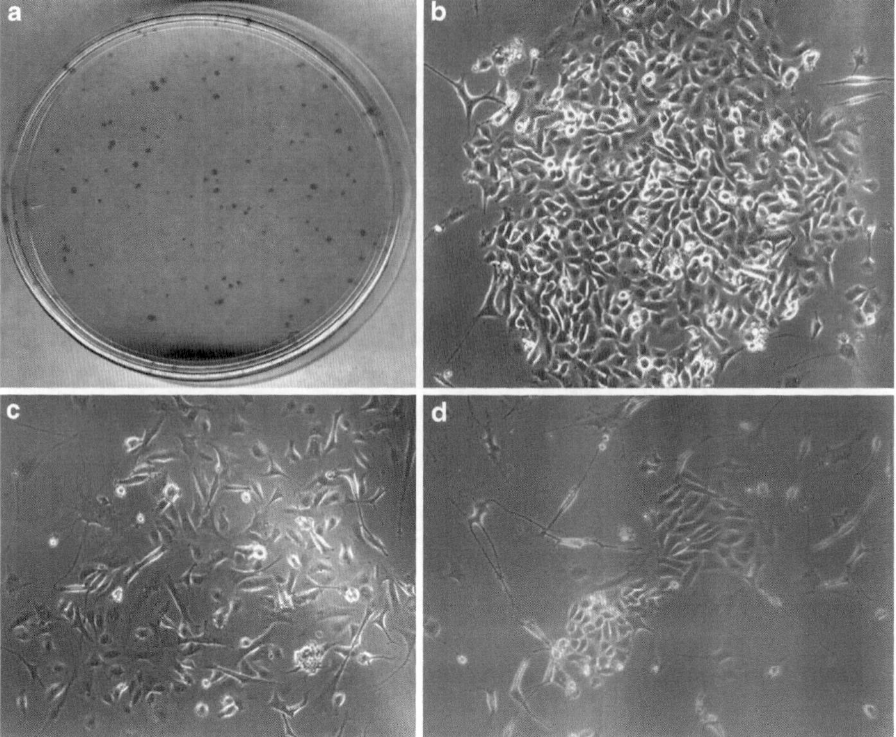

Fig. 2 Colony growth of keratinocytes from 8-week-old mice. Keratinocyte colonies from 5×10^4 primary mouse keratinocytes on MEF feeder layer in 35-mm dishes stained with 0.5% rhodamine B (**a**). Holoclone, large and round, consists of small cells (**b**). Meroclone, with irregular shape, contains large, flattened, differentiated squame-like cells (**c**). Paraclone, smaller, contains intermingled areas of small and large differentiated cells (**d**)

4. Change the medium the day after initial seeding and three times weekly thereafter for mass cultures.

5. For this assay, cells are typically grown for two or three weeks. After this time, remove medium by aspiration and fix the colonies in 10% buffered formalin overnight. After fixation, stain the colonies with 0.5% rhodamine B in distilled water for 30 min. Remove the stain and rinse dishes using cold tap water for 10 s. After drying, count the colonies (*see* Note 11) (*see* Fig. 2).

3.5 Immunofluorescent Labeling of Keratinocytes

1. Fix the cells in cold acetone (precooled to −20°C) for 10 min at −20°C.

2. Remove the acetone and wash three times with PBS. Take care not to detach cells.

3. Permeabilize the cells by incubating with 0.5% Triton X-100 in PBS for 20 min on ice.

4. Wash gently twice with 3 mL PBS.

5. Block nonspecific binding with blocking solution (5% BSA in PBS) for 30 min at room temperature.

6. Dilute the primary antibody appropriately in blocking solution and add sufficient solution to cover the cells.

7. Incubate for 30 min at room temperature (see Note 12).

8. Remove the primary antibody and wash the cells three times with 3 mL PBS.

9. Dilute Alexa Fluor488™ goat anti-rabbit IgG 1:300 in blocking solution containing DAPI and add sufficient solution to cover the cells.

10. Incubate for 30 min at room temperature. Perform all following steps in the dark.

11. Remove the antibody and wash gently three times with PBS.

12. Analyze for successful cell staining using an immunofluorescence microscope. The complete final analysis should be performed using a confocal microscope.

---

4 Notes

1. Day 0.5 is the first day that the vaginal plug can be observed

2. Ideally, the cell suspension should be essentially free of larger pieces of tissue and not too viscous (viscosity is caused by genomic DNA from lysed cells). A high level of viscosity may prevent efficient cell pelleting during centrifugation. The addition of DNase during trypsinization may solve this problem.

3. Before generating MEF feeder layers, subculture MEFs several times, allowing them to reach confluency each time (otherwise too many embryos will be wasted). Making feeder cells involves mitotically inactivating the MEFs using either γ-irradiation or mitomycin C treatment and then replating them.

4. Feeder cells can be used after one day and for up to about 1 week (changing the medium every 3 days or so). At this stage, the cells will have flattened nicely, which improves the growth of stem cells. Change to keratinocyte-SFM medium immediately before plating out the primary mouse keratinocytes for clonogenicity assays.

5. For maximum efficiency, we recommend using the epidermis from neonatal mice or embryos to harvest cells for molecular biology and biochemistry studies. Just peel the skins from embryos or neonates using sterile forceps, then carry out step 5.

6. Do not use excessive force, otherwise the cell preparation will have reduced viability. Either discard the dermis or retain it to confirm the complete removal of the hair follicles along with the epidermis.

7. We recommend the use of protease (TrypLE Select) for cell separation instead of trypsin, based on its effectiveness, price, and ease of use. In particular, unlike trypsin, the digestion does not need to be stopped.

8. For maximum efficiency, we recommend using the epidermis from neonatal mice. The epidermis from embryos between embryonic days 18 and 20 has good cell activity, and the lack of any significant hair follicle formation facilitates separation of the epidermis and dermis.

9. Coating culture dishes is very important for the attachment, spreading, and growth of epidermal cells from adult mice. We coat the culture dishes with 1 mg/mL rat-tail collagen and incubate at 37°C for 1 h. After this time, any remaining solution is removed by aspiration.

10. Cells harvested by this method can be used for molecular biology, biochemistry, and a variety of cell culture techniques (such as wound healing assays).

11. Colonies are usually visible by microscopy 5 days after seeding.

12. The dilution of primary antibodies is K14, 1:1,000; involucrin, 1:1,000; Ki67, 1:300.

Acknowledgments

This work was supported by grant 2012CB945100, 2011CB504202 from Chinese National Key Program on Basic Research, grant 31030040, 30871396, and 81123001 from National Natural Science Foundation of China, and 2012ZX10004502 from Chinese Key Project for the Infectious Diseases.

References

1. Terskikh VV, Vasiliev AV, Vorotelyak EA (2012) Label retaining cells and cutaneous stem cells. Stem Cell Rev 8(2):414–425

2. Blanpain C, Fuchs E (2006) Epidermal stem cells of the skin. Annu Rev Cell Dev Biol 22:339–373

3. Blanpain C, Lowry WE, Geoghegan A et al (2004) Self-renewal, multipotency, and the existence of two cell populations within an epithelial stem cell niche. Cell 118:635–648

4. Horsley V, Aliprantis AO, Polak L et al (2008) NFATc1 balances quiescence and proliferation of skin stem cells. Cell 132:299–310

5. Rhee H, Polak L, Fuchs E (2006) Lhx2 maintains stem cell character in hair follicles. Science 312:1946–1949

6. Watt FM, Jensen KB (2009) Epidermal stem cell diversity and quiescence. EMBO Mol Med 1:260–267

7. Inoue K, Aoi N, Sato T et al (2009) Differential expression of stem-cell-associated markers in human hair follicle epithelial cells. Lab Invest 89:844–856

8. Medina RJ, Kataoka K, Takaishi M et al (2006) Isolation of epithelial stem cells from dermis by a three-dimensional culture system. J Cell Biochem 98:174–184

9. Nowak JA, Fuchs E (2009) Isolation and culture of epithelial stem cells. Methods Mol Biol 482:215–232

10. Brownell I, Guevara E, Bai CB et al (2011) Nerve-derived sonic hedgehog defines a niche for hair follicle stem cells capable of becoming epidermal stem cells. Cell Stem Cell 8:552–565

11. Strachan LR, Ghadially R (2008) Tiers of clonal organization in the epidermis: the epidermal proliferation unit revisited. Stem Cell Rev 4:149–157

12. Greco V, Chen T, Rendl M et al (2009) A two-step mechanism for stem cell activation during hair regeneration. Cell Stem Cell 4:155–169

13. Barrandon Y, Green H (1987) Three clonal types of keratinocyte with different capacities for multiplication. Proc Natl Acad Sci USA 84:2302–2306

14. Pellegrini G, Ranno R, Stracuzzi G et al (1999) The control of epidermal stem cells (holoclones) in the treatment of massive full-thickness burns with autologous keratinocytes cultured on fibrin. Transplantation 68:868–879

15. Yang L, Mao C, Teng Y et al (2005) Targeted disruption of Smad4 in mouse epidermis results in failure of hair follicle cycling and formation of skin tumors. Cancer Res 65:8671–8678

16. Yang L, Wang L, Yang X (2009) Disruption of Smad4 in mouse epidermis leads to depletion of follicle stem cells. Mol Biol Cell 20: 882–890

Chapter 8

Monitoring the Cycling Activity of Cultured Human Keratinocytes Using a CFSE-Based Dye Tracking Approach

Loubna Chadli, Emmanuelle Cadio, Pierre Vaigot, Michèle T. Martin, and Nicolas O. Fortunel

Abstract

The development of methods and tools suitable for functional analysis of keratinocytes placed in an in vitro context is of great importance for characterizing properties associated with their normal state, for detecting abnormalities related to pathological states, or for studying the effects of extrinsic factors. In the present chapter, we describe the use of the intracellular fluorescent dye carboxyfluorescein succinimidyl ester (CFSE) to monitor cell division in mass cultures of normal human keratinocytes. We detail the preparation of CFSE-labeled keratinocyte samples and the identification by flow cytometry of cell subpopulations exhibiting different cycling rates in a mitogenic culture context. In addition, we show that the CFSE-based division-tracking approach enables the monitoring of keratinocyte responsiveness to growth modulators, which is here exemplified by the cell-cycling inhibition mediated by the growth factor TGF-β1. Finally, we show that keratinocyte subpopulations, separated according to their mitotic history using CFSE fluorescence tracking, can be sorted by flow cytometry and used for further functional characterization, including determination of clone-forming efficiency.

Key words CFSE, Keratinocytes, Cell division tracking, Mitotic history, Proliferation assay, Flow cytometry, TGF-β1

---

1 Introduction

Heterogeneity is an intrinsic characteristic of many normal cellular systems, which complicates the study of their functional and molecular characteristics. In particular, the research domain of stem and progenitor cell biology is fully concerned by this difficulty (1, 2). Firstly, the subpopulations of stem and progenitor cells present within human tissues and organs, including the epidermis, are in fact generally constituted by cells with nonequivalent immaturity statuses and metabolic activities (3–5). Secondly, when subpopulations enriched in stem or progenitor cells are placed in an in vitro context, they spontaneously tend to evolve toward an increase in

Kursad Turksen (ed.), *Skin Stem Cells: Methods and Protocols*, Methods in Molecular Biology, vol. 989, DOI 10.1007/978-1-62703-330-5_8, © Springer Science+Business Media New York 2013

heterogeneity (4, 6). In addition to many other biological aspects, the proliferation status and the responsiveness to growth modulators are important parameters that need to be monitored (7). Such investigations are of particular interest for the knowledge of cellular systems such as epidermal keratinocytes, which are originated from dynamic tissues with a naturally high cell renewal rate (8).

The principle of cell division tracking is based on the use of a vital fluorescent cellular labeling, which remains detectable at a stable level for few days in nondividing cells, and whose intensity is divided by half at each round of division. The dyes most commonly used for such purpose are PKH26 (membrane labeling) and CFSE (cytoplasmic labeling). Dye tracking approaches have been applied to the monitoring of cell division in various cell types, from bacteria (9) to different eukaryotic cells, including fibroblasts (10), lymphocytes (11, 12), and hematopoietic stem and progenitor cells (13, 14). Notably, combination of division tracking together with analysis of cell-surface markers has led to progresses in the understanding of extrinsic regulators promoting the maintenance of hematopoietic stem and progenitor cell immaturity throughout successive mitoses (13). Interestingly, the development of mathematical models and algorithms has enabled sophisticated analysis and interpretation of dye tracking flow cytometry data, integrating multiple biological parameters, including duration of the cell cycle, differentiation, cell death, or survival (11, 12, 15–19).

In this chapter, we show the feasibility of using CFSE to perform cell division tracking assays on human keratinocytes and describe the corresponding experimental procedures. CFSE division tracking should have broad interest for analyzing properties of normal or pathological keratinocyte samples, as well as specific responses to biological effectors, chemical compounds, or toxic agents.

1.1 Description of Typical CFSE Dye Tracking Experiments

As described in details in the next sections of the chapter, the CFSE-based dye tracking procedure includes four main successive steps: (1) preparation of CFSE-labeled cell samples (see Subheading 3.1); (2) short-term culture of labeled cell samples (see Subheading 3.2); (3) analysis of CFSE fluorescence profiles (see Subheading 3.3); and (4) sorting of cells with different mitotic histories for further investigations (see Subheading 3.4). The design of the experimental procedure is schematized in Fig. 1.

Fig. 1 (continued) (**a**). During this time lapse, keratinocytes can be submitted to extrinsic stimuli to visualize their impact on keratinocyte proliferation. CFSE profiles were then determined by flow cytometry (**b**). We identified three fluorescence peaks which corresponded to subpopulations of CFSE-labeled keratinocytes that have performed 0, 1, or 2 cell divisions. These three peaks could be separately sorted to further investigate keratinocyte functional properties (clone-forming efficiency (as shown in Fig. 3) or mass expansion (not shown here)) (**c**)

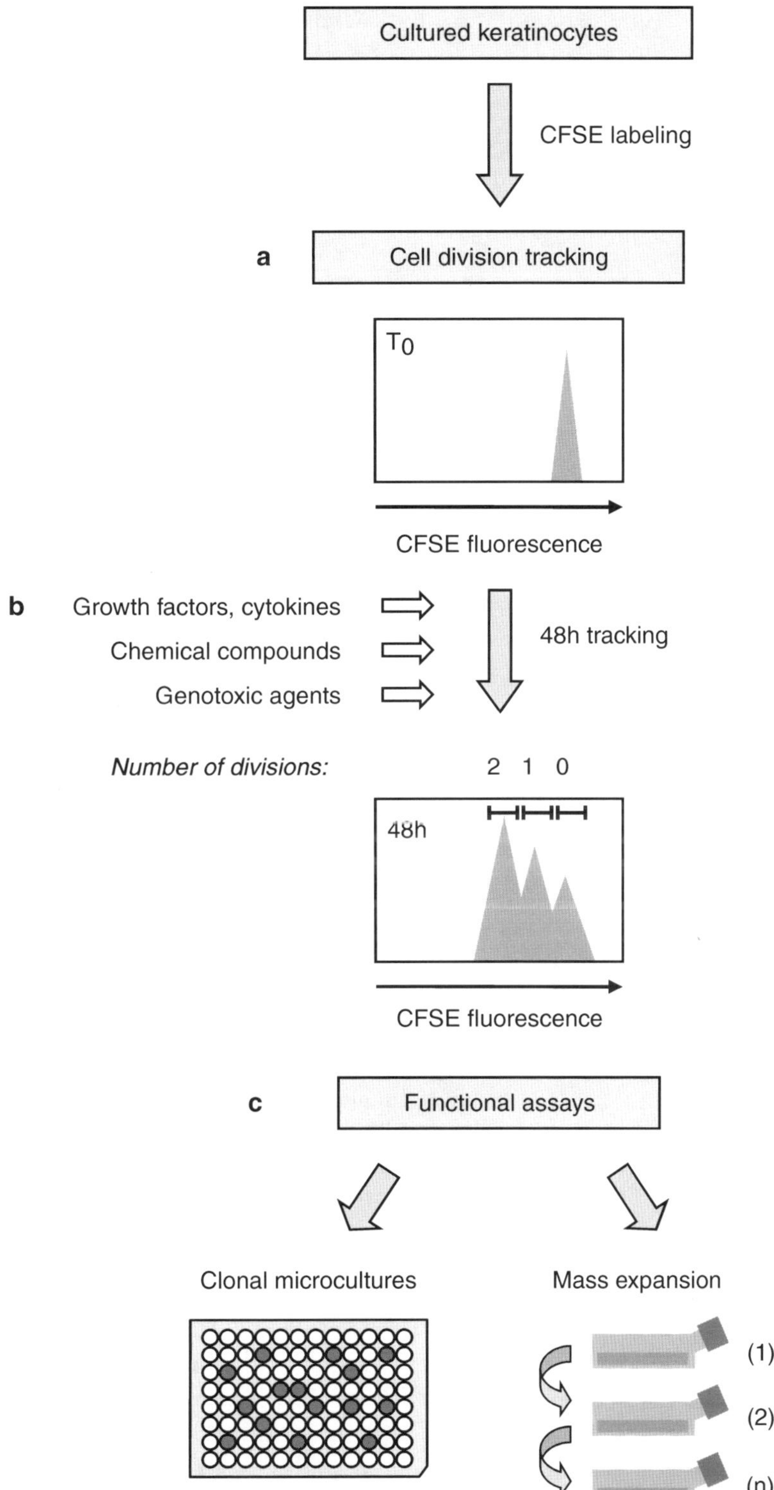

Fig. 1 Experimental procedure. Cultured keratinocytes were stained with the CFSE fluorescent dye. Labeled cells were then sorted and maintained in vitro for 48 h

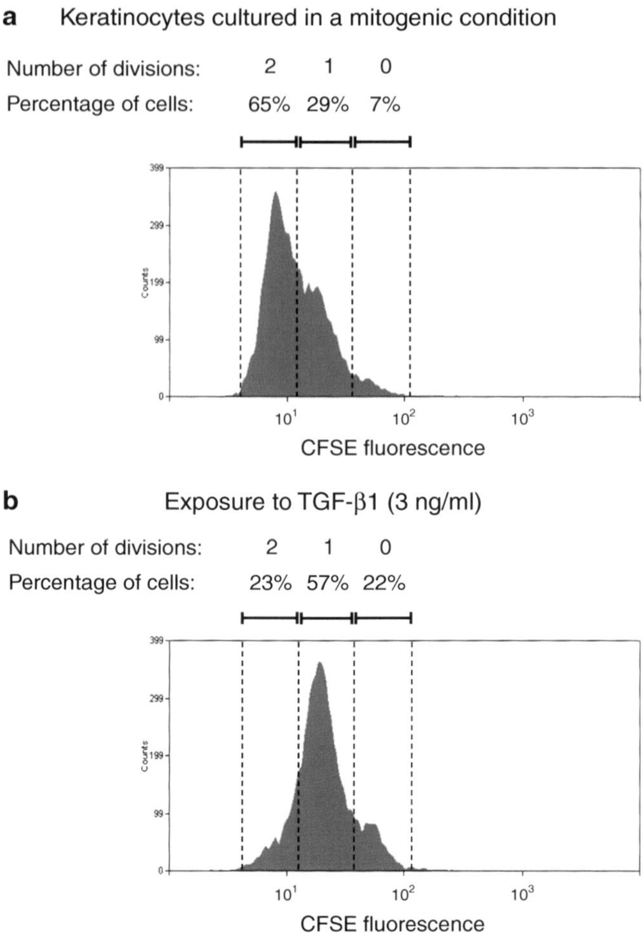

a   Keratinocytes cultured in a mitogenic condition

Number of divisions: 2   1   0
Percentage of cells: 65%   29%   7%

CFSE fluorescence

b   Exposure to TGF-β1 (3 ng/ml)

Number of divisions: 2   1   0
Percentage of cells: 23%   57%   22%

CFSE fluorescence

Fig. 2 Keratinocytes cultured for 48 h in mitogenic conditions (**a**) and exposed to an inhibitory dose of TGF-β1 (**b**). CFSE profiles are composed by three peaks of fluorescence. The CFSE peak with the lowest fluorescence intensity represents the cells that have performed 2 cell divisions in 48 h (65% of total profile). The intermediate peak corresponds to keratinocytes that have divided once (29%), while the peak with the highest fluorescence intensity, corresponds to cells that have not divided (7%) (**a**). TGF-β1 was added to the culture medium at an inhibitory concentration of 3 ng/mL, 24 h after culture initiation. One day later, CFSE profiles were determined. The frequency of actively cycling cells is markedly reduced (23% vs. 65% in the control condition), while those of the two other subpopulations (0 and 1 cell division in 48 h) are increased (respectively, 22% and 57% after TGF-β1 exposure vs. 7% and 29% in the control condition) (**b**)

Firstly, we illustrate here the use of the CFSE dye tracking approach to monitor the cycling activity of normal human skin keratinocytes placed in a culture environment promoting cell proliferation. Labeled keratinocytes, sorted as a homogeneously stained single peak, were plated onto a feeder layer of growth-arrested fibroblasts and cultured for 48 h in the presence of mitogenic factors. Cells were then harvested, and the CFSE fluorescence profile was analyzed by flow cytometry (Fig. 2a). At this stage, three Gaussian subpopulations could be identified within the CFSE profile, corresponding to keratinocytes with different mitotic histories. The Gaussian subpopulation with the lower fluorescence level had performed 2 cell divisions during the 48 h of culture (65% of total profile). The Gaussian subpopulation with intermediate fluorescence level had performed 1 cell division during the same culture time (29%). A less represented subpopulation with high fluorescence level corresponded to cells that have not divided (7%). In the present experiments, the gates used to determine percentages were not defined using algorithmic calculation.

Secondly, we have used CFSE tracking to visualize the impact of short-term exposure to TGF-β1, which is a reference growth inhibitor for keratinocytes, when present at a high concentration. The assay was performed as described above (control condition), except that TGF-β1 was added to the medium at the concentration of 3 ng/mL, 24 h after culture initiation (TGF-β1 condition). The method enabled a clear visualization of the antiproliferative effect of TGF-β1 on keratinocytes (Fig. 2b). Within the CFSE fluorescence profile, the cell subpopulation with low fluorescence level that performed 2 cell divisions during the 48 h tracking was markedly reduced (23% vs. 65% in the absence of TGF-β1). Accordingly, the subpopulations with intermediate and high CFSE fluorescence levels, corresponding respectively to 1 and 0 division, were markedly increased (57% and 22% in TGF-β1-supplemented condition vs. 29% and 7% in the control condition).

Finally, we show the feasibility of using keratinocyte subpopulations sorted according to their mitotic history (different CFSE fluorescence levels) for subsequent functional investigations. The CFSE profile corresponding to Fig. 2a was used to define three subpopulations of keratinocytes exhibiting distinct cycling activities, which were then functionally characterized using "multiparallel clonal microcultures," as described in a previous chapter of *Methods Mol Biol* (20). Briefly, cells exhibiting a low, intermediate, or a high CFSE fluorescence level after the 48 h tracking were plated individually in microculture wells using a flow cytometer equipped with an automated single-cell deposition module. The frequencies of clones obtained with these three different cell cohorts were then determined after 14 days of culture (Fig. 3). We could determine that keratinocytes with different mitotic histories exhibited nonequivalent clone-forming efficiencies (CFE). The most actively cycling subpopulation which performed 2 cell divisions during the 48 h CFSE tracking contained the highest percentage of clonogenic keratinocytes (17% CFE), as compared with the intermediate subpopulation which performed only 1 division during the same time lapse (4% CFE) and with the subpopulation which did not divide during the tracking (<1% CFE).

Together, the different experiments detailed above illustrate and validate the use of CFSE division tracking for distinguishing cells with different cycling activities within a heterogeneous

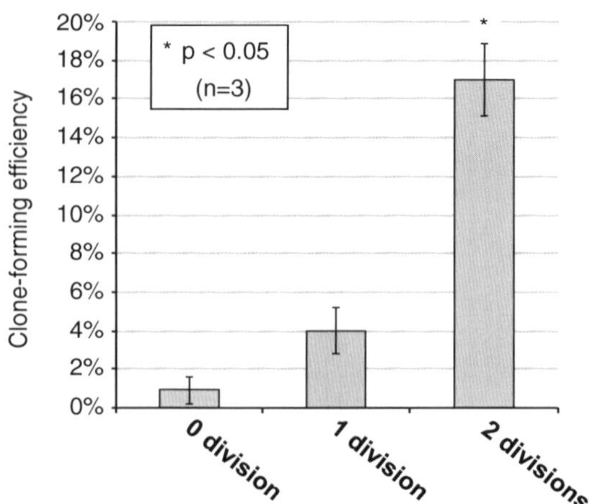

Fig. 3 Functional analysis of keratinocyte clonogenic capacity by using the CFSE dye tracking method. After 48 h of dye tracking, three keratinocyte subpopulations were identified and sorted by flow cytometry (see Fig. 2a). Cells were individually plated in 96-well plates and the clone-forming efficiency of the three subpopulations was analyzed after 14 days of clonal growth. The most actively cycling subpopulation (2 cell divisions in 48 h) contained the highest percentage of clonogenic keratinocytes (17% CFE). In the intermediate subpopulation (1 cell division) we observed a decrease of keratinocyte clonogenic capacity (4% CFE), as well as in the subpopulation which did not divide during the tracking (<1% CFE)

keratinocyte population and for analyzing their responsiveness to a growth modulator. Moreover, we could determine that CFSE labeling is compatible with the sorting of viable keratinocytes, which enables functional studies of keratinocytes separated according to their mitotic history.

2   Materials

2.1   *Cell Culture*

2.1.1   Cell Culture Devices and Reagents

1. Cell culture flasks (75 cm²), coated with type I collagen (BD BioCoat; cat. no. 354485).

2. Multi-well cell culture plates (96 wells), coated with type I collagen (BD BioCoat; cat. no. 356407).

3. DPBS (Gibco; cat. no. 14190).

4. Trypsin (Gibco; cat. no. 27250–018).

5. Trypan blue (Sigma; cat. no. T8154).

2.1.2   Components of the Keratinocyte Growth Medium and Working Concentrations

According to the present experimental procedure, keratinocytes are cultured in a base medium composed of a 3/1 mixture of DMEM 1 g/L D-glucose (Gibco; cat. no. 11880028) and Ham's F12 (Gibco; cat. no. 04195122 M), supplemented with the following components, added at the indicated final concentrations:

1. Adenine hydrochloride (Sigma; cat. no. A9795): 180 µM.

2. Recombinant human epidermal growth factor (rEGF) (Millipore; cat. no. GF001—500 µg): 10 ng/mL.

3. Hydrocortisone (Sigma; cat. no. H4881—100 mg): 0.4 µg/mL.

4. Apo-transferrin (Sigma; cat. no. T2252—100 mg): 5 µg/mL.

5. Triiodothyronine (Sigma; cat. no. T2752—100 mg): 2 nM.

6. Insulin (Sigma; cat. no. I5500—100 mg): 5 µg/mL.

7. L-glutamine (Gibco; cat. no. 25030–024): 2 mM.

8. Fetal bovine serum (FBS); research grade EU approved (Hyclone, FetalClone II, triple 0.1 µm filtered): 10% (v/v).

9. Antibiotics, penicillin/streptomycin (Gibco; cat. no. 15140–122): 100 U/mL and 100 µg/mL.

2.2 Flow Cytometry

2.2.1 Equipment

1. MoFlo™ cell sorter (Beckman Coulter, Inc.).

2. Excitation sources: 5 W water-cooled Argon Ion Laser (Coherent 70C) tuned to 488 nm, 7 W water-cooled Argon Ion Laser (Coherent 90C) turned to MLUV lines, and red laser diode.

3. Summit™ software for data analysis (MoFlo™ software).

4. Automated cell deposition unit (CyClone®) allowing single-cell deposition for cloning purpose.

5. Fluoresbrite® YG Carboxylate Microspheres, diameter 1.50 µm (Polysciences Inc., cat. no. 9719).

6. Flow cytometry tubes, 5 mL polypropylene round-bottom (Falcon; cat. no. 352063).

2.2.2 Reagents and Products

1. Sterile cells trainers 70 µm (BD Falcon; cat. no. 352350).

2. Filters 0.22 µm (Millipore; cat. no. 051246).

3. BSA (Sigma; cat. no. A9418).

4. DPBS-2% BSA filtered through a 0.22 µm filter and stored as frozen aliquots at –20°C.

5. CFSE—CellTrace™ Cell Proliferation kit (Molecular Probes®; cat. no. C34554).

3   Methods

3.1 CFSE Staining of Keratinocytes

CFSE cell staining can be performed on keratinocytes freshly extracted from tissue samples or on cultured keratinocytes (see Note 1). In this chapter, we detail the procedure for the use of CFSE dye tracking assay on cultured skin keratinocytes.

3.1.1 Harvesting Keratinocytes from Cell Cultures

NB. Keratinocyte cell cultures can be performed by using either coculture systems with growth-arrested fibroblast or feeder-free culture conditions (see Note 2).

1. Remove culture medium and rinse twice with DPBS.

2. Remove the feeder layer by adding prewarmed trypsin. Incubate for 2–5 min at room temperature until all the fibroblasts are detached (see Note 3).

3. Rinse twice with DPBS.

4. Detach the keratinocytes by adding prewarmed trypsin and incubate for 5 min at 37°C (see Note 4).

5. Harvest the cells and rinse twice with DPBS (see Note 4).

6. Centrifuge at $180 \times g$ for 15 min at 4°C.

7. Resuspend the pellet in DPBS-2% BSA and determine cell number and viability by trypan blue exclusion.

8. Centrifuge at $180 \times g$ for 15 min at 4°C. After this step, the cells will be resuspended in the CFSE staining solution.

3.1.2 Preparation of the CFSE Staining Solution

1. Do not expose the CFSE reagent and stained cells to light (see Note 5).

2. Allow solid CFSE vials to warm up for 2 min at room temperature.

3. Prepare a 5-mM CFSE stock solution immediately prior to use by adding 18 µL DMSO in each vial of CFSE (see Note 5).

4. Prewarm a volume of 1 mL DPBS-2% BSA per 10^6 of counted keratinocytes.

5. Dilute 2 µL of 5 mM CFSE/mL of prewarmed DPBS-2% BSA to obtain a final working concentration of 10 µM CFSE (see Note 6).

3.1.3 Cell Staining

1. Resuspend the keratinocytes gently in 10 µM CFSE working solution to obtain a single-cell suspension at a concentration of 10^6 cells/mL (see Note 7).

2. Incubate the cells for 15 min at 37°C in a water bath, to allow passive diffusion of the CFSE into the cells (see Note 8).

3. Centrifuge at $180 \times g$ for 15 min at room temperature.

4. Resuspend the pellet in prewarmed keratinocyte growth medium (see Subheading 2.1.2) at a concentration of 2×10^6 cells/mL.

5. Keep the stained samples at room temperature for 30 additional min to ensure a complete modification of the CFSE into a fluorescent dye (see Note 8).

6. Stop the staining reaction by diluting the cell suspension in 5–10 volumes of keratinocyte growth medium.

7. Centrifuge at $180 \times g$ for 15 min at 4°C.

8. Wash with DPBS-2% BSA and centrifuge at $180 \times g$ for 15 min at 4°C.

9. Resuspend the pellet and filtrate through a 70-μm cell strainer to eliminate residual aggregates (see Note 7).

10. Centrifuge at $180 \times g$ for 15 min at 4°C.

11. Resuspend the keratinocytes to obtain a single-cell suspension at a final concentration of 6×10^6 cells/mL of DPBS-2% BSA.

12. Filtrate through a 70-μm cell strainer and place the stained cells in flow cytometry tubes. Samples are now ready for the sorting of CFSE-labeled cells by flow cytometry.

3.2 Initiation of the Dye Tracking Assay

3.2.1 Sorting of a Homogeneously Stained Cell Population

1. Set up the cell sorter using a low sheath pressure (20 psi) and a 100 μm nozzle (see Note 9).

2. Wash and rinse the fluidic system to obtain sterile conditions.

3. Adjust the cell sorter for standard fluorescein detection. Use 488 nm laser excitation and a 530/30 band pass fluorescence filter.

4. To normalize the detection of the fluorescence intensity, use fluorescent beads for calibration. Note the channel position of the fluorescent peak.

5. Rinse the fluidic system to remove all the beads.

6. Use standard flow cytometry gating on FSC and SSC parameters to exclude cellular debris.

7. Set up the cell sorter to select CFSE-labeled cells.

8. At this stage, the distribution of the fluorescence constitutes a Gaussian single peak. Select a narrow gate (width: 20–30 channels) around the center of the peak to sort a cell population with homogeneous CFSE fluorescence (see Note 10).

9. After cell sorting, centrifuge at $180 \times g$ for 15 min at 4°C.

10. Resuspend the pellet in keratinocyte growth medium and count the cells. Cells are now ready for starting the dye tracking assay.

3.2.2 Initiation of the Dye Tracking Cultures

1. The day before culture initiation, prepare cell culture flasks by seeding feeder fibroblasts at a density of 6,000 cells/cm² in the keratinocyte growth medium (see Note 2).

2. Plate freshly sorted CFSE-labeled keratinocytes (see Note 11).

3. Place the cell cultures at 37°C in an incubator with a fully humidified atmosphere containing 5% CO_2.

4. The next day, renew the medium to remove nonattached cells. For all manipulations, avoid exposition to light.

NB. At this stage of the protocol, labeled cells can be submitted to external stimuli (growth factors, chemical compounds, genotoxic stresses) to study their impact on keratinocyte proliferation. In the experiment detailed in Subheading 1, the keratinocyte growth medium was supplemented with TGF-β1 at a concentration of 3 ng/mL (Recombinant Human TGF-β1 (R&D Systems; cat. no. 240B)).

3.3 Analysis of Cell Mitotic History

3.3.1 Harvesting CFSE-Labeled Keratinocytes from Cultures and Processing Fluorescence Analysis by Flow Cytometry

NB. The CFSE-based dye tracking method is generally used for monitoring short-term proliferation (24–72 h). In the experiments detailed here, the mitotic activity of keratinocytes was analyzed 48 h after the plating of CFSE-labeled cells.

1. Remove the culture medium and rinse twice with DPBS. At this stage, keratinocytes are at a very low confluence (1–2 cell divisions). Differential removal of feeder cells cannot be safely performed without the risk of losing the keratinocytes (see Note 12).

2. Trypsinize the cells and rinse twice with DPBS.

3. Centrifuge at $180 \times g$ for 15 min at 4°C.

4. Resuspend the pellets in DPBS-2% BSA.

5. Count the cells.

6. Add a larger volume of DPBS-2% BSA and filtrate through 70-μm cell strainers to eliminate residual aggregates that could disturb flow cytometry analysis (see Note 8).

7. Centrifuge at $180 \times g$ for 15 min at 4°C.

8. Resuspend the pellets in 1 mL of DPBS-2% BSA to obtain a single-cell suspension. CFSE cell samples are now ready for analysis of CFSE fluorescence by flow cytometry.

3.3.2 Analysis of CFSE Profile by Flow Cytometry

1. Set up the cell sorter as previously described at the steps 1–3 of Subheading 3.2.1.

2. Calibrate the fluorescence detection with the fluorescent beads. Place the peak of fluorescence obtained with the beads at the channel position previously set up at the step 4, Subheading 3.2.1.

3. Rinse the fluidic system to remove all the beads.

4. For analysis, use standard flow cytometry gating on FSC and SSC parameters to exclude cellular debris and feeder cells (see Note 12).

5. Analyze the fluorescence profile of CFSE-labeled keratinocytes. At 48 h poststaining, three peaks of fluorescence can be detected. The CFSE profile is composed by overlapping Gaussian peaks (see Note 13).

NB. Within CFSE profiles, the relative sizes of the different peaks reflect the cycling activity of the different keratinocyte subpopulations present in the analyzed samples. The parameters characterizing CFSE profiles (percentage of cells per peak, fluorescence intensity) can be used to quantify the impact of treatments or stimuli on keratinocyte cycling activity. In the experiment described in Subheading 1, the analysis of CFSE profiles is used to visualize the growth inhibitory effect of the TGF-β1.

3.4 Analysis of the Link Between Mitotic Activity and Proliferation Potential

In the present experiments, the three detected peaks correspond to keratinocyte subpopulations with different cycling rates (0, 1, and 2 cell divisions performed during the 48 h tracking). CFSE tracking is compatible with the sorting of viable cells, which enables to study their functional properties.

3.4.1 Sorting of Cells Corresponding to the Different Fluorescence Peaks

1. Set up the cell sorter (see Subheading 3.4.2 for cell cloning)

2. Identify the different keratinocytes subpopulations. For this purpose, select narrow gates (width of 5–10 channels around the center of the peak) that have to be separated enough to allow the sorting of keratinocytes with the same cycling rate.

3. Sort separately the different keratinocyte subpopulations.

4. Centrifuge the cells at $180 \times g$ for 15 min at 4°C.

5. Resuspend the pellets and count the cells.

6. Initiate cell cultures.

NB. The sorted cells can be used to achieve different functional tests, such as long-term expansion, clonal microcultures, and colony assays (as described in ref. 20).

3.4.2 Cloning Cells Corresponding to the Different Fluorescence Peaks

NB. In this chapter, we illustrate the feasibility of using the CFSE approach to determine the clonogenic capacity of the three detected cell subpopulations (see Subheading 1.1 and Fig. 3). These experiments were achieved by using clonal microcultures, as described in a previous chapter of *Methods Mol Biol* (20). The cloning procedure is summarized below.

1. The day before cloning, prepare the 96-well plates with feeder fibroblasts (6,000 cells/cm²) in the keratinocyte growth medium.

2. Set up the cell sorter for cell cloning using the CyClone® deposition system.

3. Identify the different keratinocyte subpopulations. For this purpose, select narrow gates (width of 5–10 channels around the center of the peak) that have to be separated enough to allow the sorting of keratinocytes with the same cycling rate.

4. Clone separately the different keratinocyte subpopulations.

5. Deposit a single keratinocyte per well using the CyClone® unit of the MoFlo™ cell sorter.

6. Place the culture plates immediately after cloning at 37°C in an incubator with a fully humidified atmosphere containing 5% CO_2.

7. Culture for 14 days and renew the medium three times a week.

8. Determine clone-forming efficiency.

---

4 Notes

1. The present dye tracking protocol can be applied to the study of keratinocytes from normal or pathological human skin. These cells can correspond to primary keratinocytes processed directly after extraction from tissue samples, or to their progeny, processed after a step of expansion in culture. The input cellular material can be total unfractionated keratinocytes, populations/subpopulations selected according to specific characteristics, or cell banks of clonal origin. A procedure of epidermal keratinocyte extraction from human skin samples is detailed in a previous issue of *Methods Mol Biol* (20).

2. The culture system used in the experiments detailed in this chapter consists of coculturing the keratinocytes onto a feeder layer of growth-arrested fibroblasts. In the present case, the cells used as feeders are normal fibroblasts extracted from human dermis and expanded in culture. The murine J2-3T3 fibroblast line constitutes an alternative source of feeder cells, which efficiency for supporting human keratinocyte growth is widely documented. Irreversible growth arrest of feeder cells was here induced by γ-irradiation (60Gy). An alternative method used to induce fibroblasts growth arrest is mitomycin-C treatment. A procedure of dermal fibroblast extraction from human skin samples, and preparation of feeder cell banks, is described in a previous issue of *Methods Mol Biol* (20).

3. Trypsin is prewarmed to ensure a quick and gentle removal of the feeder cells, which has to be monitored by observation under microscope. Feeder cells have to be fully detached from cell cultures to avoid the presence of labeled fibroblasts in CFSE-labeled samples of keratinocytes.

4. After trypsinization, place the detached cells in FBS to neutralize trypsin. Moreover, to avoid keratinocyte mortality, do not exceed 5 min of trypsinization. If necessary, make serial short trypsinizations: harvest the detached cells, transfer them into the tubes containing FBS, and repeat the trypsinization to fully collect the keratinocytes.

5. CFSE is photosensitive and stock solutions are unstable. Protect from light the CFSE vials and the stained cells at each step of the experiments, including medium changes. Use CFSE

stock solutions as promptly as possible and avoid repeated freezing and thawing of the reagents (solid CFSE is stable for at least 6 months).

6. The concentration of CFSE for optimal staining may vary depending on cell types and applications (flow cytometry, microscopy). The staining of rapidly dividing cells is generally performed at a concentration of 5–10 µM of CFSE working solution. We suggest that the optimal concentration of CFSE should be determined by using a standard curve in the range of 0.5–25 µM. If several concentrations give reproducible and reliable results, select the lowest concentration to preserve the normal cellular physiology and reduce artifacts from overloading. In our conditions, the optimal working concentration was 10 µM.

7. Cultured keratinocytes are more likely to constitute aggregates than freshly extracted keratinocytes. To ensure uniform labeling and avoid the obstruction of the flow cytometer or a misinterpretation of the fluorescence signal, cell pellets have to be resuspended and filtered to obtain single-cell suspensions.

8. This dye tracking method is based on the use of CFDA-SE (carboxyfluorescein diacetate succinimidyl ester) that can passively enter into the cells. CFDA-SE is not fluorescent until acetates groups are cleaved by intracellular esterases. Succinimidyl ester groups then react with intracellular amines, forming a highly fluorescent and stable compound, the CFSE (carboxyfluorescein succinimidyl ester). The CFSE is trapped in the stained cells and cannot be transferred to adjacent cells. Due to this stable linkage, the CFSE fluorescence is maintained in the cells when they do not divide, while it is split by half in daughter cells at each mitosis. Nonmodified CFDA-SE can passively diffuse out of the cells and be washed away. These chemical reactions are more likely to happen when the metabolism of the cell is optimal, at 37°C. To facilitate the comprehension of protocols, CFDA-SE is usually designated by the term "CFSE" in CFSE staining procedures.

9. Keratinocytes are fragile cells, susceptible to mechanical stress. This unusual low pressure and the choice of a large 100 µm nozzle permit to slightly increase both the viability of sorted cells and cell deposition in the case of cloning experiments.

10. Selection of a narrow gate enables the sorting of more homogeneous cell populations. The homogeneity of the sorted samples is clearly visible if sorted keratinocytes are reanalyzed. The distribution of the CFSE fluorescence then constitutes a narrow peak with a low coefficient of variation.

11. After cell sorting, cells have to be plated at an appropriate density to obtain sufficient cell numbers for the next flow cytometry

analysis. The working density has to be set up according to culture conditions and cell characteristics. For example, in our experiments, we have used densities of 2,500 cells/cm² for a 24h post-staining analysis and 2,000 cells/cm² for more than a 36h analysis (density in 75 cm² culture flasks).

12. At this stage, the presence of feeder cells in CFSE-labeled samples is not susceptible to bias the analysis of CFSE profiles. These unlabeled fibroblasts can be easily excluded from the flow cytometry analysis by appropriate gating.

13. After each round of cell division, fluorescence peaks are shifted toward lower intensities and their intensity values are divided by half. After 48 h of dye tracking, the CFSE profile is composed of three peaks of fluorescence corresponding to cells that have undergone 0, 1, or 2 cell divisions. Over time, the fluorescent peaks become less discernible.

Acknowledgments

The authors wish to thank AFM-Myobank (La Pitié-Salpêtrière Hospital, Paris) for providing biopsies of normal adult human skin. This research project was financially supported by CEA and ANR-CESA024. Loubna Chadli benefited from doctoral fellowships from CEA (direction of life sciences, International Research Training for Excellence in Life Science (IRTELIS) program), and from the Association pour la Recherche sur le Cancer (ARC).

References

1. Chadli L, Martin MT, Fortunel NO (2011) Investigating human keratinocyte stem cell identity. Eur J Dermatol 21(Suppl 2):4–11

2. Kaur P, Potten CS (2011) The interfollicular epidermal stem cell saga: sensationalism versus reality check. Exp Dermatol 20:697–702

3. Li A, Simmons PJ, Kaur P (1998) Identification and isolation of candidate human keratinocyte stem cells based on cell surface phenotype. Proc Natl Acad Sci USA 95:3902–3907

4. Fortunel NO, Cadio E, Vaigot P, Chadli L, Moratille S, Bouet S, Roméo PH, Martin MT (2010) Exploration of the functional hierarchy of the basal layer of human epidermis at the single-cell level using parallel clonal microcultures of keratinocytes. Exp Dermatol 19:387–392

5. Fortunel NO, Chadli L, Bourreau E, Cadio E, Vaigot P, Marie M, Deshayes N, Rathman-Josserand M, Leclaire J, Martin MT (2011) Cellular adhesion on collagen: a simple method to select human basal keratinocytes which preserves their high growth capacity. Eur J Dermatol 21(Suppl 2):12–20

6. Larderet G, Fortunel NO, Vaigot P, Cegalerba M, Maltère P, Zobiri O, Gidrol X, Waksman G, Martin MT (2006) Human side population keratinocytes exhibit long-term proliferative potential and a specific gene expression profile and can form a pluristratified epidermis. Stem Cells 24:965–974

7. Fortunel NO, Hatzfeld JA, Rosemary PA, Ferraris C, Monier MN, Haydont V, Longuet J, Brethon B, Lim B, Castiel I, Schmidt R, Hatzfeld A (2003) Long-term expansion of human functional epidermal precursor cells: promotion of extensive amplification by low TGF-beta1 concentrations. J Cell Sci 116:4043–4052

8. Barrandon Y, Green H (1987) Three clonal types of keratinocyte with different capacities for multiplication. Proc Natl Acad Sci USA 84:2302–2306

9. Ueckert JE, Nebe von Caron G, Bos AP, ter Steeg PF (1997) Flow cytometric analysis of *Lactobacillus plantarum* to monitor lag times, cell division and injury. Lett Appl Microbiol 25:295–299

10. Khil LY, Kim JY, Yoon JB, Kim JM, Keum WK, Kim ST, Yoon Y, Yoon MY, Moon CK, Lee JH, Ha J, Kim SS, Kang I (1997) Insulin has a limited effect on the cell cycle progression in 3T3 L1 fibroblasts. Mol Cells 7:742–748

11. De Boer RJ, Ganusov VV, Milutinović D, Hodgkin PD, Perelson AS (2006) Estimating lymphocyte division and death rates from CFSE data. Bull Math Biol 68:1011–1031

12. Hawkins ED, Hommel M, Turner ML, Battye FL, Markham JF, Hodgkin PD (2007) Measuring lymphocyte proliferation, survival and differentiation using CFSE time-series data. Nat Protoc 2:2057–2067

13. Batard P, Monier M-N, Fortunel N, Ducos K, Sansilvestri-Morel P, Phan T, Hatzfeld A, Hatzfeld JA (2000) TGF-(beta)1 maintains hematopoietic immaturity by a reversible negative control of cell cycle and induces CD34 antigen up-modulation. J Cell Sci 113:383–390

14. Bennaceur-Griscelli A, Pondarré C, Schiavon V, Vainchenker W, Coulombel L (2001) Stromal cells retard the differentiation of CD34(+) CD38(low/neg) human primitive progenitors exposed to cytokines independent of their mitotic history. Blood 97:435–441

15. Ganusov VV, Pilyugin SS, de Boer RJ, Murali-Krishna K, Ahmed R, Antia R (2005) Quantifying cell turnover using CFSE data. J Immunol Methods 298:183–200

16. Ko KH, Odell R, Nordon RE (2007) Analysis of cell differentiation by division tracking cytometry. Cytometry A 71:773–782

17. Luzyanina T, Mrusek S, Edwards JT, Roose D, Ehl S, Bocharov G (2007) Computational analysis of CFSE proliferation assay. J Math Biol 54:57–89

18. Yates A, Chan C, Strid J, Moon S, Callard R, George AJ, Stark J (2007) Reconstruction of cell population dynamics using CFSE. BMC Bioinformatics 8:196

19. Banks HT, Sutton KL, Thompson WC, Bocharov G, Roose D, Schenkel T, Meyerhans A (2011) Estimation of cell proliferation dynamics using CFSE data. Bull Math Biol 73:116–150

20. Fortunel NO, Vaigot P, Cadio E, Martin MT (2010) Functional investigations of keratinocyte stem cells and progenitors at a single-cell level using multiparallel clonal microcultures. Methods Mol Biol 585: 13–23

Chapter 9

Quantification of Cyclobutane Pyrimidine Dimers in Human Epidermal Stem Cells

M. Ruetze, S. Gallinat, H. Wenck, and A. Knott

Abstract

The common procedures that are used to quantify cyclobutane pyrimidine dimers (CPD) comprise the extraction of cellular DNA followed by the detection of this nucleic acid modification by immunoblotting or electrophoretic methods. Consequently, these approaches provide an averaged damage intensity value of a whole population of cells and are not applicable to studies where a small subgroup such as somatic stem cells are intended to be investigated and the individual cellular damage is of interest. Here, we describe a strategy to isolate epidermal stem cells from minimum human epidermis samples and a subsequent immunocytochemical quantification of cellular CPDs. Besides the determination of the DNA damage status, this technique allows for the examination of cellular CPD intensity distributions.

Key words Skin, Suction blister, Keratinocytes, Epidermal stem cells, UV irradiation, DNA damage, Cyclobutane pyrimidine dimers

1 Introduction

The human skin is continuously exposed to UV irradiation that leads to different types of DNA modifications of which cyclobutane pyrimidine dimers (CPD) are the most prevalent (1). Whereas resulting DNA mutations would pose a minor threat to epidermal keratinocytes that have withdrawn from the cell cycle, terminally differentiate, and are eventually shed from the skin surface, DNA damage in stem cells bears the potential to manifest a persistent defect in the tissue and result in dysfunctional homeostasis or in tumorigenesis (2). It has been postulated that stem cells therefore harbor specific resistance and repair mechanisms that contribute to the prevention of functional tissue defects (3). Respective characteristics could be identified in embryonic (4), hematopoietic (5) and mesenchymal (6) stem cells that were subjected to mutagenic chemical agents (5), ionizing radiation (6, 7), or reactive oxygen species (8). Whereas these studies were primarily performed to unravel mechanisms of chemoresistance in

Kursad Turksen (ed.), *Skin Stem Cells: Methods and Protocols*, Methods in Molecular Biology, vol. 989,
DOI 10.1007/978-1-62703-330-5_9, © Springer Science+Business Media New York 2013

cancer therapy, the investigation of CPDs in epidermal stem cells would be of interest to examine DNA damage under physiological conditions. However, this special experimental setting has so far been hampered by a variety of circumstances. At first, the accessibility of human epidermal stem cells is considerably lower than of those stem cells mentioned above. It is assumed that only 10% of the basal layer of skin keratinocytes are stem cells (9) and a method for in vitro expansion of this fraction has not been developed so far (10). Stem cell numbers that were sufficient for DNA isolation could possibly be isolated from adequately large skin samples that can be obtained from plastic surgery or circumcision procedures, but these biopsies are generally derived from areas that are not sun-exposed or can hardly be integrated into an in vivo irradiation study. Conversely, an adaption to small amounts of cells is difficult due to the reported impurity of epidermal stem cell preparations. The collagen IV adherence assay that is routinely used for that purpose (11) does only provide an enrichment of stem cells that is documented by an increase of colony forming efficiencies to approximately 30% compared to less than 5% of the unsorted population (12, 13). Slightly higher values can be achieved by the usage of flow cytometric sorting with various cell surface markers of epidermal stem cells (14–18), but this approach cannot be combined with CPD quantification as DNA staining requires the digestion of proteins and all stem cell markers that could be used for sorting would be lost within this step. This circumstance is also detrimental for immunostaining of CPDs on tissue sections as identification of stem cells and detection of DNA lesions cannot be performed simultaneously. Finally, an assay that is capable of detecting CPDs in human epidermal stem cells after in vivo irradiation has to be accomplished within a few hours as the repair of CPDs rapidly occurs (19). A recent approach that circumvents some of these constraints has been performed in mice where stem cells on skin sections have been identified by their slow cycling nature (20). In contrast to protein markers, this feature can be visualized by the retention of a second fluorescent DNA label which is compatible with a simultaneous CPD staining. Unfortunately, DNA intercalating agents are highly mutagenic and a corresponding study can consequently not be transferred to humans. In an attempt to deal with this range of restrictions, we here present a strategy that is based on an immunocytochemical method for CPD detection in stem cells. We developed a procedure to isolate epidermal stem cells from biological material that can be acquired with minimal invasive methods such as the suction blister technique. Small numbers of stem cells are then fixed on microscopic slides and protein digested, leading to immobilized DNA spots that can be stained with a thymidine dimer antibody. Whereas the global impact of irradiation can be represented by determining the cumulative signal intensity, the cellular distribution of DNA damage can also be displayed (Fig. 1). Initial experiments revealed a surprisingly high variation of CPD intensities in individual

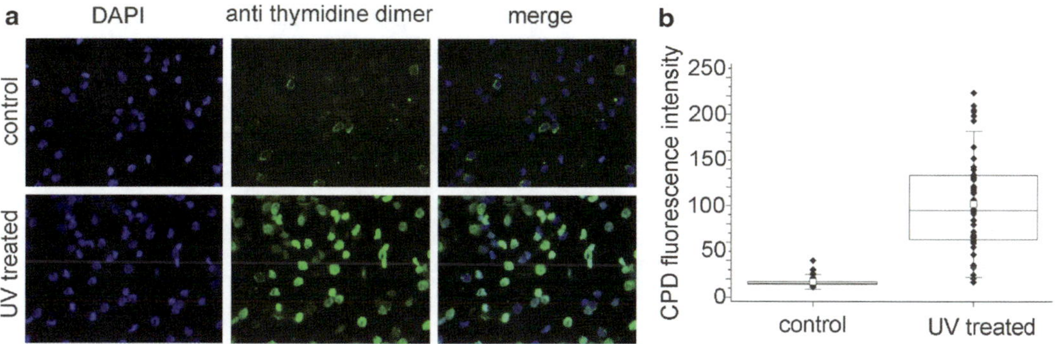

Fig. 1 Immunocytochemical staining of thymidine dimers in epidermal stem cells. (**a**) Immunofluorescence picture of stem cell nuclei that were stained using a thymidine dimer antibody (*green*) and DAPI (*blue*). Cells that were isolated from irradiated human skin sites exhibit substantially increased signal intensities. (**b**) Densitometric quantification of DNA damage in 50 stem cells within the images shown in (**a**). Cells that were derived from irradiated skin areas show an increased average damage status (*white square*), but individual cells are differentially affected by the UV exposure

stem cells of in vivo irradiated human skin (21). Applying the assay for complemental experiments will shed additional light on the induction, repair or resistance of epidermal stem cells towards UV induced DNA modifications.

2   Materials

2.1   Isolation of Human Epidermal Keratinocytes

1. Skin disinfectant (e.g., Cutasept©).

2. Adjustable vacuum pump, silicone tubes, and suction blister cups (see Note 1).

3. Double-faced adhesive tape and medical adhesive bandages (e.g., Hansaplast©).

4. Syringe with 21-gauge needle.

5. Sterilized micro-scissors.

6. Standard wound healing plasters.

7. Dulbecco´s Phosphate buffered saline (PBS, standard formulation with Mg and Ca).

8. Disinfecting solution: 70% non-denatured ethanol, 30% Dulbecco´s PBS.

9. Trypsin solution: 2.5 mg/ml trypsin in Dulbecco´s PBS without Mg and Ca.

10. Trypsin inhibitor solution: 10% fetal calf serum (FCS), 90% Dulbecco´s PBS.

11. KGM-2 medium: Keratinocyte basal medium (KBM-2, Lonza), KGM-2 Bullet Kit supplements (Lonza), 100 µM $CaCl_2$.

2.2 Isolation of Epidermal Stem Cells	1. 60 mm collagen IV coated petri dishes (BD Biocoat). 2. Cell lifter (Corning, catalog number 3008).

2.3 Immunocytochemical CPD Staining

1. Poly-L-lysine-coated microscopic slides (e.g., Polysine© slides, Thermo Scientific, catalog number P4981).

2. Fixation solution: 50% methanol, 50% acetone. Store at –20 °C.

3. Water repellant pen (DAKO pen, catalog number S2002).

4. Pepsin solution: 5,000 U/ml pepsin in 0.05 N HCl (see Note 2).

5. Humidified incubation chamber for microscopic slides.

6. Washing Buffer: 0.1% bovine serum albumin (fraction V) in Dulbecco´s PBS.

7. CPD antibody solution: 1% thymidine dimer antibody solution (clone KTM53, Abnova) in washing buffer.

8. Secondary antibody solution: 0.5% goat anti-mouse antibody (Alexa Flour© conjugated, Invitrogen catalog number A-11001) in washing buffer.

9. Mowiol/DAPI solution: 12% mowiol, 30% glycerine in 0.2 M Tris (pH 8.5), 0.1% 4′,6-diamidino-2-phenylindole (DAPI, Molecular Probes).

10. Microscopic coverslips.

3 Methods

3.1 Isolation of Human Epidermal Keratinocytes

1. Thoroughly treat the skin area of interest with disinfectant (e.g., Cutasept©) and wait until the surface is completely dry.

2. Fix the suction cup on the skin using double-faced adhesive tape on the bottom and adhesive bandages on the top.

3. Connect the cups to the vacuum pump and apply a negative pressure of 180 mbar for 30 min.

4. Increase the pressure to 300 mbar and wait until suction blister are properly formed. Depending on skin type and area, this step usually takes 2–3 h (see Note 3).

5. Carefully remove the suction cup and again disinfect the suction blisters and the surrounding area.

6. Use a syringe with 14-gauge needle to aspirate the suction blister fluid (see Note 4).

7. Carefully prepare the suction blister epidermis using microscissors and directly transfer the tissue to cold PBS (4 °C).

Individual epidermis samples can be stored at 4 °C for 1–2 h if the experiment comprises the preparation of suction blisters from additional donors or skin areas.

8. Cover the wounds with a plaster.

9. Sterilize the epidermis samples in 70% EtOH/PBS for 5 s and immediately transfer the specimen to PBS at room temperature.

10. Incubate the samples in 3 ml 0.25% trypsin for 5 min at 37 °C.

11. Cut off approximately 5 mm of the top of a 1,000 µl pipette tip and gently dissociate the epidermal keratinocytes by trituration.

12. Add 7 ml 10% FCS/PBS and pellet the cells by centrifugation at $150 \times g$ for 5 min.

13. Remove the supernatant and resuspend the cells in 1 ml KGM-2 (37 °C).

14. Proceed to the separation of epidermal stem cells as soon as possible (see Note 5).

3.2 Isolation of Epidermal Stem Cells

1. Add 2 ml of KGM-2 into two collagen IV coated petri dishes each. Make sure that the surface is completely covered with medium and equilibrate for 5 min at 37 °C.

2. Transfer the dishes to a sterile laminar flow.

3. Gently mix the keratinocytes suspension and distribute the complete volume dropwise on one dish.

4. Let the stem cells adhere for 5 min.

5. Very carefully remove the supernatant without moving or inclining the dish. Transfer the medium to the second pre-warmed collagen IV coated dish (see Note 6).

6. Check if individual adherent cells are attached to the collagen IV using a standard microscope. If more than approximately 300 cells are visible in a 20× magnification, the separation efficiency can be enhanced by gently washing the cells with 1 ml of KGM-2 (see Note 7). This additional volume should also be transferred to the second collagen IV coated dish afterwards.

7. Transfer the second collagen IV coated dish into an incubator and culture slowly-adherent cells at 37 °C with 7% $CO2$ (see Note 8).

8. Add 500 ml of pre-warmed PBS to the dish with rapidly adhering cells and manually detach those using a cell lifter.

9. Transfer the cell suspension into a 1.5 ml tube. Wash the dish twice with 400 μl of PBS and also add this volume to the suspension.

10. Pellet the rapidly adhering cells by centrifugation at $500 \times g$ for 5 min, aspirate the supernatant, and resuspend the cells in 30 μl of PBS (see Note 9).

11. Rapidly adhering cells should immediately be used for immunocytochemistry. After a culture time of at least 24 h, the slowly adhering cells on the second collagen IV coated dish can also be analogously detached and harvested. As their cell number is expected to be considerably higher, an increase of the resuspension volume to 50–100 μl is advisable.

3.3 Immunocyto-chemical CPD Staining

1. Place a 20 μl drop of the cell suspension on a poly-L-lysine-coated microscopic slide. Spread the fluid using a pipette tip to an area of 5–7 mm in diameter.

2. If the experiment includes a comparison of cells that were derived from different areas or individuals, the corresponding probes should be placed next to each other on the same microscopic slide. The size of standard slides should be sufficient for analyzing at least four different probes in parallel.

3. Let the fluid evaporate in a drying oven at 40 °C. After 25–35 min, a cloudy layer of PBS salt crystals indicates that drying is complete.

4. Fix the cells by incubating the microscopic slide in –20 °C cold methanol/acetone for 10 min.

5. Air dry the slide for 10 min at RT.

6. Draw a circle around the spot with a water repellant pen and let the fluid dry for additional 5 min (see Note 10).

7. Rehydrate the spot within the circle with 200 μl of PBS for 5 min at RT.

8. Exchange the PBS with 100 μl of pepsin solution and transfer the slide into an humidified incubator for 30 min at 37 °C (see Note 11).

9. Aspirate the pepsin solution and wash the slide two times by adding 200 μl 0.1% BSA/PBS for 5 min at RT.

10. Exchange PBS with 100 μl freshly prepared CPD antibody solution. Incubate in a humidified chamber for 1 h at RT.

11. Aspirate the primary antibody solution and wash the slide three times by adding 200 μl 0.1% BSA/PBS for 5 min at RT.

12. Incubate with 100 μl of secondary antibody solution for 30 min at RT in a humidified chamber in the dark (see Note 12).

13. Repeat the washing step with 200 μl of 0.1% BSA/PBS three times for 5 min in a darkened environment.

14. Place a 10–20 µl drop of mowiol/DAPI solution in the middle of each spot and lay a cover slip on top. Lead any air inclusions to the edge of the slide by gentle pushing using a pipette tip. Take care that the cover slip does not move during this step (see Note 13).

15. Depending on the stability of the secondary antibody fluorophor, the slides can be stored in a dark and dry environment for a few hours although immediate analysis is preferable.

3.4 Analysis of Signal Intensities

1. Use a standard fluorescence microscope for the visualization of CPD stainings. A 40× magnification should be suitable to analyze 50 cells per image. Use the DAPI staining to select an area where enough intact nuclei are visible and acquire a multi-channel image (see Note 14).

2. The quantitative analysis can be performed by any program that supports the extraction of grayscale values from multi-channel images (see Note 15). In a 40× magnification and a picture size of 1,360 × 1,024 pixels, the diameters of nuclei are typically in the range of 40–60 pixels.

3. Select a circular region of interest (ROI) size that is considerable smaller than the smallest nuclei in the image to prevent that background pixels are included into the analysis. At the mentioned resolution, 30 pixels is a reasonable ROI diameter that corresponds to an area of approximately 700 pixels. Do not change the ROI size or shape when analyzing different cells or specimen.

4. Place the ROI into the center of the first nucleus to analyze. Measure the average grayscale value within the ROI of the DAPI and the CPD channel. Repeat the measurement for a sufficient number of nuclei (see Note 16).

5. Divide the average grayscale value of the CPD channel by the corresponding value of the DAPI channel. The resulting data can be regarded as CPD intensities that are normalized to the amount of immobilized DNA of the slide.

6. For the comparison of cells that were derived from different donors or skin sites, the normalized CPD values of all cells that were analyzed from an individual probe can be averaged to determine the global DNA damage status (see Note 17). Alternatively, the cellular intensity distribution of each probe can be visualized in histogram charts.

4   Notes

1. To our knowledge, a suction blister apparatus is not commercially available. Usual systems work with a standard vacuum pump and costum-made plexiglass cups that contain 1–5

cylindrical holes with diameters of 3–10 mm. The original technique is based on a publication of U. Kiistala (22) and adaptions to specific dermatological questions have been frequently reported (23, 24). Our data were generated using a 6 mm system but we also expect other diameters to be applicable.

2. Take care that the pepsin solution is always freshly prepared as the enzymatic activity rapidly decreases when lyophilized pepsin is resolved. Pepsin powder is available at activities of approximately 1,000 U/mg leading to a final concentration of about 5 mg/ml.

3. The time until a suction blister is formed largely depends on individual parameters. We frequently observe that the formation of blisters is for example substantially slower when young donors are investigated. Using transparent plexiglass cups enables to visually control when the epidermis beneath the cup holes is completely detached. The negative pressure values can be slightly adjusted to fasten or slow down the blister formation.

4. Insert the needle and pull the piston of the syringe simultaneously as the wound fluid rapidly flows out of the blister after injection. Suction blister fluid can be stored at –20 °C and used for additional analyzes such as antibody based protein quantifications.

5. Keratinocytes should only be kept in suspension when the experimental design requires additional steps prior to the collagen IV adhesion assay. In that case, we recommend to maintain the cells at 37 °C to avoid any additional stress induced defects. Note that the cells will nevertheless start to differentiate in suspension and the cellular characteristics may change within this timeframe.

6. This step is only necessary if the CPD content of normal basal keratinocytes should also be investigated.

7. Additional washing steps will increase the fraction of rapidly adhering stem cell on the dish. However, if the initial amount of keratinocytes that has been plated is already quite low, it is recommended to resign this step in favor of a sufficient cell number. Make sure that extend of aspiration and washing is equal between cell populations that should be compared in the analysis afterwards.

8. Note that the comparison of CPDs in stem cells that rapidly adhere and normal keratinocytes that slowly adhere will be falsified by the fact that the analysis can only be carried out at different time points after plating. Due to the prolonged adherence time, normal keratinocytes will exhibit an increased CPD repair after irradiation. However, the comparison of probes

from different donors or skin areas is possible with both fractions.

9. In general, the amount of stem cells is too low to obtain a visible cell pellet. To prevent the loss of cells, the supernatant should not be removed completely.

10. It is important that the repellant is entirely dry as it will otherwise partially be dissolved in the washing solution and subsequently impair the fluorescence image acquisition.

11. Do not carry out the digestion step in a dry environment since the small volume will quickly evaporate otherwise.

12. An incubation chamber made of black plastic is most suitable for the prevention of fluorophore bleaching.

13. Any movement of the coverslip after mounting will result in damage to the nuclear structure. Fluorescence values may be nevertheless determined, but the possibility of comparison to other probes will presumably be hindered.

14. Image analysis is generally possible irrespective of the format in which the picture is saved. However, acquiring TIFF-files will ensure that the most comprehensive data will be accessible.

15. For our purposes, we used ImageJ to measure grayscale intensities in multi-image stacks.

16. Depending on the available number of cells and the size of the study, the analysis of 50 cells per sample is a reasonable starting point. ImageJ offers the possibility to develop macros for the simplification of multiple data extractions. Using this option enables the data acquisition for 50 cells in approximately 2 min in our hands.

17. Note that the determination of an average CPD status across all cells is only sensible if a normal distribution of the intensities can be observed. A discrete occurrence of damaged and unaffected cells would indicate the existence of different populations what would be inconsistent with taking the mean. We therefore recommend the visualization of the damage distribution as the initial step of analysis.

References

1. Cadet J, Douki T, Pouget JP et al (2001) Effects of UV and visible radiations on cellular DNA. Curr Probl Dermatol 29:62–73
2. Molho-Pessach V, Lotem M (2007) Ultraviolet radiation and cutaneous carcinogenesis. Curr Probl Dermatol 35:14–27
3. Frosina G (2010) The bright and the dark sides of DNA repair in stem cells. J Biomed Biotechnol: 845396
4. Maynard S, Swistowska AM, Lee JW et al (2008) Human embryonic stem cells have enhanced repair of multiple forms of DNA damage. Stem Cells 26:2266–2274
5. Bracker TU, Giebel B, Spanholtz J et al (2006) Stringent regulation of DNA repair during human hematopoietic differentiation: a gene expression and functional analysis. Stem Cells 24:722–730

6. Chen MF, Lin CT, Chen WC et al (2006) The sensitivity of human mesenchymal stem cells to ionizing radiation. Int J Radiat Oncol Biol Phys 66:244–253

7. Rachidi W, Harfourche G, Lemaitre G et al (2007) Sensing radiosensitivity of human epidermal stem cells. Radiother Oncol 83:267–276

8. Saretzki G, Armstrong L, Leake A et al (2004) Stress defense in murine embryonic stem cells is superior to that of various differentiated murine cells. Stem Cells 22:962–971

9. Sun X, Fu X, Sheng Z (2007) Cutaneous stem cells: something new and something borrowed. Wound Repair Regen 15:775–785

10. Kaur P (2006) Interfollicular epidermal stem cells: identification, challenges, potential. J Invest Dermatol 126:1450–1458

11. Jones PH, Watt FM (1993) Separation of human epidermal stem cells from transit amplifying cells on the basis of differences in integrin function and expression. Cell 73:713–724

12. Li J, Miao C, Guo W et al (2008) Enrichment of putative human epidermal stem cells based on cell size and collagen type IV adhesiveness. Cell Res 18:360–371

13. Jones PH, Harper S, Watt FM (1995) Stem cell patterning and fate in human epidermis. Cell 80:83–93

14. Ohyama M, Terunuma A, Tock CL et al (2006) Characterization and isolation of stem cell-enriched human hair follicle bulge cells. J Clin Invest 116:249–260

15. Nijhof JG, Braun KM, Giangreco A et al (2006) The cell-surface marker MTS24 identifies a novel population of follicular keratinocytes with characteristics of progenitor cells. Development 133:3027–3037

16. Wan H, Stone MG, Simpson C et al (2003) Desmosomal proteins, including desmoglein 3, serve as novel negative markers for epidermal stem cell-containing population of keratinocytes. J Cell Sci 116:4239–4248

17. Webb A, Li A, Kaur P (2004) Location and phenotype of human adult keratinocyte stem cells of the skin. Differentiation 72:387–395

18. Li A, Simmons PJ, Kaur P (1998) Identification and isolation of candidate human keratinocyte stem cells based on cell surface phenotype. Proc Natl Acad Sci U S A 95:3902–3907

19. Moriwaki S, Takahashi Y (2008) Photoaging and DNA repair. J Dermatol Sci 50:169–176

20. Nijhof JG, van Pelt C, Mulder AA et al (2007) Epidermal stem and progenitor cells in murine epidermis accumulate UV damage despite NER proficiency. Carcinogenesis 28:792–800

21. Ruetze M, Dunckelmann K, Schade A et al (2011) Damage at the root of cell renewal–UV sensitivity of human epidermal stem cells. J Dermatol Sci. doi:10.1016/j.jdermsci.2011.06.010

22. Kiistala U (1968) Suction blister device for separation of viable epidermis from dermis. J Invest Dermatol 50:129–137

23. Gupta S, Shroff S (1999) Modified technique of suction blistering for epidermal grafting in vitiligo. Int J Dermatol 38:306–309

24. Kim HU, Yun SK (2000) Suction device for epidermal grafting in vitiligo: employing a syringe and a manometer to provide an adequate negative pressure. Dermatol Surg 26:702–704

Chapter 10

Analysis of Gene Expression in Skin Using Laser Capture Microdissection

Briana Lee, Mikhail Geyfman, Bogi Andersen, and Xing Dai

Abstract

Gene expression analysis is a useful tool to study the molecular mechanisms underlying skin development and homeostasis. Here we describe a method that utilizes laser capture microdissection (LCM) to isolate RNAs from localized areas of skin, allowing the characterization of gene expression by RT-PCR and microarray technologies.

Key words Laser capture microdissection, Epidermis, Hair follicle, Stem/progenitor cells, mRNA, Reverse transcription, Gene expression

1 Introduction

Mammalian skin is a complex organ with a multitude of epithelial and stromal cell types and harbors various appendages such as hair follicles which themselves are "miniorgans." As the skin epithelial compartment develops, it acquires multiple cell types with specific cellular functions. The interfollicular epidermis continuously self-renews throughout life, owing to the activity of epidermal stem/progenitor cells (1, 2). In contrast, hair follicles undergo periodic cycles of proliferation and apoptosis, a process referred to as the hair cycle in which regeneration is fueled by stem cells located in distinct anatomic locales (e.g., bulge, secondary hair germ, and isthmus) (3). A thorough understanding of the molecular mechanisms governing epidermal/hair follicle development and regeneration will benefit from a systematic analysis of gene expression in the constituent cell types at different developmental and regeneration stages.

Physical separation, especially when used in combination with enzyme digestion, provides a useful way for isolating distinct skin cell populations. A time-tested method is to separate epidermis from its underlying dermis using trypsin/dispase followed by

Kursad Turksen (ed.), *Skin Stem Cells: Methods and Protocols*, Methods in Molecular Biology, vol. 989,
DOI 10.1007/978-1-62703-330-5_10, © Springer Science+Business Media New York 2013

Table 1
Summary of markers used to identify skin epithelial stem/progenitor cell populations

Marker	Location	Reference
α6 Integrin	Basal epidermis	(10)
Sca1	Basal epidermal cells	(11)
CD34	Bulge	(12)
Lgr5	Lower bulge, hair germ	(13)
Lgr6	Isthmus	(14)
Lrig1	Infundibulum	(15)
MTS24	Infundibulum	(16)
Blimp1	Infundibulum	(17)
Gli1	Upper bulge	(18)
Ephrin β1	Transient amplifying cells of the hair matrix	(19)

physical peeling. However, this method does not allow the separation of distinct epidermal cell types. Moreover, epidermis of earlier developmental stages (e.g., E13.5) is extremely thin and difficult to peel off. The culture of primary keratinocytes from skin offers a valid substitute for analysis; however, it is unavoidably accompanied by a deviation from the true in vivo gene expression program due to altered growth conditions and lack of normal epithelial–mesenchymal interactions.

In recent years, fluorescence activated cell sorting (FACS) based on various surface markers has emerged as a powerful strategy to isolate distinct populations of skin epithelial stem cells, leading to the accumulation of tremendous amount of knowledge about adult stem cell behavior (4) (Table 1). However, the utility of this methodology is limited by the availability of markers. FACS has also been applied to cases where the desired cell types are labeled with green fluorescent protein (GFP) using cell type-specific promoters or based on characteristic cell behaviors (e.g., slow-cycling of bulge stem cells) (5, 6). However, these approaches typically involve complex breeding of transgenic mouse lines and sometimes drug administration to achieve precisely controlled GFP expression in a spatiotemporal manner.

For reasons discussed above, laser capture microdissection (LCM), which allows the collection of cells from histological sections, offers a useful alternative approach to compare gene expression in various skin cell types during development as well as in adulthood. Unlike FACS, this method enables one to directly identify and recover cell populations without reliance on known cell-surface markers or undertaking of complicated and expensive transgenic experiments.

LCM offers the added advantage of permitting the isolation of cells in their normal in vivo settings. This is particularly relevant for study of stem cells, which are known to reside in niches which include, in addition to stem cells themselves, nearby signaling cells and extracellular matrix (7). LCM allows one to collect an intact niche without disrupting its components as is required by FACS. Additionally, it is possible to use immunofluorescence to label a subset of cells within small regions, thus guiding the capture of specific cells (8).

Currently methods are available for using the small amount of material obtained from LCM to perform molecular/biochemical assays such as RT-PCR, microarray gene expression studies, and even proteomic analyses (9). The shortcomings of LCM include a relatively high initial cost of the LCM microscope and contamination from adjacent tissue. However, experienced users report minimal contamination, and multiple tissue-specific genes can be used to accurately evaluate the relative amount of contaminating material in laser captured tissues.

2 Materials

For RNA isolation and cDNA synthesis, modified protocols and reagents from Qiagen RNeasy and Sensiscript kits, respectively, were utilized.

2.1 Preparing Skin Cryoblocks

1. Petri dishes (Bacterial grade).
2. Tissue-TekCryomold (15 mm × 15 mm × 5 mm), Electron Microscopy Sciences, Cat. No. 62352–15.
3. Tissue TekCryo-OCT, Fisher, Cat. No. NC9695545.
4. Dry ice.
5. Ice-cold PBS.
6. Fine scissors and forceps, RNAse Away-treated.

2.2 Sectioning Skin Cryoblocks

1. PEN-membrane slides, Leica, Cat. No. 11505158.
2. UV crosslinker (e.g., Stratagene UV Stratalinker 2400).
3. Tissue TekCryo-OCT.
4. Cryostat (e.g., Leica CM1850).
5. RNase Away, Fisher, Cat. No. 7003.
6. Ethanol Gold Shield, 200 proof.

2.3 Fixation and H&E Staining

1. 70% and 100% Ethanol.
2. Xylene, Fisher, Cat. No. X3S-4.
3. Acetone, Fisher, Cat. No. A929-4.
4. DEPC-treated water, Sigma, Cat. No. D5758 or home-made.

	5. Meyer's hemotoxylin solution, Sigma, Cat. No. MHS1.
	6. Slide mailers, Fisher, Cat. No. HS15986.

2.4 LCM

1. Flat cap 0.5 ml PCR tubes, Fisher, Cat. No. 14-222-265.
2. RNase away.
3. Buffer RLT, Qiagen, Cat. No. 79216.
4. Beta-mercaptoethanol (BME), Sigma, Cat. No. M3148.
5. Laser microdissection microscope (e.g., Leica LS-AMD).

2.5 RNA Isolation

1. RNeasyMicro kit, Qiagen, Cat. No. 74004.
   - Buffer RLT and RW1.
2. 20 ng/µl proteinase K, Fisher, Cat. No. BP1700-100.
3. 100% ethanol.
4. Microcentrifuge.

2.6 RNA Quality Control

1. Spectrophotometer.
2. Agilent 2100 Bioanalyzer.

2.7 Reverse Transcription

1. Sensiscript RT kit, Qiagen, Cat. No. 205221.
2. Random hexamer primers, Applied Biosystems, Cat. No. N8080127.
3. 37°C water bath.
4. Eppendorf tubes.

3 Methods

3.1 Preparing Skin Cryoblocks

1. Always use skin samples that have been harvested immediately before placing in cryoblock and keep the tissues on ice until freezing in OCT.
2. Excise a piece of back skin using fine scissors and forceps (pretreated with RNase Away) and briefly wash with ice-cold PBS (see Note 1).
3. Place a cryoblock on a piece of dry ice and cover the bottom with a few drops of OCT medium. Allow it to freeze.
4. Place skin sample with epidermis facing up into the cryoblock and completely cover the tissue with OCT medium, being careful to avoid bubbles (see Note 2).
5. Wait until the OCT is frozen thoroughly and store the samples at −80°C or proceed immediately to sectioning step.

3.2 Sectioning

1. Clean the inside of the cryostat with ethanol and RNase away.
2. Incubate slides in UV crosslink chamber on maximum power for 30 min.

3. Adjust cryostat chamber temperature to –25°C and place OCT tissue samples into the chamber for 20 min to allow the tissue to equilibrate to chamber temperature.

4. Cut 8 μm sections and place no more than four slices onto one membrane slide (see Note 3).

5. Keep slides at –80°C for up to 1 week in a tightly closed slide box until they are ready to be fixed and stained.

3.3 Fixation and H&E Staining

All materials, including slide containers and forceps should be treated with RNase away prior to use.

1. Prepare the following slide holders containing: ice-cold acetone, ice-cold 70% ethanol, DEPC water (x2), Meyer's hematoxylin, 70% ethanol, 100% ethanol (x2), and xylene (x2).

2. Incubate slides in the following order:

Fixation

- Cold acetone for 5 min.

- Cold 70% ethanol for 1 min.

- Wash with DEPC water (RNase-free) for 10 s.

- H&E staining

- Meyer's hematoxylin for 30 s.

- Wash with DEPC water for 10 s.

Dehydration

- 70% ethanol for 10 s.

- 100% ethanol for 30 s.

- 100% ethanol for another 30 s.

- Xylene for 30 s.

- Xylene for 5 min (see Note 4).

3. Proceed immediately to LCM.

3.4 LCM (Refer to Leica LMD Manual for Details and Illustrations of Cap Loading and Basic Operation)

1. Add 30 μl BME per 1 ml RLT buffer.

2. Insert a flat cap PCR tube into the tube holder on the sample collection tray.

3. Load the cap of PCR tubes with 75 μl BME/RLT buffer.

4. Insert a LCM membrane slide facing down.

5. Align the sample tray according to operational instructions.

6. Cut the desired area by tracing a line around the outside region of the area so that the laser is cutting the cells directly adjacent to your desired cells (see Note 5).

7. Collect at least 2,000 epidermal keratinocytes (see Fig. 1a) or hair follicle keratinocytes or dermal papillae cells in one tube (see Fig. 2a). Make sure that a total dissection time does not

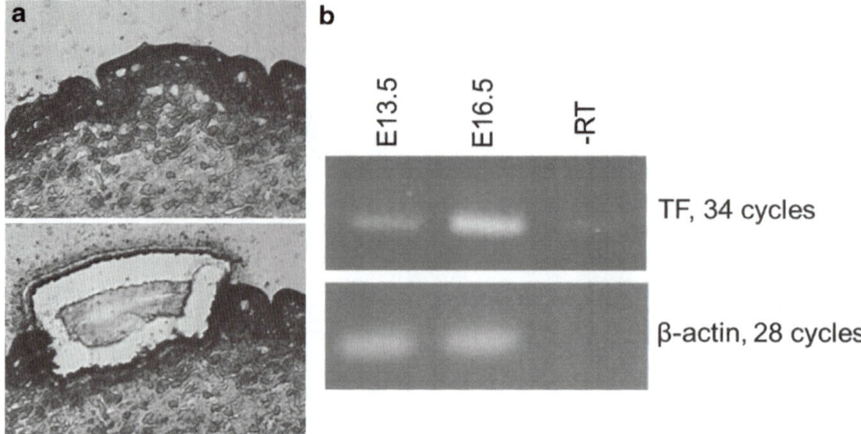

Fig. 1 LCM of embryonic epidermal cells. (**a**) E16.5 skin sections before and after LCM. Cutting was performed with attempt to minimize burning of desired epidermal cells while avoiding the capture of dermal cells. (**b**) Semi-quantitative RT-PCR was performed on RNAs prepared from E13.5 and E16.5 laser captured epidermal cells as shown in (**a**), using primers against a transcription factor (TF) and housekeeping gene β-actin

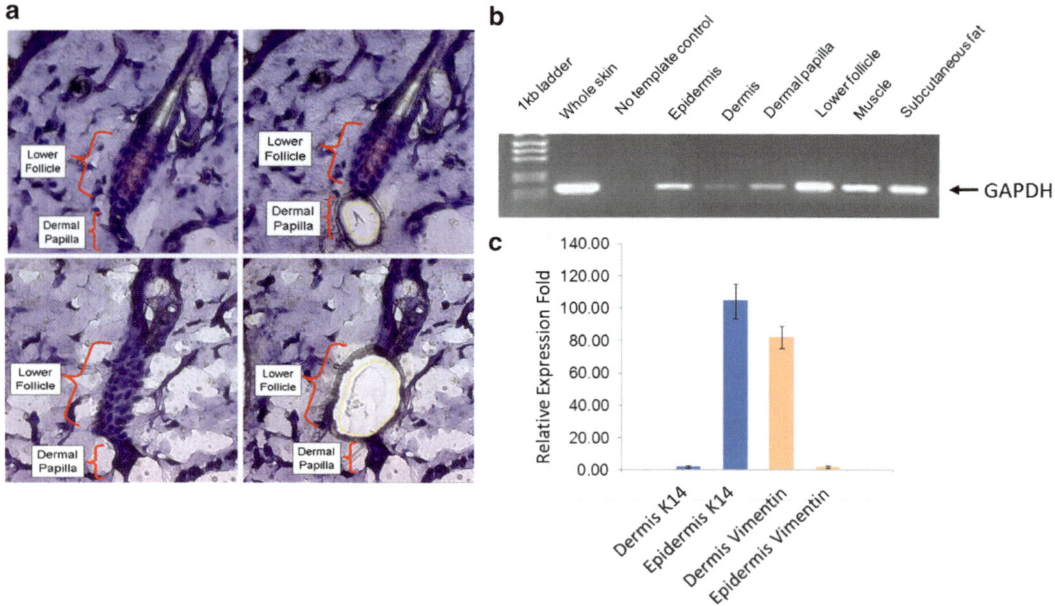

Fig. 2 LCM of adult hair follicle cells. (**a**) Adult skin sections before and after LCM. *Top*, collection of dermal papilla cells. *Bottom*, collection of the lower part of the follicle containing the bulge area. Cutting was performed with attempt to minimize contamination while avoiding UV laser induced thermal damage of RNA. (**b**) RT-PCR analysis of RNAs prepared from laser captured cells of epidermis, dermis, dermal papilla, lower follicle, muscle, and subcutaneous layer. A 1/10 of cDNA generated was used to PCR amplify a ~100-bp segment of GAPDH (35 cycles). (**c**) qRT-PCR analysis of the expression of Keratin 14 (basal epidermis-specific gene) and Vimentin (dermis-specific gene) in laser captured dermis and epidermis

exceed 1 h. Therefore we recommend processing no more than two slides at a time.

8. Vortex tissue in buffer for 2 min at full speed.

9. Process samples immediately or store at –80°C for up to a week (see Note 6).

3.5 RNA Isolation

1. If possible, carry out all pipetting steps using filter tips in a sterile laminar flow hood and clean equipment with RNase away.

2. RNA extraction using RNeasy Micro Kit.

   (a) Vortex for a few minutes at full speed if the samples have been stored at –80°C.

   (b) Adjust volume of tissue harvests to 150 μl (you can combine two 75 μl tubes or just add another 75 μl RLT buffer to one tube).

   (c) Vortex at full speed for 30 s.

   (d) Add 295 μl RNase-free water.

   (e) Add 5 μl proteinase K (20 ng/μl). Mix by pipetting.

   - Incubate at 55°C for 10–15 min.

   (f) Microfuge for 3 min at 12,000 rpm ($10,000 \times g$).

   (g) Pipette supernatant into a new tube provided with the kit.

   (h) Add 225 μl of 100% RNase-free ethanol.

   - Mix by pipetting.

   (i) Transfer to RNeasy MiniElute spin column with 2 ml collection tube.

   (j) Close the lid of the microfuge gently and microfuge for 15 s at 10,000 rpm ($8,000 \times g$).

   - Optional: run the flow-through on the column again. This may increase the amount of RNA that binds to the column; however, you also run the risk of increasing salt contamination.

   - Discard the flow-through.

   (k) Add 700 μl Buffer RW1 to the spin column.

   - Microfuge 15 s at 10,000 rpm ($8,000 \times g$).

   - Discard the flow-through and collection tube.

   (l) Place spin column in a new 2 ml collection tube provided with the kit.

   - Add 500 μl Buffer RPE (ethanol added).

   - Microfuge 15 s at 10,000 rpm ($8,000 \times g$).

   - Discard the flow-through.

(m) Add 500 μl fresh 80% ethanol (use RNase-free water) to the spin column.

- Microfuge 15 s at 10,000 rpm (8,000×g).

- Discard the flow-through and the collection tube. Do not let spin column contact the flow-through when removing the column.

(n) Place spin column in a new 2 ml collection tube provided with the kit.

- Microfuge at full speed (14,000×g) for 5 min. Leave the spin column lid open (evaporates ethanol).

- Discard the flow-through and the collection tube.

(o) Place spin column in a new 1.5 ml collection tube.

- Add 20 μl RNase-free water directly to the center of the spin column membrane.

- Microfuge for 1 min at full speed to elute the RNA.

- Run the eluate through the spin column again to increase the RNA yield.

3.6 Quality Control

1. Maintain RNase-free working conditions as described above.

2. Determine RNA concentration with a spectrophotometer measuring 260 nm absorption and Agilent Bioanalyzer to check RNA integrity (see Note 7).

3. Proceed to reverse transcription.

3.7 Reverse Transcription

1. Maintain RNase-free working conditions as described above.

2. Use the Sensiscript RT kit and random hexamer primers for reverse transcription.

3. Aqueous solutions of cDNA can be stored at −20°C.

4. Perform RT-PCR with housekeeping and tissue-specific genes in order to validate lack of contamination from unintended tissue and to estimate cDNA yield (see Figs. 1b and 2b, c).

4   Notes

1. For adult mouse skin, make sure to clip hair after sacrificing and prior to removal of mouse skin.

2. Since the specimen is no longer visible after OCT embedding, make a note of which way the epidermis is facing.

3. Be careful not to overlap the OCT sections, since that leads to loss of sections in subsequent staining steps.

4. Slides can be kept in xylene for longer than 5 min during transport to the LCM. We have not noted deleterious effects to the

     tissue or purified material when slides had been kept in xylene for longer than 5 min.

5. If the line is drawn too close to the cells to be collected, the laser will burn them away. Draw a line just outside the area of interest (refer to Fig. 1a, 2a).

6. Once all samples are cut and captured, immediately process or store at −80°C. Do not leave samples out at room temperature for longer than 30 min.

7. For each 2,000-cell harvest, typical yield of RNA is 15–20 ng/μl for embryonic epidermis, 10–20 ng/μl for hair germ, 10–20 ng/μl for dermal papilla, and 20–30 ng/μl for the bulge.

References

1. Clayton E et al (2007) A single type of progenitor cell maintains normal epidermis. Nature 446(7132):185–189
2. Jensen KB et al (2009) Lrig1 expression defines a distinct multipotent stem cell population in mammalian epidermis. Cell Stem Cell 4(5):427–439
3. Woo WM, Oro AE (2011) SnapShot: hair follicle stem cells. Cell 146(2):334–334e2
4. Fuchs E (2009) The tortoise and the hair: slow-cycling cells in the stem cell race. Cell 137(5):811–819
5. Rhee H, Polak L et al (2006) Lhx2 maintains stem cell character in hair follicles. Science 312(5782):1946–1949
6. Fuchs E, Horsley V (2011) Ferreting out stem cells from their niches. Nat Cell Biol 13(5):513–518
7. Spradling A, Drummond-Barbosa D, Kai T (2001) Stem cells find their niche. Nature 414(6859):98–104
8. Murakami H, Liotta L et al (2000) IF-LCM: laser capture microdissection of immunofluorescently defined cells for mRNA analysis rapid communication. Kidney Int 58(3):1346–1353
9. Xu BJ et al (2009) Combining laser capture microdissection and proteomics: methodologies and clinical applications. Proteomics Clin Appl 4(2):116–123
10. Blanpain C et al (2004) Self-renewal, multipotency, and the existence of two cell populations within an epithelial stem cell niche. Cell 118(5):635–648
11. Albert MR, Foster RA, Vogel JC (2001) Murine epidermal label-retaining cells isolated by flow cytometry do not express the stem cell markers CD34, Sca-1, or Flk-1. J Invest Dermatol 117(4):943–948
12. Trempus CS et al (2003) Enrichment for living murine keratinocytes from the hair follicle bulge with the cell surface marker CD34. J Invest Dermatol 120(4):501–511
13. Jaks V et al (2008) Lgr5 marks cycling, yet long-lived, hair follicle stem cells. Nat Genet 40(11):1291–1299
14. Snippert HJ et al (2011) Lgr6 marks stem cells in the hair follicle that generate all cell lineages of the skin. Science 327(5971):1385–1389
15. Jensen KB, Collins CA et al (2009) Lrig1 expression defines a distinct multipotent stem cell population in mammalian epidermis. Cell Stem Cell 4(5):427–439
16. Nijhof JG et al (2006) The cell-surface marker MTS24 identifies a novel population of follicular keratinocytes with characteristics of progenitor cells. Development 133(15):3027–3037
17. Horsley V et al (2006) Blimp1 defines a progenitor population that governs cellular input to the sebaceous gland. Cell 126(3):597–609
18. Brownell I et al (2011) Nerve-derived sonic hedgehog defines a niche for hair follicle stem cells capable of becoming epidermal stem cells. Cell Stem Cell 8(5):552–565
19. Lien WH et al (2011) Genome-wide maps of histone modifications unwind in vivo chromatin states of the hair follicle lineage. Cell Stem Cell 9(3):219–232

Chapter 11

Qualitatively Monitoring Binding and Expression of the Transcription Factor Sp1 as a Useful Tool to Evaluate the Reliability of Primary Cultured Epithelial Stem Cells in Tissue Reconstruction

Manon Gaudreault, Danielle Larouche, Lucie Germain, and Sylvain L. Guérin

Abstract

Electrophoretic mobility shift assay and Western blot are simple, efficient, and rapid methods for the study of DNA–protein interactions and expression, respectively. Primary cultures and subcultures of epithelial cells are widely used for the production of tissue-engineered substitutes and are gaining popularity as a model for gene expression studies. The preservation of stem-cells through the culture process is essential to produce high quality substitutes. However as such cells are passaged in culture, they often lose their ability to proliferate, a process likely to be determined by the altered expression of nuclear-located transcription factors such as Sp1, whose expression has been documented to be required for cell adhesion, migration, and differentiation. Our recent studies demonstrated that reconstructed tissues exhibiting poor histological and structural characteristics are also those that were produced with epithelial cells in which expression and DNA binding of Sp1 was reduced in vitro. Therefore, monitoring both the expression and DNA binding of this transcription factor in human skin and corneal epithelial cells might prove a particularly useful tool for selecting which cells are to be used for tissue reconstruction.

Key words Stem cells, Epidermis, Cornea, Corneal epithelial cells, Tissue engineering, Transcription factor, Sp1, EMSA, Gel shift, Mobility shift assay, Western blot

1 Introduction

Several considerations must be taken into account when developing engineered tissues for transplantation. The preservation of stem cells into the implanted tissues is of the utmost importance to warrant survival and long-term persistence after grafting. Consequently, it has become particularly important to develop molecular tools that validate the cell culture method in order to make sure that stem cells are present and preserved. Even if the absence of allogeneic feeder layer cells would be desirable during the production of autologous tissues for clinical purpose, yet culturing epithelial cells in the presence

Kursad Turksen (ed.), *Skin Stem Cells: Methods and Protocols*, Methods in Molecular Biology, vol. 989, DOI 10.1007/978-1-62703-330-5_11, © Springer Science+Business Media New York 2013

of fibroblasts and/or their secreted products greatly improve their morphological characteristics (1–4). We have previously shown that a feeder layer of irradiated Swiss 3T3 (i3T3) cells increases the number of cell passages that primary cultured human epithelial cells can sustain in vitro (5). In human skin keratinocytes, the presence of such a feeder layer delays keratinocytes terminal differentiation by preventing extinction of the positive transcription factor Sp1 within the nucleus of these cells. Furthermore, the progressive loss of Sp1 expression during the last passages of cultured human skin keratinocytes or corneal epithelial cells correlates well with their growth arrest and terminal differentiation (5, 6). Monitoring both the expression and DNA-binding properties of Sp1 can therefore be very useful as predictors of epithelial cell culture's quality and reliability before they are to be used for the production of engineered tissues.

This chapter details the methods for harvesting and processing nuclear and total protein extracts of epithelial cells that have been cultured on an irradiated i3T3 feeder layer as described (for a detailed protocol, see (7)), as well as the further analysis of Sp1 by Western blotting and electrophoretic mobility shift assay (EMSA).

2 Materials

2.1 Culture Media

2.1.1 Tissue Transport Medium

1. Components used for the preparation of the tissue transport medium (for final concentration see Table 1):

   (a) *DME-Ham.* Dulbecco's modified Eagle's medium (DME) (cat. no. 12800, Invitrogen, Oakville, Ont.): Dissolve Ham's F12 medium (cat. no. 21700, Invitrogen), 3:1, 3.07 g/L NaHCO$_3$ (36.5 mM) (cat. no. S233, Fisher Scientific, Ottawa, Ont.), 24.3 mg/L adenine (0.18 mM) (first dissolve adenine in 0.02N HCL to make a 1 M solution) (cat. no. A2786, Sigma Chemicals, St-Louis, MO), in apyrogenic ultrapure water. Adjust pH to 7.1. Sterilize by filtration through a 0.22-μm low-binding disposable filter. Aliquot and store at 4°C.

   (b) *Fetal calf serum* (cat. no. SH30396, HyClone, Logan, UT). Thaw in cold water. Inactivate in hot water (56°C) for 30 min. Distribute in single use aliquots (50 mL) and store at –20°C.

   (c) *Penicillin G and Gentamicin* (cat. no. P3032 and G1264, respectively, Sigma Chemicals). Dissolve 50,000 IU/mL of penicillin G and 12.5 mg/mL of gentamicin sulfate in apyrogenic ultrapure water to make a 500× stock solution. Sterilize by filtration through a 0.22 μm low-binding disposable filter, distribute in single use aliquots (2 mL), and store at –80°C.

   (d) *Fungizone* (amphotericin B, cat. no. A9528, Sigma Chemicals). Dissolve 0.25 mg/mL of amphotericin B

Table 1
Tissue transport medium

Component	Quantity (mL)	Final concentration
DME-Ham	900	90% (v/v)
Fetal calf serum	100	10% (v/v)
Penicillin G–gentamicin 500×	2	Penicillin G 100 IU/mL Gentamicin 25 µg/mL
Fungizone 500×	2	0.5 µg/mL

(0.27 mM), in apyrogenic ultrapure water to make a 500× stock solution. Sterilize by filtration through a 0.22-µm low-binding disposable filter, distribute in single use aliquots, and store at –80°C.

2. Thaw all components at 4°C. To make 1 L, refer to Table 1.

3. Store the transport medium at 4°C.

2.1.2 Complete Human Keratinocyte Culture Medium (Complete hkDME-Ham)

1. Components used for the preparation of the human keratinocyte culture medium (for final concentration see Table 2):

(a) *DME-Ham* (see Subheading 2.1.1; item 1(a)).

(b) *Fetal clone II serum* (cat. no. SH30066, HyClone). Thaw in cold water. Inactivate in hot water (56°C) for 30 min. Distribute in single use aliquots (25 mL) and store at –20°C.

(c) *Insulin* (cat. no. I1882, Sigma Chemicals). Dissolve 250 mg in 50 mL 0.005N HCl (125 µL 2N HCl/50 mL apyrogenic ultrapure water) to make a 1,000× stock solution (5 mg/mL). Sterilize by filtration through a 0.22-µm low binding disposable filter, distribute in single use aliquots, and store at –80°C.

(d) *Hydrocortisone* (cat. no. 386698, Calbiochem, San Diego, CA). Dissolve 25 mg in 5 mL of 95% ethanol (4.8 mL 99% ethanol (Les Alcools de Commerce Inc., Brampton, Ont.)/0.2 mL apyrogenic ultrapure water). Complete to 125 mL with DME-Ham (see Subheading 2.1.1; item 1(**a**)) to make a 500× stock solution (0.53 mM). Sterilize by filtration through a 0.22-µm low-binding disposable filter, distribute in single use aliquots (1 mL), and store at –80°C.

(e) *Cholera toxin* (cat. no. C8052, Sigma Chemicals). Dissolve 1 mg in 1 mL of apyrogenic ultrapure water. Complete to 118 mL with DME-Ham (see Subheading 2.1.1; item 1(a)) supplemented with 10% (v/v) Fetal clone II (see Subheading 2.1.2; item 1(b)) to make a 1,000× stock solution. Sterilize by filtration through a 0.22-µm low-binding

Table 2
Complete human keratinocyte culture medium (complete hkDME-Ham)

Component	Quantity (mL)	Final concentration
DME-Ham	950	95% (v/v)
Fetal clone II	50	5% (v/v)
Insulin 1,000×	1	5 µg/mL
Hydrocortisone 500×	2	0.4 µg/mL
Cholera toxin 1,000×	1	10^{-10} M
Epidermal growth factor 1,000×	1	10 ng/mL
Penicillin G–gentamicin 500×	2	Penicillin G 100 IU/mL Gentamicin 25 µg/mL

disposable filter, distribute in single use aliquots and store at –80°C.

(f) *Epidermal growth factor* (cat. no. GF-010-8, Austral Biologicals, San Ramon, CA). Dissolve 500 µg in 2.5 mL of 10 mM HCl. Complete to 50 mL with DME-Ham (see Subheading 2.1.1, item 1(a)) supplemented with 10% (v/v) fetal clone II (see Subheading 2.1.2, item 1(b)) to make a 1,000× stock solution. Sterilize by filtration through a 0.22-µm low-binding disposable filter, distribute in single use aliquots, and store at –80°C.

(g) *Penicillin G–Gentamicin* (see Subheading 2.1.1, item 1(c)).

2. Thaw all components at 4°C. To make 1 L, refer to Table 2 (see Note 1).

3. Complete hkDME-Ham can be stored at 4°C for 10 days.

2.1.3 Complete Human Corneal Epithelial Cell Culture Medium (Complete hcDME-Ham)

1. Components used for the preparation of the human corneal epithelial cell culture medium (for final concentration, see Table 3):

(a) *Transferrin/T3*. Transferrin (vial of 600 mg/20 mL, cat. no. 652202, Roche Diagnostics, Laval, QC, Canada)/3,3′,5′-triiodo-L-thyronine (T3) (cat. no. T2752, Sigma Chemicals). Solution #1: Dissolve 6.8 mg of T3 in a small volume (about 3 mL) of 0.02N NaOH (10 µL 10N NaOH/4.99 mL apyrogenic ultrapure water). Complete to 50 mL with apyrogenic ultrapure water. Sterilize by filtration through a 0.22-µm low-binding disposable filter, distribute in single use aliquots (1.2 mL), and store at 4°C. Solution #2: Mix the transferrin vial with 1.2 mL of solution #1 and complete to 120 mL with apyrogenic ultrapure water to make a 1,000× stock solution. Sterilize by filtration through a 0.22-µm low-binding disposable filter, distribute in single use aliquots (0.5 mL), and store at –80°C.

Table 3
Complete human corneal epithelial cell culture medium (complete hcDME-Ham)

Component	Quantity (mL)	Final concentration
DME-Ham	900	90% (v/v)
Fetal clone II	100	10% (v/v)
Insulin 1,000×	1	5 µg/mL
Hydrocortisone 500×	2	0.4 µg/mL
Cholera toxin 1,000×	1	10^{-10} M
Epidermal growth factor 1,000×	1	10 ng/mL
Transferrin/T3 1,000×	1	Transferrin: 5 µg/mL T3: 2×10^{-9} M
Penicillin G–gentamicin 500×	2	Penicillin G 100 IU/mL Gentamicin 25 µg/mL

Table 4
Freezing medium

Component	Quantity (mL)	Final concentration
Fetal calf serum	9	90% (v/v)
DMSO	1	10% (v/v)

2. Thaw all components described in Subheading 2.1.2, item 1, with the addition of Transferrin/T3, at 4°C. To make 1 L, refer to Table 3 (see Note 1).

3. Complete hcDME-Ham can be stored at 4°C for 10 days.

2.1.4 Freezing Medium

1. Preparation of freezing medium components.

    (a) Fetal calf serum (see Subheading 2.1.1, item 1(b)).

    (b) DMSO (see Note 2).

2. Thaw all components at 4°C. To make 10 mL, refer to Table 4.

2.2 Monolayer Culture of Human Keratinocytes

2.2.1 Extraction and Culture of Human Keratinocytes

1. Source of human keratinocytes: normal adult skin or newborn foreskin removed by surgery.

2. Sterile container.

3. Transport medium (see Subheading 2.1.1).

4. 50 mL centrifuge tubes (cat. no. 352070, BD Falcon, Franklin Lakes, NJ).

5. Phosphate buffered saline–Penicillin G/Gentamicin/Fungizone (PBS-P/G/F): 137 mM NaCl (cat. no. S271, Fisher Scientific),

2.7 mM KCl (cat. no. P285, Fisher Scientific), 6.5 mM Na_2HPO_4 (cat. no. S369, Fisher Scientific), 1.5 mM KH_2PO_4 (cat. no. P217, Fisher Scientific). Dissolve in apyrogenic ultrapure water to make a 10× stock solution. Store at room temperature. To dilute 10× PBS to 1×, add apyrogenic ultrapure water. Verify pH is 7.4. Store at room temperature. Before use, add Penicillin G–Gentamicin 500× stock solution (see Subheading 2.1.1, item 1(c)) and Fungizone 500× stock solution (see Subheading 2.1.1, item 1(d)) by diluting these additives to 1×.

6. Petri dishes (size: 100×15 mm, cat. no. 08-757-12, Fisher Scientific).

7. HEPES (cat. no. 194549, MP Biomedicals Inc., Montreal, QC, Canada). Make a 5× stock solution in apyrogenic ultrapure water: 0.05 M HEPES, 33.5 mM KCl (cat. no. P217, Fisher Scientific), 0.71 M NaCl (cat. no. S271, Sigma Chemicals). Adjust pH to 7.3. Protect from light and store at 4°C. HEPES 1×: dilute the 5× stock solution to 1× with apyrogenic ultrapure water and add 1 mM $CaCl_2$ (cat. no. C7902, Sigma Chemicals). Adjust pH to 7.45. Protect from light (see Note 3) and store at 4°C.

8. Thermolysin (cat. no. T7902, Sigma Chemicals). Dissolve 500 µg/mL in HEPES 1× (see Subheading 2.2.1, item 7). Sterilize by filtration through a 0.22-µm low-binding disposable filter and store at 4°C (see Note 4).

9. Trypsin/EDTA. 2.8 mM D-glucose (cat. no. DX0145, EMD Chemicals Inc., Gibbstown, NJ), 0.05% (w/v) trypsin 1-500 (cat. no. 7003, Intergen, Toronto, ON, Canada), 0.00075% (v/v) phenol red (Phenol red solution 0.5%, sterile-filtered, cat. no. P0290, Sigma Chemicals), 100,000 IU/L penicillin G (cat. no. P3032, Sigma Chemicals), 25 mg/L gentamicin (cat. no. G1264, Sigma Chemicals), 0.01% (w/v) EDTA (EDTA, Disodium salt, cat. no. 8993, J.T. Baker, Phillipsburg, NJ). Dissolve in 1× PBS. Adjust pH to 7.45. Sterilize by filtration through a 0.22-µm low-binding disposable filter, distribute in single use aliquots (10 mL), and store at −20°C.

10. Dissecting curved forceps (cat. no. 08-953F, Fisher Scientific).

11. Scalpel (#4 cat. no. 08-917-5, Fisher Scientific) and blade (size: 22, cat. no. 73-0422, Personna Medical, Verona, VA).

12. Trypsinization unit (Celstir suspension culture flask, cat. no. 356875, Wheaton Sciences Products, Millville, NJ).

13. Parafilm (cat. no. 13-374, Fisher Scientific).

14. Tissue culture flask, 75 cm^2 (cat. no. 35310, BD Falcon).

15. Complete hkDME-Ham (see Subheading 2.1.2).

16. Irradiated i3T3 (ATCC, #CCL-92). To obtain about $8–10 \times 10^6$ cells, seed 1×10^6 cells in a 75 cm^2 culture flask (Tissue culture flask, cat. no. 35310, BD Falcon) with 20 mL of DME. Incubate for 4 days in 8% CO$_2$, 100% humidity atmosphere at 37°C. Irradiate at 6,000 rads with a Gammacell irradiator (^{60}Co source), see Note 5.

2.2.2 Subculture of Human Keratinocytes (Cell Passaging)

1. Complete hkDME-Ham (see Subheading 2.1.2).
2. i3T3 (see Subheading 2.2.1, item 16).
3. Trypsin/EDTA (see Subheading 2.2.1, item 9).
4. 50 mL centrifuge tube (cat. no. 352070, BD Falcon).
5. Tissue culture flask, 75 cm^2 (cat. no. 35310, BD Falcon).

2.2.3 Cryopreservation of Human Keratinocytes

1. Freezing medium (see Subheading 2.1.4).
2. Sterile cryogenic vials (cat. no. 5000-020, Nalgene Labware, Rochester, NY).
3. Freezing container (cat. no. 5100-0001, Nalgene Labware), filled with 99% ethanol and pre-cooled at –20°C.

2.2.4 Thawing of Human Keratinocytes

1. Complete hkDME-Ham (see Subheading 2.1.2).
2. Tissue culture flask, 75 cm^2 (cat. no. 35310, BD Falcon).
3. i3T3 (see Subheading 2.2.1, item 16).

2.3 Monolayer Culture of Human Corneal Epithelial Cells

2.3.1 Extraction and Culture of Human Corneal Epithelial Cells

1. Source of human corneal epithelial cells: corneas from human post-mortem donors unsuitable for transplantation (Banque nationale d'yeux du CHUQ, Quebec, Canada).
2. 35 and 100-mm tissue culture Petri dishes (size: 35×10 mm, cat. no. 351008, BD Falcon, and size: 100×15 mm, cat. no. 08-757-12, Fisher Scientific, respectively).
3. Sterile 3 in. $\times$ 3 in. gauze (cat. no. A3120, AMD-Ritmed, Lachine, QC, Canada).
4. Scalpel (#4 cat. no. 08-917-5, Fisher Scientific) and blades (size: 11, cat. no. 73-0411, and size: 22, cat. no. 73-0422, Personna Medical, Verona, VA).
5. Dissecting curved forceps (cat. no. 11272-30, Fine Science Tools, North Vancouver, BC, Canada).
6. Dissecting curved scissor (cat. no. E3220, Storz, St-Louis, MO).
7. 15 and 50 mL tubes (cat. no. 352096 and cat. no. 352070, respectively, BD Falcon).
8. PBS-P/G/F: (see Subheading 2.2.1, item 5).
9. 7.5 mm diameter trephine (cat. no. 9711, Medtronic of Canada, Mississauga, ON, Canada).
10. Dispase II (cat. no. 165-859, Roche Diagnostics). Dissolve 2 mg/mL dispase in HEPES 1$\times$ (see Subheading 2.2.1,

item 7). Sterilize by filtration through a 0.22-μm low-binding disposable filter. Store at 4°C (see Note 3).

11. Complete hcDME-Ham (see Subheading 2.1.3).

12. Tissue culture flask, 75 cm² (cat. no. 35310, BD Falcon).

13. i3T3 (see Subheading 2.2.1, item 16).

14. Dissecting microscope (Nikon, SMZ-800).

2.3.2 Subculture of Human Corneal Epithelial Cells (Cell Passaging)

1. Complete hcDME-Ham (see Subheading 2.1.3).

2. i3T3 (see Subheading 2.2.1, item 16).

3. Trypsin/EDTA (see Subheading 2.2.1, item 9).

4. 50 mL tubes (cat. no. 352070, BD Falcon).

5. Tissue culture flask, 75 cm² (cat. no. 35310, BD Falcon).

2.3.3 Cryopreservation of Human Corneal Epithelium Cells

1. Freezing medium (see Subheading 2.1.4).

2. Sterile cryogenic vials (cat. no. 5000-020, Nalgene Labware).

3. Freezing container (cat. no. 5100-0001, Nalgene Labware), filled with 99% ethanol and pre-cooled at –20°C.

2.3.4 Thawing of Human Corneal Epithelial Cells

1. Complete hkDME-Ham (see Subheading 2.1.2).

2. Tissue culture flask, 75 cm² (cat. no. 35310, BD Falcon).

3. i3T3 (see Subheading 2.2.1, item 16).

2.4 Preparation of Crude Nuclear Extracts

1. Three nuclear extract (NE) solutions: NE1 (0.14 M NaCl, 10 mM Tris–HCl pH 7.5), NE2 (25 mM KCl, 2 mM $Mg(CH_3COO)_2$, 1 mM DTT, 10 mM Tris–HCl pH 7.5) and NE3 (50 mM KCl, 1 mM $MgCl_2$, 20 mM K_3PO_4 pH 7,4, 1 mM β-mercaptoethanol, 20% (v/v) glycerol).

2. Homogenizer potter (size: 2 mL or 5 mL, cat. no. 14559038, Fisher Scientific).

3. A solution of protease inhibitor cocktail (cat. no. P8340 Sigma-Aldrich) see Note 6.

4. Beckman Optima TLX, Optima TL or TL-100 Ultracentrifuge with rotor TLA-120.2 (10×2.0 mL, 8×34 mm).

5. Dialysis molecular porous membrane tubing and universal closures: Spectra/Por Dialysis membrane number 3, MWCO 3,500. Nominal flat width: 18 mm, diameter: 11.5 mm, vol/length: 1.1 mL/cm, length 15 m/50 ft (recorder no.: 132720, Spectrumlabs.com).

6. A solution of dialysis buffer (50 mM KCl, 1 mM $MgCl_2$, 20 mM K_3PO_4 pH 7.4, 1 mM β-mercaptoethanol, 20% (v/v) glycerol, 1 mM DTT, and 0.5 mM PMSF).

7. Bio-rad protein assay for measuring total protein concentration (cat. no. 500-0006, Bio-rad Laboratories, Inc.) see Note 7.

Table 5
Sequence of oligonucleotides bearing the high affinity binding site for the transcription factor Sp1

Strand	Sequence
Top (5′–3′)	GATCATATCTGCGGGGCGGGGCAGACACAG
Bottom (5′–3′)	CTGTGTCTGCCCCGCCCCGCAGATATGATC

2.5 Labeling and Purification of a Double-Stranded Sp1 Oligonucleotide

1. $(\gamma^{32}P)$ dATP (see Note 8).

2. 50 ng of a synthetic double-stranded oligonucleotide (refer to Table 5) bearing the high-affinity binding site for the transcription factor Sp1 (see Note 9).

3. T4 polynucleotide kinase (PNK) and 10× kinase buffer (0.5 M Tris–HCl pH 7.5, 0.1 M $MgCl_2$, 40 mM DTT, 1 mM spermidine, 1 mM EDTA).

4. For purification: DNA purification columns like the Elutip-D columns (item no: 10462617, Whatman Inc., UK).

5. A stock of wash solution (0.2 M NaCl, 20 mM Tris–HCl pH 7.5, 1 mM EDTA).

6. A stock of elution solution (1 M NaCl, 20 mM Tris–HCl pH 7.5, 1 mM EDTA).

7. For DNA precipitation, a preparation of 2 µg/µL linear polyacrylamide (stock solution: 1 g acrylamide, 20 mL H_2O, 200 µL ammonium persulfate 10% (w/v), 80 µL TEMED), a solution of 3 M NaOAc (pH 5.2).

2.6 Electrophoretic Mobility Shift Assay

1. Standard vertical electrophoresis apparatus for polyacrylamide gels.

2. A stock solution of 40% (w/v) acrylamide prepared in a 39:1 (w/w) ratio of acrylamide and N',N'-methylene bis-acrylamide (see Note 10).

3. A stock solution of 5× Tris–glycine (250 mM Tris–HCl, 12.5 mM EDTA, 2 M glycine) (see Note 11).

4. Extract (crude or enriched) containing cell or tissue nuclear proteins (see Subheading 2.4 and Note 12).

5. A stock solution of 2× binding buffer (20 mM Hepes pH 7.9, 20% glycerol, 0.2 mM EDTA, 1 mM tetrasodium pyrophosphate, and 0.5 mM PMSF).

6. A stock solution of 6× loading buffer (0.25% bromophenol blue, 0.25% xylene cyanol, and 40% sucrose).

7. A stock solution of 1 mg/mL poly(deoxyinosinic-deoxycytidylic) acid sodium salt (poly(dI:dC)) (cat. no. P4929, Sigma-Aldrich).

8. Whatman chromatographic paper (Grade No. 3MM, size: 10.2×13.3 cm, cat. no.: 3030-6189, Whatman Inc.)

9. Autoradiography cassette (electrophoresis systems, FBXC 810, Fisher Scientific).

10. Scientific imaging films (KodacBioMax XAR Film, 20.3×25.4 cm, cat. no. 165-1454).

2.7 Detection of Sp1 Protein in Western Blot Analyses

2.7.1 Sample Preparation

1. 2× SDS PAGE sample buffer (for 100 mL: mix 10 mL 1.5 M Tris–HCl (pH 6.8), 6 mL 20% SDS, 30 mL glycerol, 15 mL β-mercaptoethanol, and 1.8 mg bromophenol blue. Adjust volume to 100 mL with H_2O and aliquot in 10 mL stock solution and store at –20°C).

2.7.2 Electrophoresis (Stacking and Running Gels) and Transfer

1. Mini-protean cell with mini Trans-Blot module (cat. no. 165-8061, Bio-rad).

2. 30% acryl–bisacrylamide mix (for 100 mL: dissolve 29 g acrylamide and 1 g N,N'-methylenebisacrylamide in 60 mL H_2O and adjust volume to 100 mL with H_2O).

3. 1.5 M Tris–HCl pH 8.8 and 1.0 M Tris–HCl pH 6.8 solutions.

4. 10% ammonium persulfate (APS) (for 10 mL: dissolve 1 g APS in 8 mL H_2O and adjust volume to 10 mL with H_2O).

5. 10× SDS PAGE running buffer (for 1 L: dissolve 10 g SDS, 30.3 g Tris–HCl and 144.1 g glycine in 800 mL H_2O and adjust volume to 1 L with H_2O).

6. 10× transfer buffer (for 1 L: dissolve 30.3 g Tris–HCl and 144.1 g glycine in 800 mL H_2O and adjust volume to 1 L with H_2O).

7. Prestained molecular weight markers (Precision Plus protein Standards, Biorad cat. no. 161-0376).

8. Nitrocellulose membrane (Amersham™Hybond™-ECL, 0.45 μm, cat. no. RPN303D, GE Healthcare).

2.7.3 Membrane Blocking and Antibody Incubations

1. A solution of 1× Tris-buffered saline (TBS): for 1 L, dissolve 5.05 g Tris–HCl (50 mM), 8.76 g NaCl (150 mM), and 0.2% Triton X-100 in 800 mL and complete at 1 L.

2. Blocking solution: a solution of 5% (w/v) nonfat milk in TBS (see Subheading 2.7.3, item 1).

3. Anti-Sp1 (Rabbit polyclonal IgG Sp1 (clone PEP2), cat. no. Sc-59, Santa Cruz Biotechnology).

4. Peroxidase-conjugated AffiniPure Goat Anti-Rabbit IgG (H+L), cat. no. 111-035-003, Jackson ImmunoResearch Laboratories Inc.

<table>
<tr><td>

*2.7.4 Detection of Sp1
Protein*

</td><td>

1. ECL Plus Western blotting detection system (cat. no. RPN2132, Amersham™).

2. Autoradiography cassette (electrophoresis systems, FBXC 810, Fisher Scientific).

3. Scientific imaging films (KodacBioMax XAR Film, 20.3×25.4 cm, cat. no. 165-1454).

</td></tr>
</table>

3 Methods

<table>
<tr><td>

3.1 Monolayer Culture of Human Skin Epithelial Cells (Keratinocytes)

*3.1.1 Extraction
and Culture of Human
Keratinocytes*

</td><td>

1. Human keratinocytes are isolated from normal adult or newborn skin (see Note 13).

2. Transport and conservation: In the surgery room, put the skin specimen into a sterile container filled with cold (4°C) transport medium.

   All further manipulations are performed under a sterile laminar flow hood cabinet.

3. Wash the skin specimen in a 50-mL centrifuge tube containing 30 mL PBS-P/G/F. Agitate vigorously. With sterile forceps, transfer the skin specimen in another PBS-P/G/F tube. Repeat this step three times.

4. Spread out the skin specimen, epidermis on the top, into a 100-mm tissue culture Petri dish.

5. Cut the skin in 3×100 mm pieces with a scalpel (blade 22).

6. Add 10 mL of cold (4°C) thermolysin. Seal the Petri dish with parafilm.

7. Incubate overnight at 4°C.

8. With two curved forceps, separate the epidermis from the dermis. Put epidermal pieces within a trypsinization unit containing 20 mL of warm (37°C) trypsin/EDTA. Incubate under agitation during 15–30 min at 37°C.

9. Collect the cell suspension. Put into a 50-mL centrifuge tube and add 20 mL of warm (37°C) DME-Ham supplemented with 5% (v/v) fetal clone II. Wash the trypsinization unit with 10 mL of DME-Ham 5% (v/v) fetal clone II and add it to the tube (complete the volume to a total of 40 mL).

10. Count the cells and measure the viability by trypan blue staining. The cell viability should be superior to 80%.

11. Centrifuge cell suspension at $300 \times g$ for 10 min at room temperature.

12. Seed 1×10^6 keratinocytes and 1.5×10^6 i3T3 by 75 cm² culture flask with 20 mL of complete hkDME-Ham. Incubate in

</td></tr>
</table>

8% CO_2, 100% humidity atmosphere at 37°C. Change culture medium three times a week.

13. When the keratinocytes reach 80% confluence, subculture (see Subheading 3.1.2) or freeze (see Subheading 3.1.3) cells.

3.1.2 Subculture of Human Keratinocytes (Cell Passaging)

For a 75-cm² tissue culture flask of keratinocytes:

All further manipulations are performed under a sterile laminar flow hood cabinet.

1. Remove medium.

2. Wash the culture flask with 2 mL of warm (37°C) Trypsin/EDTA and remove it.

3. Add 8 mL of Trypsin/EDTA. Incubate at 37°C until the cells are detached from the flask.

4. Add 8 mL of complete hkDME-Ham (37°C). Collect the cell suspension. Put into a 50-mL centrifuge tube. Wash the flask with 2 mL of complete hkDME-Ham, collect the cell suspension and add it to the 50-mL tube (total 18 mL).

5. Count the cells and measure the viability by trypan blue staining. The cell viability should be superior to 80%.

6. Centrifuge cell suspension at $300 \times g$ for 10 min at room temperature.

7. Resuspend cell pellet in a given volume of complete hkDME-Ham.

8. Seed 2×10^5 to 7×10^5 keratinocytes and 1.5×10^6 i3T3 by 75 cm² culture flask with 20 mL of complete hkDME-Ham. Incubate in 8% CO_2, 100% humidity atmosphere at 37°C. Change culture medium three times a week.

3.1.3 Cryopreservation of Human Keratinocytes

All further manipulations are performed under a sterile laminar flow hood cabinet.

1. Resuspend cells in a given volume of cold (4°C) freezing medium in order to obtain 2×10^6 cells/mL (see Note 14).

2. On ice, aliquot in cryogenic vials. Put them in a freezing container filled with 99% ethanol that has previously been cooled at −20°C.

3. Freeze overnight at −80°C, in the freezing container.

4. Store in liquid nitrogen.

3.1.4 Thawing of Human Keratinocytes

1. Put the cryogenic vial in 37°C water until only a small cluster of ice remains.

All further manipulations are performed under a sterile laminar flow hood cabinet.

2. Using a pipette, put 1 mL of cold (4°C) complete hkDME-Ham in the cryogenic vial. When the ice is melted, put the

cryogenic vial content in a sterile centrifuge 15 mL tube containing 9 mL of cold complete hkDME-Ham.

3. Count the cells and measure the viability by trypan blue staining.

4. Centrifuge cell suspension at $300 \times g$ for 10 min at room temperature.

5. Resuspend cell pellet in a given volume of warm (37°C) complete hkDME-Ham.

6. Seed 2×10^5 to 7×10^5 keratinocytes and 1.5×10^6 i3T3 by 75 cm^2 tissue culture flask with 20 mL of complete hkDME-Ham. Incubate in 8% CO_2, 100% humidity atmosphere at 37°C. Change culture medium three times a week.

3.2 Monolayer Culture of Human Corneal Epithelial Cells

3.2.1 Extraction and Culture of Human Corneal Epithelial Cells

1. Human corneal epithelial cells are obtained from the cornea of human post-mortem donors unsuitable for transplantation (see Note 13).

2. Wash the eye in a 50-mL centrifuge tube containing 30 mL PBS-P/G/F. Agitate gently 1–2 min. With sterile forceps, transfer the eye in another tube filled with PBS-P/G/F. Repeat this step three times.

3. Place the eye into a 100-mm Petri dish.

4. Surround the eye with a folded sterile gauze. This helps in holding the eye without having to touch it.

5. With the #11 scalpel blade, make a small opening of 2–3 mm in the sclera.

6. With curved scissors, cut-out the cornea in order to obtain only the limbus and central cornea. Avoid leaving sclera to eliminate conjunctival epithelial cell contamination.

7. With two curved forceps, peel-off the iris. Do this step while holding the cornea in the air to avoid any damage to the epithelium during the procedure.

8. Place the limbal ring into a 35-mm tissue culture Petri dish. The limbal ring is obtained by separating the limbus from the central cornea with a 7.5-mm diameter trephine. Add 5 mL of cold (4°C) dispase II. Seal the Petri dish with parafilm.

9. Incubate overnight at 4°C.

10. With two curved forceps, mechanically separate the epithelium from the stroma under a dissecting microscope.

11. Put the epithelium in a 35-mm tissue culture Petri dish and cut into small pieces with a #22 scalpel blade (see Note 15).

12. With a plastic pipette, collect the pieces with 5 mL of warm (37°C) complete hcDME-Ham and put into a 15-mL tube. Wash with another 5 mL of warm (37°C) complete hcDME-Ham and put in the same 15 mL tube.

13. Centrifuge at $300 \times g$ for 10 min at room temperature.

14. Seed 1×10^6 epithelial cells and 1.5×10^6 i3T3 by 75 cm² tissue culture flask with 20 mL of complete hcDME-Ham. Incubate in 8% CO_2, 100% humidity atmosphere at 37°C. Change culture medium three times a week.

15. When the human corneal epithelial cells reach 80% confluence, subculture (see Subheading 3.2.2), or freeze (see Subheading 3.2.3) cells.

3.2.2 Subculture of Human Corneal Epithelial Cells (Cell Passaging)

For a 75-cm² tissue culture flask of corneal epithelial cells: Follow steps 1–8 from Subheading 3.1.2 but use complete hcDME-Ham instead hkDME-Ham.

3.2.3 Cryopreservation of Human Corneal Epithelial Cells

Follow steps 1–4 from Subheading 3.1.3.

3.2.4 Thawing of Human Corneal Epithelial Cells

Follow steps 1–6 from Subheading 3.1.4 but use complete hcDME-Ham instead of complete hkDME-Ham.

3.3 Preparation of Crude Nuclear Extracts

1. Resuspend the cell pellet (10–50 millions of cells) in 5 mL NE1 in a 15-mL centrifuge tube (see Subheading 2.4, item 1 for NE solutions). Centrifuge at $900 \times g$ for 5 min at 4°C, used a table centrifuge.

2. Remove the supernatant and resuspend the cells pelleted in 5 mL NE1 and centrifuge 5 min at 4°C

3. Remove the supernatant, resuspend cells in 5 mL NE2, and centrifuge 5 min at 4°C.

4. Remove the supernatant, resuspend cells in 2 mL NE2, and incubate 5 min on ice.

5. Transfer the cellular suspension in the homogenizer potter (see Subheading 2.4, item 2) and push down and up the piston 20 times.

6. Centrifuge at $500 \times g$ for 3 min at 4°C and remove the supernatant.

7. Resuspend the pellet in 1 mL NE3. Add 1/10 the cell pellet volume of KCL 4 M and 10 μL of protease inhibitor cocktail (see Subheading 2.4, item 3).

8. Transfer the cell suspension in the homogenizer potter (see Subheading 2.4, item 2) and push down and up the piston ten times.

9. With an ultracentrifuge, centrifuge in a 1.5-mL tube at maximal speed ($16,000 \times g$) 5 min at 4°C.

10. Transfer the supernatant in a conical tube for the TL 100.2 rotor and centrifuge 2 h at $20,000 \times g$ at 4°C (see Subheading 2.4, item 4).

11. Transfer the supernatant in a dialysis molecular porous membrane tubing (see Subheading 2.4, item 5), close the tube with two universal closures prior to dialyzing 1 h against the dialysis buffer (see Subheading 2.4, item 6).

12. Determine total protein concentration of the nuclear extract by the Bio-Rad protein assay (see Subheading 2.4, item 7). The quality of the proteins from the crude nuclear extracts has been evaluated by coomassie blue staining of the proteins following their separation by gel electrophoresis (SDS-PAGE; see example in Fig. 1a).

3.4 Labeling and Purification of a Double-Stranded Sp1 Oligonucleotide

1. Anneal equal amounts of the complementary Sp1 strands (see Subheading 2.5, item 2 and Table 5) (example: for 1 µg/µL stock: 25 ng of 1 µg/µL of each strands in 50 µL total volume). Heat the resulting DNA mix at 4°C over the specific melting temperature (T_M) of the oligonucleotides (alternatively, you may heat the oligonucleotide mix at 90°C) for 90 s and let cool at room temperature. When DNA reaches RT, store at –20°C prior to use.

2. Use 50 ng (5 µL of 10 ng/µL solution) of the Sp1 double-stranded oligonucleotide prepared in Subheading 3.4, step 1 and perform DNA labeling with 5 µL T4 PNK buffer, 34 µL H_2O, 1 µL T4 PNK and 5 µL of 10 µCi/mL $(\gamma^{32}P)$ dATP (see Subheading 2.5, items 1 and 3). Let incubate for 2 h at 37°C and add 950 µL TBE 1× to obtained a total volume of 1 mL.

3. The Sp1 labeled probe is then separated from free nucleotides by purification on a Elutip-D column (see Subheading 2.5, item 4). 5 mL of wash solution (see Subheading 2.5, item 5) is passed thought the column prior to loading of the Sp1 probe. Wash the column containing the probe with 5 mL of wash solution. Elute Sp1 labeled probe with 300 µL of elution solution (see Subheading 2.5, item 6) in a fresh tube.

4. Precipitate the Sp1 labeled probe with 30 µL NaOAc 3 M pH 5.2, 2.5 µL linear polyacrylamide at 2 µg/µL (see Subheading 2.5, item 7) and 750 µL 95–100% ethanol. Allow labeled DNA to precipitate on dry ice or –80°C for 30 min.

5. Centrifuge for 15 min at top speed in a microcentrifuge and carefully remove the supernatant, wash the pellet with 70% ethanol, and dry.

6. Count the labeled Sp1 probe using a beta counter by Cerenkov counting or resuspend Sp1-labeled probe in a small volume (100 µL) and count a 1-µL aliquot.

7. Resuspend labeled Sp1 probe to 30,000 cpm/µL.

3.5 Sp1DNA Interaction in EMSA

1. Rigorously clean and dry the electrophoresis apparatus and its accessories prior to use. Gel plates should be cleaned using any good quality commercial soap and then rinsed with 95% ethanol.

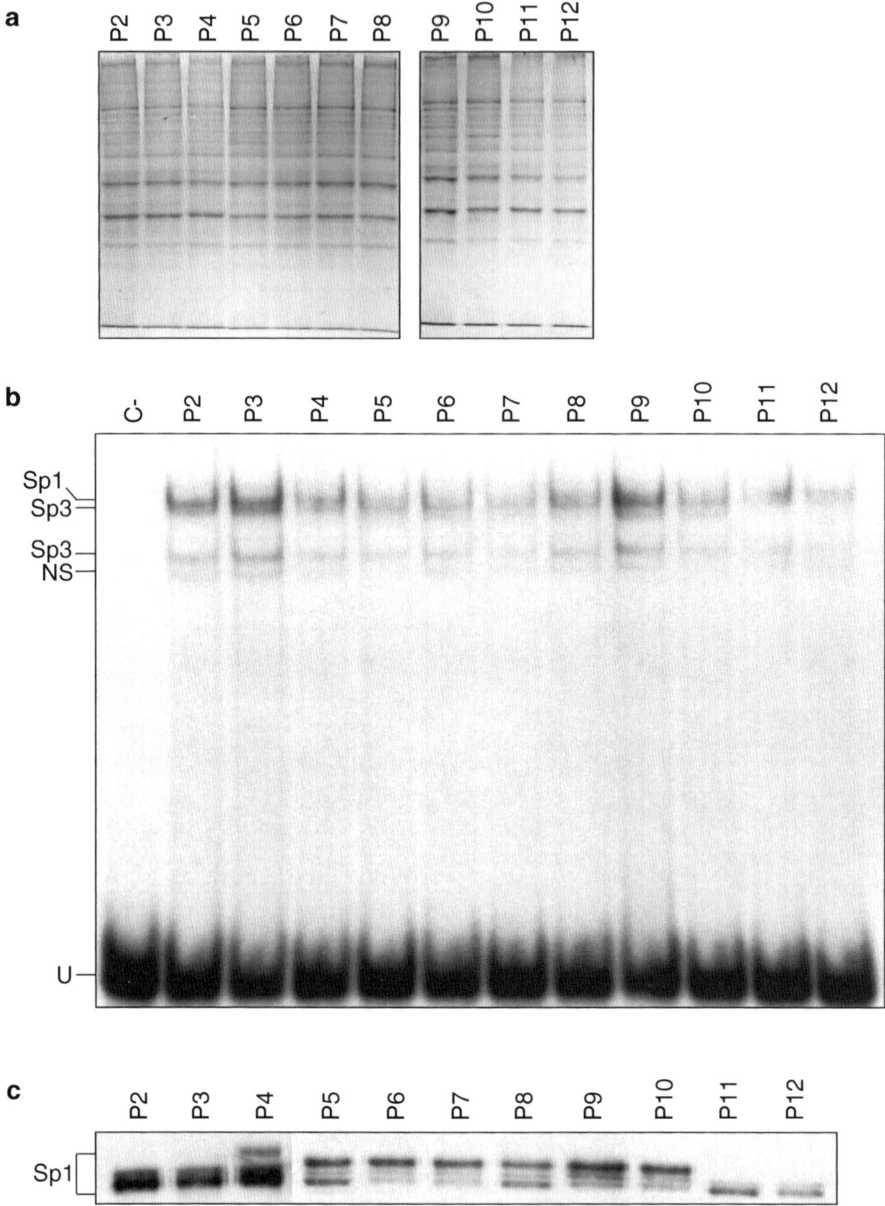

Fig. 1 Binding and expression of the transcription factor Sp1 by EMSA and Western blot. Crude nuclear extracts were prepared from newborn keratinocytes (NbKs) grown at passage 3 (P3) to passage 12 (P12) with i3T3. (**a**) 10 μg nuclear proteins from each cell passage were separated by gel electrophoresis (SDS-PAGE) and then stained with Coomassie blue in order to ensure that no protein degradation is occurring from one sample to another. (**b**) A synthetic double-stranded oligonucleotide bearing the high-affinity binding site for the transcription factor Sp1 was 5′-end labeled and incubated with or without (C−) crude nuclear extracts prepared from NbKs at each cell passage (P3–P12). The EMSA was conducted at 4°C on a 6% native polyacrylamide gel for 5 h at 120 V. The position corresponding to Sp1 and Sp3, as well as that of the free probe (U) is indicated. NS: non-specific signal. (**c**) Expression of Sp1 was monitored on the crude nuclear extracts (10 μg from each) used in panel (**b**) by Western blot using the Sp1 antibody. The results presented on this figure indicate clearly that expression of Sp1 vary greatly in human skin keratinocytes between cell passages, reaching high levels between passages P2–P4 and P9–P10

One plate can be treated with a coat of Sigmacote (chlorinated organopolysiloxane in heptane, Sigma-Aldrich) to facilitate gel removal from the plates after running.

2. Prepare a 6% polyacrylamide gel as follows; mix 5 mL of 5× Tris–glycine, 3.75 mL of 40% acrylamide (39:1) stock solution, and 16.25 mL H_2O to obtain 25 mL (see Subheading 2.6, items 2 and 3). Add 180 µL of 10% ammonium persulfate (APS) and 25 µL of TEMED. Carefully stir and pour the acrylamide solution between the plates. Place a comb and allow the gel to set for 30 min then mount the gel in the electrophoresis tank and fill the chamber with 1× Tris–glycine buffer.

3. Pre-run the gel at 120 V at 4°C until the current becomes invariant (this usually takes 60 min on average). This pre-running step ensures that the gel will be at a constant temperature at the moment of sample loading.

4. When the gel is ready for loading, prepare samples as follows. For each sample, mix 12 µL of 2× binding buffer (see Subheading 2.6, item 5), 1 µL of 1 mg/mL poly(dI:dC) (see Note 16) and 0.6 µL of 2 M KCl (see Note 17) then add 30,000 cpm of labeled probe. When possible, pool together invariant components in one microcentrifuge tube and then redistribute equal amounts into different tubes accounting for the different experimental conditions. Finally, add 5 µg nuclear proteins enriched (*see* Subheading 3.3) and add H_2O to a final volume of 24 µL. Mix gently each tube and incubate at room temperature for 3 min. As a control, prepare a sample without nuclear extract and add 2 µL of 6× loading buffer containing bromophenol blue and xylene cyanol (see Subheading 2.6, item 6).

5. Load samples by changing the pipet tip for each sample and run the gel at 120 V for 5 h at 4°C (see Note 18).

6. After the gel is run, disassemble the apparatus and remove one of the glass plates, place a Whatman paper over the gel and carefully lift the gel off the remaining plate. Make sure that the gel is well fixed on the Whatman before lifting the gel to avoid gel breakage. Place a Saran Wrap over the gel and dry at 80°C for 30 min.

7. Place an X-ray film over the gel in an autoradiography cassette and expose at −70°C overnight or less, depending on strength of signal (see examples in Figs. 1b and 2).

3.6 Detection of the Sp1 Protein in Western Blot

3.6.1 Sample Preparation

1. Typically use 20 µg of nuclear proteins (see Subheading 3.3) per lane. For each sample, add the desired amount of proteins to a microcentrifuge tube followed by 2× SDS PAGE sample buffer (see Subheading 2.7.1, item 1), dilute sample if necessary in H_2O to get the same volume for each sample (see Note 19).

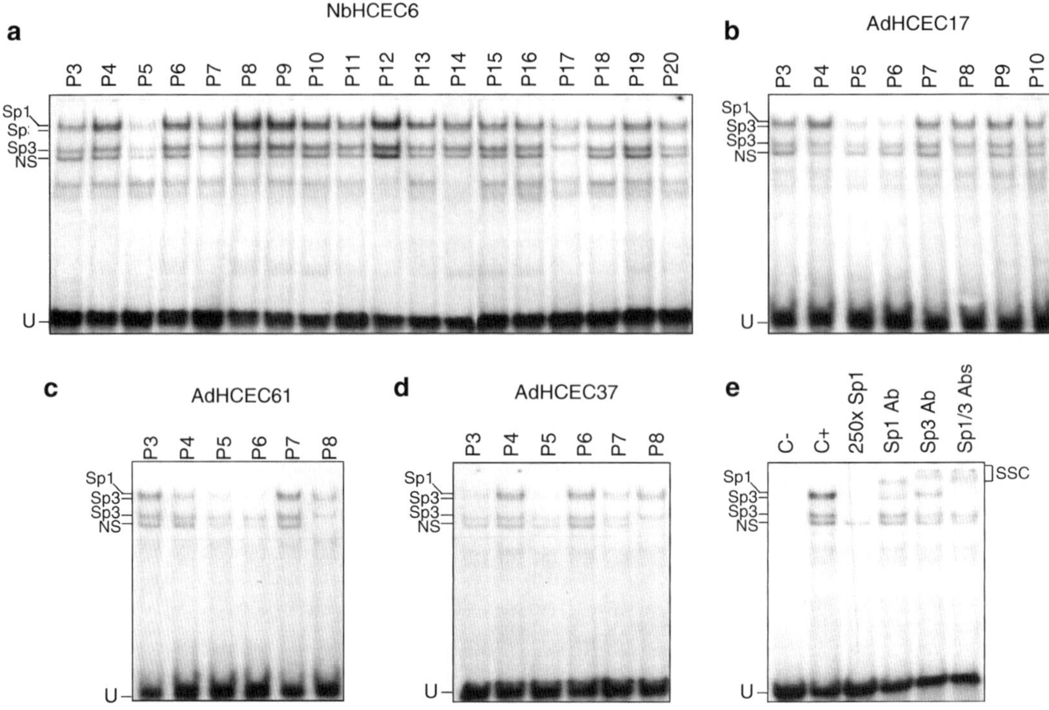

Fig. 2 Analysis of Sp1 binding in human corneal epithelial cells by EMSA analyses. A synthetic double-stranded oligonucleotide bearing the high-affinity binding site for Sp1 was 5′-end labeled and incubated with 5 μg of crude nuclear extracts prepared from four distinct primary cultured newborn (Nb) or Adult (Ad) human corneal epithelial cells (HCEC) grown with i3T3. (**a**) NbHCEC6 grown from P3 to P20, (**b**) AdHCEC17 grown from P3 to P10, (**c**) AdHCEC61 grown from P3 to P8, and (**d**) AdHCEC37 grown from P3 to P8. Formation of DNA–protein complexes was evaluated on a native 6% polyacrylamide gel. (**e**) The Sp1 probe was incubated either alone (C−) or with 5 μg nuclear extract from NbHCEC6 grown at passage P3 (C+). To validate the specificity for the formation of the Sp1/Sp3 DNA–protein complexes, competition analyses in EMSA were performed. Approximately 5 μg nuclear proteins from NbHCEC6 cells at passage P3 were incubated along with a 250-fold molar excess of an unlabeled double-stranded oligonucleotide bearing the high affinity binding site for Sp1 (250× Sp1) or in the presence of polyclonal antibodies directed against Sp1 and Sp3 (added individually (Sp1 Ab or Sp3 Ab) or in combination (Sp1/3 Abs)). Formation of DNA–protein complexes was monitored on a native 6% polyacrylamide gel. Formation of the Sp1 and Sp3 (Sp1/Sp3) complexes, as well as that of their corresponding supershifted complexes (SSC), was then monitored by EMSA U: free probe. NS: non-specific signal. The results presented on this figure indicates that as with human skin keratinocytes, DNA binding of Sp1 vary greatly between cell passages in HCECs. Furthermore, the cell cultures that sustain the highest and more prolonged expression/DNA binding of Sp1 from one passage to another (such as with NbHCEC6 cells) are also those that can sustain a higher number of cell passages in culture

2. Incubate at 95–100°C for 5 min.

3. Quick spin the samples.

4. Load immediately on gel previously prepared (see Subheading 3.6.2).

Table 6
Solutions for Tris–glycine SDS-polyacrylamide gel electrophoresis

10% running gel	10 mL	20 mL
H_2O	4.0	7.9
30% acrylamide mix	3.3	6.7
1.5 M Tris–HCl pH 8.8	2.5	5.0
10% SDS	0.1	0.2
10% APS	0.1	0.2
TEMED	0.006	0.008

Stacking gel	5 mL	10 mL
H_2O	3.4	6.8
30% acrylamide mix	0.83	1.7
1.0 M Tris–HCl pH 6.8	0.63	1.25
10% SDS	0.05	0.1
10% APS	0.05	0.1
TEMED	0.005	0.01

3.6.2 Electrophoresis (Stacking and Running Gels) and Transfer

1. Place a short plate on top of the 1.5-mm spacer plate.

2. Slide the two plates into the casting frame, keeping the short plate facing front and insure both plates are flush at the bottom on a level surface.

3. Lock the pressure cams to secure the glass plates.

4. Engage the spring-loaded lever and place the gel cassette assembly on the casting strand gasket. Insure the horizontal ribs on the back of the casting frame are flush against the face of the casting stand and the glass plates are perpendicular to the level surface.

5. Make a 10% running gel (refer to Table 6). The concentration of the gel may vary according to the separation that you need and the protein under study. For the detection of Sp1, a 10% gel is recommended.

6. Take up 7.5 mL of running gel. Put pipet on edge of glass plate, hold at a 90° angle, and pour the 7.5 mL between glass plates.

7. Add 500–1,000 µL of butanol saturated in H_2O. Let sit for 30–45 min.

8. When gel is polymerized, invert apparatus and empty the butanol. Blot with Whatman paper.

9. Make stacking gel (refer to Table 6).

10. Pour-in stacking gel and place comb into gel. Let sit for 20–30 min.

11. Remove the gel cassette sandwich from the casting frame and place it into the electrode assembly with the short plate facing inward.

12. Slide the gel cassette sandwich and electrode assembly into the clamping frame.

13. Press down the electrode assembly while closing the two cam levers of the clamping frame.

14. Lower into the mini tank and add 400 mL of 1× running buffer (see Subheading 2.7.2, item 5).

15. Prepare and load samples as described in Subheading 3.6.1 and load 5–10 μL of marker in the first well (see Subheading 2.7.2, item 7).

16. Run at 75–100 V for 1–2 h (depending on the extend of protein separation that you need).

17. Before the run is completed, get out a pyrex dish and fill with transfer buffer (see Subheading 2.7.2, item 6).

18. Soak one nitrocellulose membrane (see Subheading 2.7.2, item 8), four Whatman papers, and two fiber pads per gel for at least 5 min.

19. Stop electrophoresis and remove gel. Separate plates and put gel off plate into the transfer buffer.

20. Open sandwich with black side (or anode side) down and stack in this order: fiber pad, two Whatman papers, gel, membrane, two Whatman papers, and fiber pad. Make sure to roll out bubbles.

21. Close and put into transfer chamber in the good side (take care at the anode and cathode sides). Put in frozen block supplied with the unit. Fill transfer chamber with transfer buffer (see Subheading 2.7.2, item 6).

22. Transfer at 100 V 1 h or at 30 V overnight in the cold room (4°C).

23. Open sandwich and put membrane in clean container (pipet tip box for example) containing 1× TBS (see Subheading 2.7.3, item 1).

3.6.3 Membrane Blocking and Antibody Incubations

1. Rinse membrane twice in 1× TBS (see Subheading 2.7.3, item 1).

2. Block with 25 mL of blocking solution (see Subheading 2.7.3, item 2) for 1 h at room temperature.

3. Remove the blocking solution and add primary antibody at desired concentration. Diluted the antibody in 1% (w/v) nonfat milk in TBS. For the anti-Sp1: put 2.5 μL of the stock in 12.5 mL (1:5,000) of 1% (w/v) nonfat milk in TBS (see Subheading 2.7.3, item 3), let incubate in shaker overnight at 4°C or 1 h at room temperature.

4. Wash in TBS four times 5 min or more at room temperature.

5. Add secondary antibody (see Subheading 2.7.3, item 4 for Sp1 detection) at 1:2,000 in 1% (w/v) nonfat milk in TBS. Let incubate 1 h in shaker at room temperature.

6. Wash in TBS four times 5 min or more at room temperature.

3.6.4 Detection of Sp1 Protein

1. Drips dry the membrane onto a paper towel and lie flat on a piece of Saran Wrap with the protein side facing up.

2. Mix the ECL Reagents 1 and 2 into a 15-mL conical tube and pour onto the membrane. Let incubate 5 min at room temperature (see Subheading 2.7.4, item 1).

3. Dump solution off the membrane and cover membrane with saran wrap.

4. In a darkroom, expose to film for various periods of time (5, 15, 30, and 60 s or more, depending on strength of signal) in autoradiography cassette (see Subheading 2.7.4, items 2 and 3). A few different exposures will likely be needed. The band corresponding to Sp1 should be visible between 95 and 105 kDa (see example in Fig. 1c). Actin expression may be also monitored by Western blot for normalization purpose.

4 Notes

1. Serum must be added first followed by insulin. Insulin must be added with a new sterile plastic pipette.

2. Dimethyl sulfoxide (DMSO) (cat. no. D5879). Distribute the stock solution (99.7%) in single use aliquots and store at −20°C.

3. When exposed to light, HEPES buffer may undergo degradation and become toxic.

4. This solution must be prepared the day of its use.

5. Irradiated 3T3 should be kept 1 week in 8% CO_2, 100% humidity atmosphere at 37°C. However, the cell yield may fall by approximately 10–15% per day.

6. This protease inhibitor cocktail is a mixture of protease inhibitors with broad specificity for the inhibition of serine, cysteine, aspartic proteases, and aminopeptidase. It contains 4-(2-aminoethyl)benzenesulfonyl fluoride (AEBSF), pepstatinA, E-64,

bestatin, leupeptin, and aprotinin. It contains no metal chelators. Use at 1:100 for cell lysates or tissue extracts.

7. The Bio-Rad protein assay is a simple colorimetric assay for measuring total protein concentration and is based on the Bradfort dye-binding method (8). Based on the color change of Coomassie brilliant blue G-250 dye in response to various concentrations of protein, the dye binds to primarily basic (especially arginine) and aromatic amino acid residues. Many detergents and basic protein buffers interfere with the assay, interference may be caused by chemical–protein or chemical–dye interactions.

8. (γ^{32}P) dATP emits high-energy beta radiations. Refer to the rules established by your local control radioactivity agency for handling and proper disposal of radioactive materials and waste.

9. The protein Sp1 is a ubiquitous transcription factor usually reported to activate gene transcription in response to physiological and pathological stimuli. Binds with high affinity to GC-rich motifs (see Table 5) and regulates the expression of a large number of genes involved in a variety of processes such as cell growth, apoptosis, differentiation, and immune responses. Highly regulated by post-translational modifications (phosphorylation, sumoylation, proteolytic cleavage, glycosylation, and acetylation) (for review see refs. 9–11).

10. Acrylamide is a potent neurotoxic compound that is easily absorbed through skin. Wearing gloves and a mask to avoid direct contact with the skin or inhalation is therefore required when manipulating acrylamide dried or in solution. The acrylamide solution is light-sensitive and should be kept away from direct light.

11. The concentration of the polyacrylamide gel used in EMSA is primarily dictated by both the size of the labeled probe selected and the resolution of the DNA–protein complexes obtained. It can vary from 4% up to 12%.

12. When using crude nuclear extracts for detecting DNA–protein complexes in EMSA, their quality is obviously very critical. Whenever possible, nuclei purification procedures using sucrose pads (12) is preferred in order to eliminate contamination by cytosolic proteins, which most often also contain substantial amounts of proteases.

13. Protocols must be approved by the institution's committee for the protection of human subjects.

14. DMSO is an oxidative agent toxic for cells, especially at temperature above 10°C. Thus, it must be used at 4°C.

15. Change scalpel blade and clean other instruments (forceps) between eye specimens.

16. Nonspecific DNA–protein interactions are usually prevented by the addition to the reaction mix of 1–5 µg of a non specific competitor DNA. Although this is clearly very effective when crude nuclear extracts are used, such high concentrations of nonspecific competitor DNA were found to compete even for specific DNA–protein complexes when enriched preparations of nuclear proteins are used in EMSA (13). The more enriched the nuclear protein of interest, the lower the amount of nonspecific competitor required. For example, we routinely use 1–2 µg poly(dI:dC) with crude nuclear proteins, 250 ng when the nuclear extract is enriched on heparin–Sepharose column, and no more than 25–50 ng with purified or recombinant proteins.

17. The signal strength of a shifted DNA–protein complex can be substantially increased by favoring the odds for the interaction between the protein of interest and its target sequence. This can easily be achieved with enriched preparations of nuclear proteins either by increasing the amount of the labeled probe selected or by decreasing the concentration of poly(dI:dC), or both. Furthermore, the DNA binding ability of some nuclear proteins proved to be highly dependent on the amount of salt (usually provided by KCl) that is present in the reaction mix. Transcription factors such as NFI and Sp1 best interact with their respective target sequence in the presence of 100 mM and 150 mM KCl, respectively (14). It is therefore useful to evaluate which KCl concentration best allows binding of nuclear proteins to a specific DNA target probe.

18. Formation of DNA–protein complexes is highly dependent on the voltage selected for their migration into the polyacrylamide gel (15). We have found that reducing the migration time by running the EMSA at voltage lower than 120 V (usually corresponding to 10 mA for a single gel) on a 6% polyacrylamide gel rendered very unstable the formation of most DNA–protein complexes and consequently resulted in our inability to detect them.

19. All sample volumes and total protein amounts must be the same.

Acknowledgments

The authors would like to thank current and former members of the LOEX and LOEX/CUO-Recherche laboratories who have contributed to develop the foregoing protocols.

References

1. Green H, Rheinwald JG, Sun TT (1977) Properties of an epithelial cell type in culture: the epidermal keratinocyte and its dependence on products of the fibroblast. Prog Clin Biol Res 17:493–500
2. McLoughlin CB (1961) The importance of mesenchymal factors in the differentiation of chick epidermis. J Embryol Exp Morphol 9:370–384
3. Melbye SW, Karasek MA (1973) Some characteristics of a factor stimulating skin epithelial cell growth in vitro. Exp Cell Res 79:279–286
4. Wessells NK (1964) Substrate and nutrient effects upon epidermal basal cell orientation and proliferation. Proc Natl Acad Sci U S A 52:252–259
5. Masson-Gadais B, Fugere C, Paquet C, Leclerc S, Lefort NR, Germain L, Guerin SL (2006) The feeder layer-mediated extended lifetime of cultured human skin keratinocytes is associated with altered levels of the transcription factors Sp1 and Sp3. J Cell Physiol 206:831–842
6. Gaudreault M, Carrier P, Larouche K, Leclerc S, Giasson M, Germain L, Guerin SL (2003) Influence of sp1/sp3 expression on corneal epithelial cells proliferation and differentiation properties in reconstructed tissues. Invest Ophthalmol Vis Sci 44:1447–1457
7. Larouche D, Paquet C, Fradette J, Carrier P, Auger FA, Germain L (2009) Regeneration of skin and cornea by tissue engineering. Methods Mol Biol 482:233–256
8. Bradford MM (1976) A rapid and sensitive method for the quantitation of microgram quantities of protein utilizing the principle of protein-dye binding. Anal Biochem 72:248–254
9. Li L, Davie JR (2010) The role of Sp1 and Sp3 in normal and cancer cell biology. Ann Anat 192:275–283
10. Zhao C, Meng A (2005) Sp1-like transcription factors are regulators of embryonic development in vertebrates. Dev Growth Differ 47:201–211
11. Li L, He S, Sun JM, Davie JR (2004) Gene regulation by Sp1 and Sp3. Biochem Cell Biol 82:460–471
12. Leclerc S, Eskild W, Guerin SL (1997) The rat growth hormone and human cellular retinol binding protein 1 genes share homologous NF1-like binding sites that exert either positive or negative influences on gene expression in vitro. DNA Cell Biol 16:951–967
13. Roder K, Schweizer M (2001) Running-buffer composition influences DNA-protein and protein-protein complexes detected by electrophoretic mobility-shift assay (EMSA). Biotechnol Appl Biochem 33:209–214
14. Gaudreault M, Vigneault F, Leclerc S, Guerin SL (2007) Laminin reduces expression of the human alpha6 integrin subunit gene by altering the level of the transcription factors Sp1 and Sp3. Invest Ophthalmol Vis Sci 48:3490–3505
15. Vossen KM, Fried MG (1997) Sequestration stabilizes lac repressor-DNA complexes during gel electrophoresis. Anal Biochem 245:85–92

Chapter 12

Genetic Manipulation of Keratinocyte Stem Cells with Lentiviral Vectors

Masahito Yasuda, David J. Claypool, Erika Guevara, Dennis R. Roop, and Jiang Chen

Abstract

The epidermis of the skin and its appendages, such as the hair follicles, are formed and maintained by keratinocyte stem cells. Highly efficient and permanent genetic modifications are valuable tools to examine the multipotency and regenerative capacity of keratinocyte stem cells in skin and hair follicle development, homeostasis, and regeneration. Herein, we describe an ex vivo approach by which primary mouse keratinocytes can be permanently manipulated by lentiviral vectors at the genetic level. This protocol can be used to permanently express a gene-of-interest or selectively silence the expression of an endogenous gene, which can be used in preclinical development of gene-based therapies for skin and systemic disorders.

Key words Keratinocyte, Skin, Mouse, Gene transfer, Lentivirus

1 Introduction

Primary epidermal keratinocytes can be cultured in vitro (1–3). Cultured keratinocytes contain a heterogeneous population of stem cells and transit amplifying cells, which maintain high proliferative capability and can be used to reconstitute fully functional skin and skin appendages (4, 5). Primary keratinocytes of mouse skin can be isolated and maintained in low calcium medium for up to 10 days (6, 7) before being grafted onto recipient mice as skin grafts. Therefore, these keratinocytes can be genetically modified ex vivo and subsequently used in vivo to assay for functions.

Commonly used transfection methods for delivering genetic material to mammalian cells, including physical, chemical, and viral approaches, can be used to genetically modify epidermal keratinocytes. The lipid-mediated transfection method is one of the most efficient nonviral gene delivery methods with an efficiency up to 40% (8). In contrast, viral transduction can achieve up to 100% transfection efficiency. Among various viral vectors, adenoviral and adeno-associated viral vectors can only transiently deliver genetic

Kursad Turksen (ed.), *Skin Stem Cells: Methods and Protocols*, Methods in Molecular Biology, vol. 989, DOI 10.1007/978-1-62703-330-5_12, © Springer Science+Business Media New York 2013

material (9, 10); whereas retroviral vectors can integrate the transgene into the genome, thus achieving permanent gene expression (11). Lentiviral vectors, a variation of retroviral vectors derived from human immunodeficiency virus (HIV), can infect both dividing and nondividing cells (12), making them ideally suited for introducing genetic materials to stem cells which are often quiescent. Here, we describe a procedure through which mouse keratinocyte stem cells can be isolated, cultured, and infected by lentiviral vectors with up to 100% transduction efficiency in vitro. This procedure can be utilized to permanently express a gene product or suppress the expression of an endogenous gene in keratinocytes, which can be subsequently evaluated in vitro or grafted onto immunocompromised mice.

2   Materials

2.1   Isolation and Culture of Mouse Primary Keratinocytes

1. Mouse embryos (E18.5) or newborn (up to 3-day old) pups.

2. Dispase II solution: Dissolve 1 g (1,000 U) lyophilized Dispase II (Neutral protease, Grade II, Roche #04942078001) in 33 ml phosphate buffer saline (PBS, pH 7.4) to make 30 U/ml stock solution. Filter through a 0.22 μm filter to sterilize. 1 ml aliquots of this stock solution can be stored at –20°C for up to 1 year. Before using, dilute 1 ml of stock solution (30 U/ml) in 20 ml of serum-free medium (e.g., DMEM or EMEM) to a working concentration of 1.5 U/ml.

3. 0.25% trypsin without EDTA (Gibco, #15050-065).

4. Keratinocyte growth medium: Fibroblast-conditioned medium consisting of EMEM (BioWhittaker, #06-174G), 8% chelexed fetal bovine serum (FBS), penicillin/streptomycin (100 U/ml), and 0.05 mM $CaCl_2$ (see Note 1), or serum-free medium (see Note 2).

5. Collagen IV or collagen I-coated tissue culture dish: Collagen IV (Sigma-Aldrich, #C-5533), 0.25% acetic acid (see Note 3), or BioCoat™ Collagen I or IV 100 mm Culture Dish (BD Biosciences, #354450 or #354453, respectively).

6. Filters: Millex-MP 0.22 μm filter unit (Millipore, #SLMP025SS), Stericup-HV 0.45 μm filter unit (Millipore, #SCHVU05RE), and cell strainer, 40 μm (BD biosciences, #352340).

2.2   Lentiviral Vector Production

1. 293T cells.

2. 293T growth medium: Dulbecco's Modified Eagle Medium (DMEM; 4,500 mg/l glucose, 110 mg/l sodium pyruvate and 584 mg/l l-glutamine) (Sigma-Aldrich, #D6429), 10% heat-inactivated FBS, and 2 mM Glutamine.

3. 0.25% trypsin–EDTA.

4. Transferring lentiviral vector and packaging system: pLVTHM (expressing GFP and shRNA) lentiviral transferring vector, psPAX2 (gag/rev) packaging plasmid, and pMD2.G (VSV-G) envelope plasmid, or other lentiviral vector and compatible packaging system.

5. 2.5 M $CaCl_2$: Dissolve 183.7 g $CaCl_2$ $2H_2O$ in 500 ml of ddH_2O. Filter through 0.45 µm filter to sterilize. Store indefinitely at –20°C or –80°C in 50 ml aliquots. Once thawed, $CaCl_2$ solution can be stored at 4°C for up to 3 months.

6. 2× HeBS: Mix 28 ml 5 M NaCl, 50 ml 0.5 M HEPES, 7.5 ml 0.1 M Na_2HPO_4, 50 ml 0.1 M KCl, 0.1 M Glucose with 250 ml of ddH_2O. Adjust pH to 7.05–7.12. Add ddH_2O to 500 ml and filter through 0.22 µm filter to sterilize. Store at –20°C in 50 ml aliquots for up to 1 year.

7. HEPES buffered H_2O: Add 125 µl of 1 M HEPES, pH 7.3, into 50 ml sterile ddH_2O. Store at 4°C.

8. 1 M sodium butyrate: Dissolve 1 g sodium butyrate (Sigma-Aldrich, #B-5887) in 9.1 ml ddH_2O. Adjust pH to 7.0 with 1N HCl. Filter through 0.22 µm filter to sterilize. Store at 4°C.

2.3 Transduction of Primary Keratinocytes In Vitro

1. 8 mg/ml polybrene: Dissolve 0.1 g polybrene (hexadimethrine bromide, Sigma-Aldrich, #H9268) in 12.5 ml ddH_2O. Filter through 0.22 µm filter to sterilize. Store at 4°C.

---

3 Methods

3.1 Isolation and Culture of Mouse Keratinocytes

The isolation of keratinocyte can be completed within 1 day or 2 half days. A relatively pure battery of keratinocytes can be isolated and propagate in vitro for 1 week (see Note 4).

1. Strip off skin from E18.5 embryos or newborn pups (see Notes 5 and 6).

2. Float the skin on Dispase II solution in 100 mm dish or 6-well plate with the surface of the skin (epidermis) facing up. Make sure that the skin is spread out without curling or folding, and that the skin is not submerged under the solution. Incubate at 37°C for approximately 1 h, room temperature for 4 h, or 4°C overnight. Dispase II will disrupt the epidermal–dermal junction (e.g., type IV collagen and fibronectin) (Fig. 1a).

3. Lay skin on the dry surface of a tissue culture dish, epidermis facing down. Spread out the entire skin so that the edges of the skin are firmly attached to the surface of the dish.

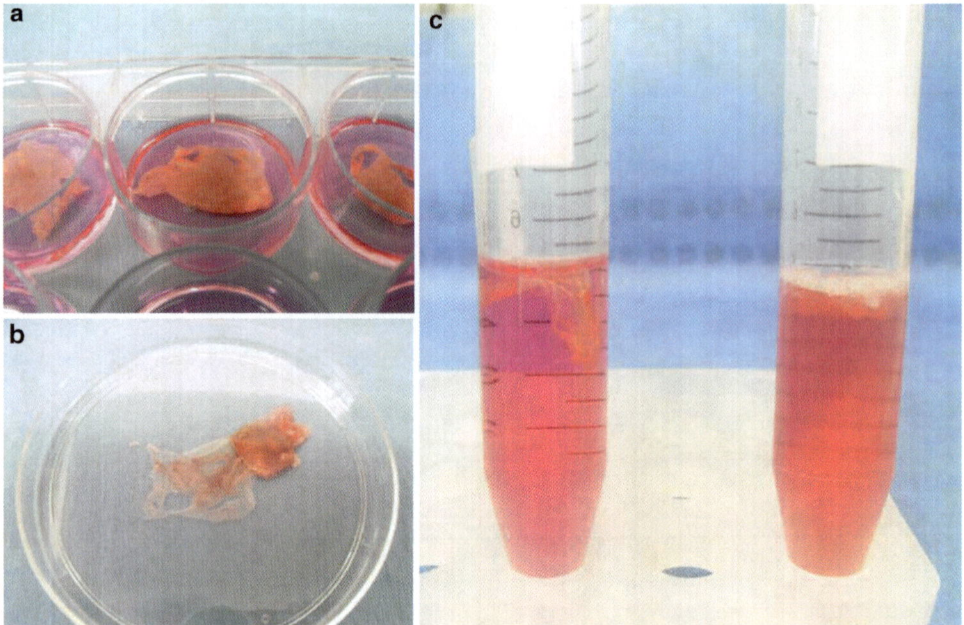

Fig. 1 Isolation of keratinocytes from embryonic mouse skin. (**a**) Skin is floated on Dispase II solution. (**b**) Dermis is lifted up and removed from the underlying epidermis. (**c**) Epidermis is incubated in 0.25% trypsin before (*left*) and after (*right*) agitation. Note that the solution becomes cloudy when keratinocytes are released from the epidermis

4. Grab the dermis, which is facing up, with a pair of forceps at the edge of the skin and carefully pull dermis away from the underlying epidermis. The dermis should be readily separated from the epidermis in its entirety when the skin is sufficiently digested by Dispase II, leaving behind an intact epidermal sheet (now the basal keratinocytes are facing up) (Fig. 1b).

5. Carefully and immediately lift the entire epidermis and transfer to 5 ml of prewarmed 0.25% trypsin in a 15-ml tube (see Note 7).

6. Tap the tube forcefully and repeatedly with finger for 2–5 min to release keratinocytes from epidermis. When sufficiently agitated, the epidermis will turn pale and float whereas the solution becomes cloudy (Fig. 1c).

7. Remove the epidermal membrane. Add 5 ml of serum-containing medium to neutralized trypsin. Pipet to disrupt clumping and filter cells through a 40-µm cell strainer into a 50-ml conical centrifugation tube.

8. Spin down cells at $200 \times g$ (approximately 1,200 rpm in an Eppendorf centrifuge 5804 fitted with an A-4-44 wing-bucket rotor) for 5 min.

9. Suspend cells thoroughly to minimize clumping. Seed cells in a 100 mm dish containing 10 ml keratinocyte growth medium

(see Note 8), and incubate overnight at 37°C in a humidified incubator with 5% CO_2.

10. The next morning, wash the cells with keratinocyte growth medium or PBS before replacing with fresh growth medium. Thereafter, change medium every other day.

3.2 Production of Lentiviral Vectors

The following protocol utilizes calcium precipitation to transfect 293T cells. Lentiviral vectors can be produced in five working days. Various quantities of lentiviral vectors can be produced by scaling up (or down) the following protocol. A lentivector, pLVTHM (Addgene plasmid 12247) (13, 14), was used in this protocol to simultaneously exemplify the capabilities of lentiviral vectors in expressing an exogenous gene or knocking down an endogenous gene. Please see Note 9 for important safety precautions.

Day 0

1. Grow 293T cells in a 150-mm tissue culture dish to near confluency with 293T growth medium in a humidified incubator containing 5% CO_2 at 37°C.

Day 1

2. Harvest 293T cells by trypsinizing with 3 ml of 0.25% trypsin–EDTA and plate the cells into three 150 mm dishes, each of which contains 15 ml 293T growth medium without antibiotics, in the afternoon before transfection.

Day 2

3. 293T cells should reach approximately 50% confluency in the morning.

4. Prepare DNA:$CaCl_2$ mixture by combining 48 µg pLVTHM, 12 µg pMD2.G, and 36 µg psPAX2 in 1,100 µl of HEPES·H_2O with 300 µl of 2.5 M $CaCl_2$. The total volume of the DNA:$CaCl_2$ mixture is 1.5 ml (for transfecting one 150 mm dish). Mix well by pipetting or brief vortexing.

5. In a separate tube, prepare 1.5 ml of 2× HeBS.

6. Add the DNA:$CaCl_2$ mixture (1.5 ml) to the above 2× HeBS (1.5 ml) dropwisely, while bubbling the latter by blow air with an electric pipette.

7. Transfect 293T cells by dropping the above transfection-ready mixture evenly to the 293T culture. Rock the plate gently to mix. The total volume is 18 ml.

8. 8 h later, add 180 µl of 1 M sodium butyrate to the 293T culture to a final concentration of 10 mM; mix well by gentle swirling.

Day 3

9. Approximately 24 h after transfection, wash 293T cells carefully with PBS for three times. Replace with 15 ml of 293T growth medium containing 10 mM HEPES (see Note 10).

10. Continue the culture for 2 days.

Day 5

11. Collect the virus-containing supernatant carefully. Pellet debris by centrifugation at $800 \times g$ for 5 min. Filter with a 0.22 μm filter unit.

12. The virus-containing medium can be used to infect keratinocytes directly or concentrated by ultracentrifugation at $50,000 \times g$ in a Beckman SW28 rotor for 2 h or centrifiltration with Centricon columns (e.g., Centricon Plus-70, Millipore, #UFC710008). Store temporarily at 4°C for up to 5 days or freeze single-use aliquots at –80°C (see Note 11).

3.3 Transduction of Keratinocytes with Lentiviral Vectors

1. Grow keratinocytes to approximately 30% confluency (normally in 1–3 days after plating) (Fig. 2a).

2. Add an appropriate amount of lentiviral solution and 5 μl of 8 mg/ml polybrene to the keratinocyte culture containing 5 ml of medium. The final concentration of polybrene is 8 μg/ml (see Notes 12 and 13).

3. Incubate overnight.

4. Replace virus-containing medium with 10 ml of normal keratinocyte growth medium and continue the culture until harvest.

5. Infected keratinocytes will reach confluency in approximately 2–4 days (Fig. 2b). Cells can be harvested for protein (Fig. 2b) or RNA (Fig. 2d) analyses or grafted with fibroblasts on immunocompromised mice to form genetically modified skin (epidermis) and appendages (hair follicles) (see Note 14).

4   Notes

1. The preparation of fibroblast-conditioned keratinocyte growth medium is described previously (2). Chelexed FBS can be prepared as the following. Fill 500 ml ddH$_2$O in an Erlenmeyer flask placed on a stirrer. Add 100 g chelex 100 resin (Bio-Rad, #142-2842) while stirring slowly. Adjust pH to 7.35–7.4 with HCl. Turn off the stirrer and allow the resin to settle for 30 min. Pour off ddH$_2$O. Wash the resin three times by stirring in 500 ml ddH$_2$O for 5 min, letting the resin settle for 30 min, and pouring off ddH$_2$O. Wash with 250 ml of PBS, twice. At the end of the last wash, cool to 4°C for 10 min; adjust to pH 7.35–7.4. Let the resin settle for 30 min and pour off PBS. Add 500 ml of FBS and stir for 1 h at 4°C. Let the resin settle for 30 min. Collect the chelexed FBS and filter through a 0.45 μm filter. Store the chelexed FBS in 45 ml aliquots at –20°C.

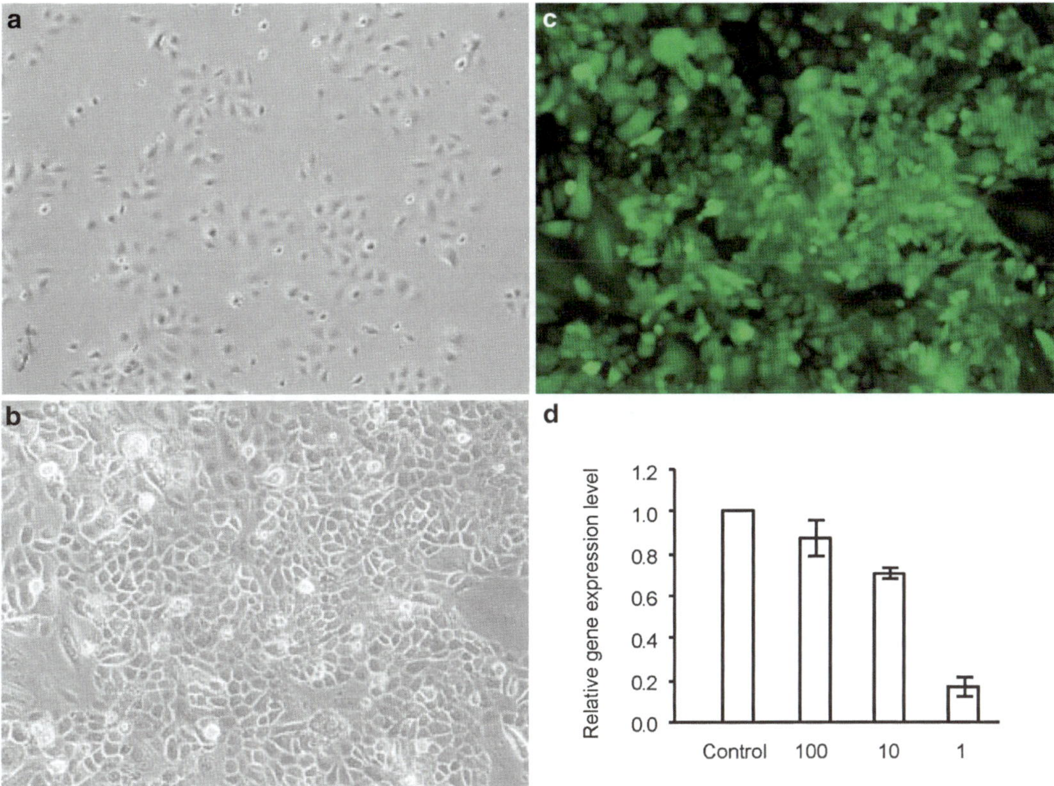

Fig. 2 Lentiviral transduction efficiency. (**a**) Primary murine keratinocytes cultured 24 h after seeding reach a confluency ideal for transduction. (**b**) Three days after transduction, keratinocytes grow to confluency. (**c**) Fluorescent microscopy demonstrating that almost all keratinocytes express GFP. (**d**) Quantitative RT-PCR demonstrates the ability of lentiviral vectors in silencing the expression of an endogenous gene. 100, 10, and 1 represent the dilution factors of the viral supernatant used for infection. Control, non-transduced cells

2. Defined Keratinocyte-Serum Free Medium (DK-SFM; Gibco, #10744-019) or CnT-07 (CELLnTEC) can be used as a substitute for fibroblast-conditioned keratinocyte growth medium.

3. Reconstitute collagen powder to 1 mg/ml in 0.25% acetic acid. Store at –20°C. Before using, thaw the frozen collagen stock on ice; mix by vortexing for 15 s. Dilute to 7 μg/ml with 0.25% acetic acid. Apply the recommended volume of diluted collagen to tissue culture plates (approximately 7.8 ml for one 100 mm dish) and let stand at 4°C overnight. Aspirate the solution. Rinse the plates with equal volume of sterile PBS, then, sterile ddH$_2$O. Aspirate the solution; air dry in the hood. Seal coated plates with parafilm and store at 4°C for up to 6 months.

4. Isolated keratinocytes may contain a small percentage of dermis fibroblasts. However, dermal fibroblasts will be selected out in keratinocyte growth medium.

5. Details for stripping skin off mouse embryos or newborn pups are described by Lichti et al. (3).

6. If the genotypes of donor embryos or pups need to be determined, genomic DNA can be extracted from the tails and used for genotyping.

7. Multiple skins can be combined and processed at the same time. It is important to protect the basal cells from drying up by folding the epidermis before all skins are ready.

8. Keratinocytes will reach 30% confluency in approximately 1–2 days.

9. As a safety precaution, all virus-related work should be handled with personal protection equipment, including gloves, mask, and laboratory coat, and in a biosafety cabinet approved for use with virus. Dry and liquid waste should be decontaminated with fresh concentrated bleach before disposal. Equipment in contact with virus needs to be cleaned with 75% ethanol.

10. Keratinocyte growth medium may be used so that the virus-containing supernatant can be used directly on keratinocytes.

11. If needed, titers of lentiviral vectors can be determined by ELISA, qRT-PCR, and flow cytometry. An example of titration based on GFP expression is described by Schroers et al. (15).

12. Polybrene (hexadimethrine bromide) is a cationic polymer known to enhance viral transduction (16). Polybrene inhibits keratinocytes proliferation at concentrations higher than 8 μg/ml (17).

13. Virus and polybrene can be premixed in keratinocyte growth medium before being added to cells. When using serum-free medium (DK-SFM or Cnt-07), supplement with 1% chelexed FBS which will enhance viral stability, hence the transduction efficiency. In addition, more than one round of lentivirus infection can be performed to enhance the transduction efficiency.

14. Grafting keratinocytes with fibroblasts on immunocompromised mice is a valuable approach in examining the effect of lentiviral vector-mediated genetic modification. The grafting procedure is described elsewhere by Lichti et al. (3).

Acknowledgments

We are grateful to colleagues in our laboratories who have modified and improved the procedure described herein. J.C. is supported by a grant from NIH/NIAMS (AR061485).

References

1. Yuspa SH, Harris CC (1974) Altered differentiation of mouse epidermal cells treated with retinyl acetate in vitro. Exp Cell Res 86:95–105
2. Hager B, Bickenbach JR, Fleckman P (1999) Long-term culture of murine epidermal keratinocytes. J Invest Dermatol 112:971–976
3. Lichti U, Anders J, Yuspa SH (2008) Isolation and short-term culture of primary keratinocytes, hair follicle populations and dermal cells from newborn mice and keratinocytes from adult mice for in vitro analysis and for grafting to immunodeficient mice. Nat Protoc 3:799–810
4. Green H, Kehinde O, Thomas J (1979) Growth of cultured human epidermal cells into multiple epithelia suitable for grafting. Proc Natl Acad Sci U S A 76:5665–5668
5. Lichti U, Weinberg WC, Goodman L, Ledbetter S, Dooley T, Morgan D, Yuspa SH (1993) In vivo regulation of murine hair growth: Insights from grafting defined cell populations onto nude mice. J Invest Dermatol 101:124S–129S
6. Hennings H, Michael D, Cheng C, Steinert P, Holbrook K, Yuspa SH (1980) Calcium regulation of growth and differentiation of mouse epidermal cells in culture. Cell 19:245–254
7. Yuspa SH, Kilkenny AE, Steinert PM, Roop DR (1989) Expression of murine epidermal differentiation markers is tightly regulated by restricted extracellular calcium concentrations in vitro. J Cell Biol 109:1207–1217
8. Dickens S, Van den Berge S, Hendrickx B, Verdonck K, Luttun A, Vranckx JJ (2010) Nonviral transfection strategies for keratinocytes, fibroblasts, and endothelial progenitor cells for ex vivo gene transfer to skin wounds. Tissue Eng Part C Methods 16:1601–1608
9. Benihoud K, Yeh P, Perricaudet M (1999) Adenovirus vectors for gene delivery. Curr Opin Biotechnol 10:440–447
10. Khare R, Chen CY, Weaver EA, Barry MA (2011) Advances and future challenges in adenoviral vector pharmacology and targeting. Curr Gene Ther 11:241–258
11. Eglitis MA, Kantoff P, Gilboa E, Anderson WF (1985) Gene expression in mice after high efficiency retroviral-mediated gene transfer. Science 230:1395–1398
12. Naldini L, Blömer U, Gallay P, Ory D, Mulligan R, Gage FH, Verma IM, Trono D (1996) In vivo gene delivery and stable transduction of nondividing cells by a lentiviral vector. Science 272:263–267
13. Zufferey R, Nagy D, Mandel RJ, Naldini L, Trono D (1997) Multiply attenuated lentiviral vector achieves efficient gene delivery in vivo. Nat Biotechnol 15:871–875
14. Wiznerowicz M, Trono D (2003) Conditional suppression of cellular genes: Lentivirus vector-mediated drug-inducible RNA interference. J Virol 77:8957–8961
15. Schroers R, Chen S-Y (2004) Lentiviral transduction of human dendritic cells. Methods Mol Biol 246:451–459
16. Toyoshima K, Vogt PK (1969) Enhancement and inhibition of avian sarcoma viruses by polycations and polyanions. Virology 38:414–426
17. Seitz B, Baktanian E, Gordon EM, Anderson WF, LaBree L, McDonnell PJ (1998) Retroviral vector-mediated gene transfer into keratocytes: in vitro effects of polybrene and protamine sulfate. Graefes Arch Clin Exp Ophthalmol 236:602–612

Chapter 13

Using 3D Culture to Investigate the Role of Mechanical Signaling in Keratinocyte Stem Cells

Lee Wallace and Julia Reichelt

Abstract

The ability to grow keratinocyte stem cells (KSCs) in 3D culture is an important step forward for investigating the physiological properties of these cells. In the epidermis, KSCs are subject to various types of mechanical stress. To study the effects of mechanical stress on KSCs, monolayer cultures are limited as the KSCs can only form cell–cell contacts in one plane and to prevent differentiation, KSCs are grown in low (0.05 mM) calcium, which impairs formation of calcium-dependent adhesion structures such as desmosomes. This is in contrast to how KSCs are found in the epidermis in vivo, where they are connected on all sides by other cells, allowing them to form a more organized cytoskeleton. The cytoskeleton is essential for transducing mechanical signals between cells, and this cannot be accurately reproduced in monolayer cultures, where the cells do not have the same level of organization or connections.

We describe a technique which allows the generation of large numbers of uniformly sized cell aggregates using cultured murine KSCs. These aggregates are produced using physiological calcium concentrations (1.2 mM), allowing the cells within the aggregates to form calcium-dependent contacts with other cells on all sides, resulting in the reorganization of the cytoskeleton, integrating the cells within each aggregate. Within the aggregates, KSCs retain stem cell properties, such as p63 expression, despite the increased calcium concentration and show activation of the mitogen-activated protein kinase ERK upon stretch. KSC aggregates can be manipulated further and provide a more physiologically relevant model for studying mechanical signaling in KSCs.

Key words Murine, Keratinocytes, Stem cells, 3D culture, Mechanical signaling

1 Introduction

The establishment of murine keratinocyte stem cell (KSC) culture (1) has paved the way for investigations which allow better understanding of the epidermis and epidermal regeneration. KSC culture allows the study of the signaling involved in the proliferation, differentiation, and migration of normal KSCs and KSCs derived from epidermal tissue from transgenic mice.

The interactions of cells and their signaling in monolayers is not necessarily a good representation of the in vivo conditions.

Kursad Turksen (ed.), *Skin Stem Cells: Methods and Protocols*, Methods in Molecular Biology, vol. 989,
DOI 10.1007/978-1-62703-330-5_13, © Springer Science+Business Media New York 2013

In vivo, multipotent KSCs reside in a niche termed "the bulge," located at the site of each individual hair follicle. KSCs in the bulge are surrounded by other KSCs and, on the outside of this structure, by connective tissue. KSCs found in the bulge region are slow cycling, only being activated to migrate and proliferate when needed to sustain the skin after injury or maintain the hair growth cycle (2). Also connected to the bulge region is the arrector pili muscle (APM). This muscle uses the bulge as its insertion site, yet the potential role of the stresses incurred by the KSCs when the APM contracts have not been studied. Mechanical stress has been shown to induce signaling pathways in other types of stem cells. For example, mesenchymal stem cells in bones differentiate along the osteogenic lineage upon exposure to mechanical stress, which subsequently influences the rates of bone production and resorption, leading to bone remodeling (3, 4). Mechanical stresses applied to keratinocytes are known to activate the mitogen-activated protein kinase ERK, resulting in an increase in cell proliferation (5, 6). Mechanical signals can also mediate effects leading to cell death, growth, and differentiation in other cell types (7).

For mechanical stresses to be applied to, and have an affect on cells, an organized cytoskeleton is required (8). In the hair follicle bulge region, KSCs exist in a 3D environment with organized cytoskeletons within the cells and cell–cell contacts on all sides with other KSCs and with the extracellular matrix allowing sensing and transduction of mechanical stimuli. In order to study mechanical signaling in cell culture, the KSC culture method needs to be developed from 2D to 3D, as the cells within monolayers do not have the same cytoskeletal structure as those in the epidermis (9, 10). The disorganization of the cytoskeleton of cells in monolayers (11) will change the way in which mechanical signals are transduced between and within cells (12) from what would be seen in vivo. The resulting cellular signaling generated under mechanical stress will not be the same in monolayers as would be seen in cells with an organized cytoskeleton in a 3D structure. For a cell culture model to be representative of the in vivo KSCs, the effects created by mechanical stress on the cytoskeleton need to be comparable.

We describe a 3D culture method for murine KSCs based on the formation of cell aggregates which works in a similar fashion to the "hanging drop" method (13). Specialized culture plates, containing microwells with a repellent surface (Aggrewell plates) to prevent the adhesion of cells, allow the number of KSC aggregates formed to be scaled up. KSCs derived from neonatal mouse skin (1) are seeded into the Aggrewell plates, and the repellent surface leaves the cells in suspension with only other cells to adhere to, resulting in the formation of spherical aggregates when seeded in

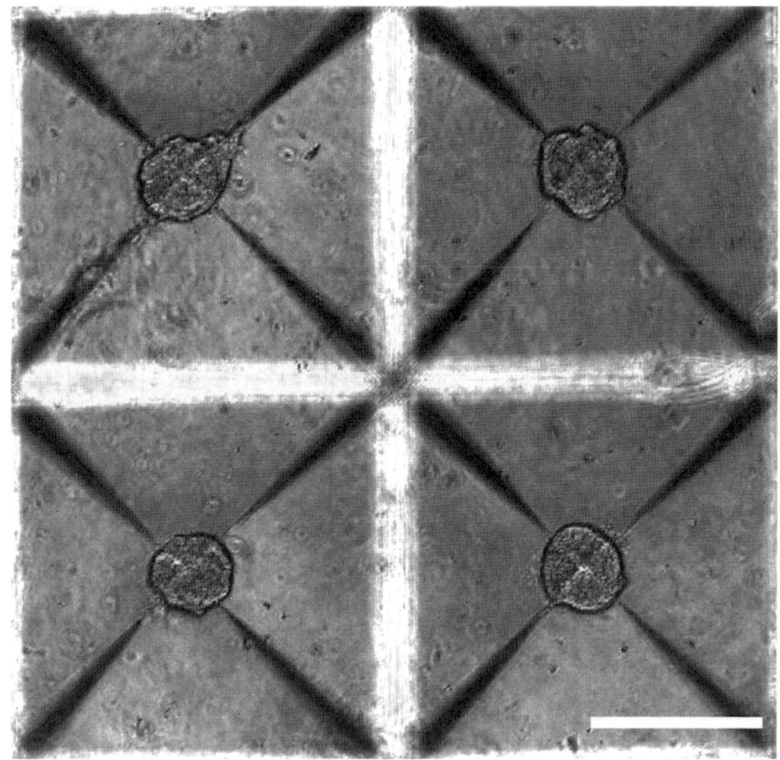

Fig. 1 Brightfield image of KSC aggregates within four microwells of an Aggrewell plate. Aggregates are uniform in size (100 cells per aggregate) and will only form when seeded in medium containing 1.2 mM calcium. Scale bar represents 200 µm

medium with a physiological calcium concentration (1.2 mM) (Fig. 1). The higher calcium concentration does not induce the differentiation of KSCs in 3D culture as it would in 2D culture (1) (Fig. 2), yet it is important as it allows the cells to form calcium-dependent cell junctions, such as adherens junctions and desmosomes, which are strongly reduced in the 2D cultures grown in low calcium (0.05 mM) medium (14). A higher calcium concentration also initiates a change in the distribution of actin and intermediate filaments, as well as microtubules within the cells, altering the overall cytoskeletal structure (14, 15).

Once formed, the KSC aggregates can be used to investigate the response of KSCs to mechanical stress. Here, we present a method which allows studying mechanosignaling in response to defined stretch using the Flexcell system (Flexcell Technologies) in combination with Western blotting and immunofluorescence analysis (Figs. 3 and 4).

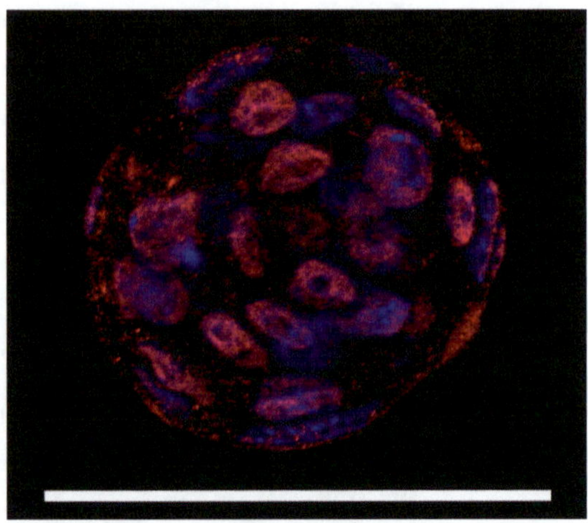

Fig. 2 KSC aggregate sections (5 μm) were fixed in 4% formaldehyde in PBS and then stained for the KSC marker p63 (17). The cells within the aggregate expressed p63, showing that even though the medium contained 1.2 mM calcium, KSC retained their stem cell-like properties. Further characterization of KSC aggregates can be found in ref. 1. Anti-p63 (Santa Cruz Biotechnology, Heidelberg, Germany), Alexa-594-conjugated anti-rabbit (Invitrogen), *red*; DAPI was used for nuclear staining, *blue*. Scale bar represents 100 μm

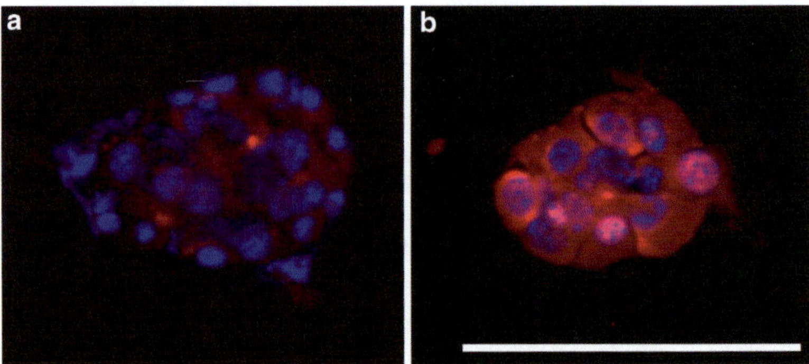

Fig. 3 Immunofluorescent staining for phospho-ERK (P-ERK) of KSC aggregates after stretch using the Flexcell system. Compared to an unstretched control **(a)**, KSC aggregates which were subjected to mechanical stretch showed increased P-ERK expression **(b)**. This experiment was performed using cyclic stretching of 10% at 0.1 Hz for 10 min. Anti-P-ERK (Cell Signalling Technology, Danvers, MA, USA), Alexa-594-conjugated anti-rabbit (Invitrogen), *red*; DAPI was used for nuclear staining, *blue*. Scale bar represents 100 μm

2 Materials

2.1 Keratinocyte Cultivation

1. Rat tail collagen I (Becton Dickinson, Oxford, UK), 50 μg/ml 0.02 M acetic acid, stored at 4°C (see Note 1).

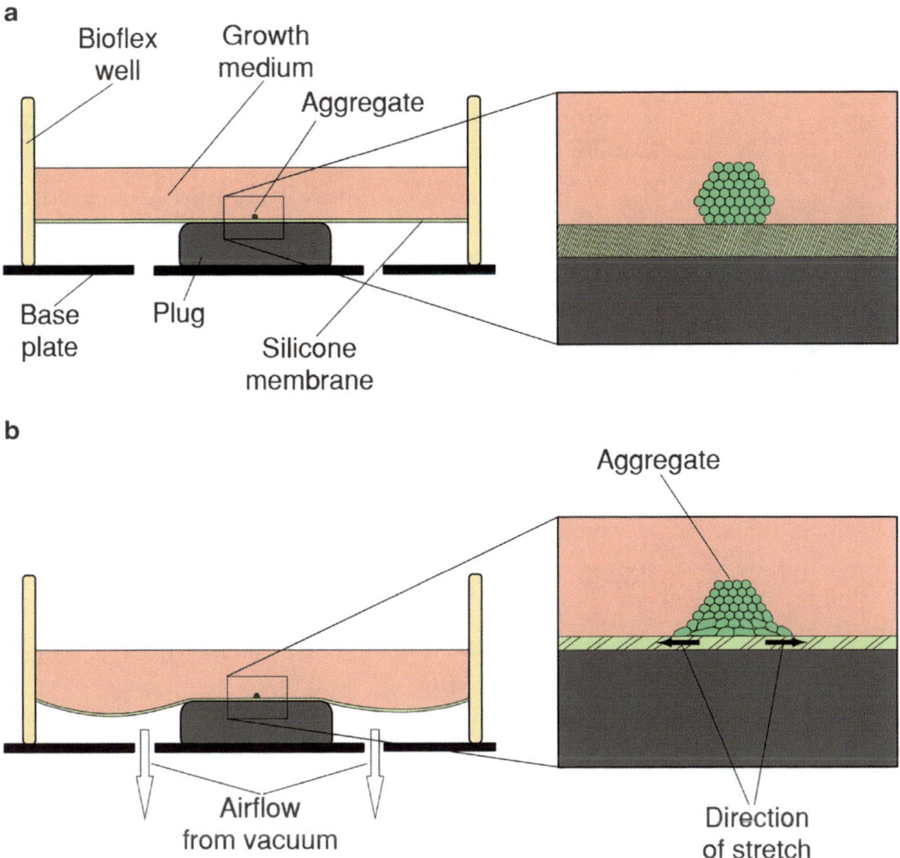

Fig. 4 Stretching of KSC aggregates using the Flexcell system. KSC aggregates are seeded into a Bioflex well and left to adhere to the silicone membrane. The aggregates adhere using only the cells at the bottom which are directly in contact with the silicone (**a**). When subjected to stretch by applying a vacuum from underneath the silicone membrane, only the cells at the bottom of the aggregate are directly affected by the mechanical stress (**b**). Any cells higher up in the aggregate would only experience stretch, or initiate signaling pathways, from the signaling and pull of cells attached to them. Figure created by Matthew D. Alexander

2. Chelex 100 resin (Bio Rad Laboratories, Hercules, CA, USA).

3. Fetal calf serum (FCS) Gold (PAA, Coelbe, Germany).

4. FAD medium: Dulbecco's Modified Eagles medium (DMEM)/ HAM's F12 3.5:1.1, low calcium (0.05 mM Ca^{2+}), 460 ml/ bottle (custom-made by Biochrom, Berlin, Germany).

5. Complete FAD medium: FAD medium, 10% Chelex-treated FCS Gold (20 g Chelex 100 resin/500 ml FCS, rotated overnight at 4°C before paper filtration and then sterile filtration to be stored in 50 ml aliquots at −20°C), 0.18 mM adenine (a), 0.5 µg/ml hydrocortisone (b), 5 µg/ml insulin (c), 10^{-10} M cholera toxin (d) (all from Sigma Aldrich, Dorset, UK), 10 ng/ ml epidermal growth factor (EGF) (e) (Invitrogen, Carlsbad,

CA, USA), 2 mM glutamine, PSA (100 U/ml penicillin, 100 µg/ml streptomycin and 0.25 µg/ml amphotericin; supplied as 100× stock solution, Invitrogen).

Stock solutions (stored at –20°C with the exception of cholera toxin which is stored at 4°C):

(a) Adenine (250×): 45 mM in 50 mM HCl, 2 ml added per 460 ml of FAD medium.

(b) Hydrocortisone (500×): 250 µg/ml in ethanol, 1 ml stock solution added per 460 ml of FAD medium.

(c) Insulin (1,000×): 5 mg/ml in 5 mM HCl, 0.5 ml added per 460 ml FAD medium.

(d) EGF (1,000×): 10 µg/ml in FAD medium, 0.5 ml added per 460 ml FAD medium.

(e) Cholera toxin (10^{-5} M): 1 mg/1.18 ml in sterile Milli-Q water, 5 µl added per 460 ml FAD medium.

6. 0.05% trypsin and 0.02% ethylenediaminetetraacetic acid (EDTA) (Lonza, Basel, Switzerland).

7. Phosphate buffered saline (PBS) (see Note 2).

8. 0.02% EDTA in PBS.

2.2 Feeder Cell Cultivation

1. 3T3-J2 fibroblasts.

2. DMEM (Sigma-Aldrich, Poole, UK) supplemented with 10% FCS and PSA (100 U/ml penicillin, 100 µg/ml streptomycin, and 0.25 µg/ml amphotericin; supplied as 100× stock solution, Invitrogen).

3. Mitomycin C (Applichem, Lancaster, UK). 100× stock solution prepared as 0.4 mg/ml in PBS. Aliquots stored at –20°C. Working concentration is 10 µl/ml medium.

4. 0.05% trypsin and 0.02% EDTA (Lonza).

2.3 Keratinocyte Aggregate Formation and Harvesting

1. High calcium (1.2 mM) complete FAD medium, made up as for keratinocyte cultivation (Subheading 2.1) with the addition of 1.2 mM $CaCl_2$.

2. Trypan blue staining solution (Sigma-Aldrich).

3. Aggrewell 400Ex culture plates (Stem Cell Technologies, Grenoble, France).

4. 40 µm cell strainer (Stem Cell Technologies).

5. Aggrewell rinsing solution (Stem Cell Technologies).

6. Hemocytometer.

2.4 Keratinocyte Aggregate Mounting for Cryosectioning

1. PBS.

2. 1.2% alginate solution: alginic acid sodium salt (Sigma-Aldrich) (see Note 3).

3. 102 mM $CaCl_2$ solution in Milli-Q water.

4. Optimal cutting temperature compound (OCT) (Fisher Scientific, Loughborough, UK).

5. Neutral Red staining solution (Sigma-Aldrich).

2.5 Keratinocyte Aggregate Stretching

1. FX-4000T Tension plus Flexcell system (Dunn, Asbach, Germany).

2. Bioflex 6-well culture plates, untreated (Dunn).

3   Methods

3.1 Keratinocyte Cultivation

1. For KSC culture, culture vessels are coated at room temperature for 1 h with rat tail collagen I and then the collagen solution is removed and the vessels are rinsed twice with PBS.

2. To subcultivate KSCs, the medium is removed and the cells are rinsed with PBS.

3. The cells are incubated with 2 ml of 0.02% EDTA in PBS, per 25 cm² surface area, for 5 min at room temperature.

4. After 5 min, the EDTA solution is removed and 1 ml of 0.05% trypsin/0.02% EDTA is added per 25 cm² surface area and incubated for 8–10 min, or until cells start to detach at 37°C (see Note 4).

5. The trypsinized cells are collected in complete FAD medium and centrifuged at $220 \times g$ for 5 min before resuspending the cell pellet in complete FAD medium.

6. For routine culture, the cells are split every 2–3 days at densities of 1:2 to 1:3. Cells are grown at 32°C and 5% CO_2.

3.2 Feeder Cell Cultivation

1. For subcultivating feeder cells, the medium is removed and they are rinsed with PBS.

2. The feeder cells are then incubated for 2 min with 0.05% trypsin/0.02% EDTA at 37°C.

3. The trypsinized cells are collected in DMEM containing 10% FCS and PSA and centrifuged at $220 \times g$ for 5 min. The feeder cells are resuspended in DMEM with 10% FCS and PSA.

4. For routine culture, feeder cells are split 1:3 every 2–3 days and incubated at 37°C, 5% CO_2.

3.3 Mitomycin C-Treatment of Feeder Cells

In order to increase KSC aggregate adhesion to the silicone growth surface of Bioflex plates, they are seeded in the presence of feeder cells. For cocultivation with KSC aggregates, feeder cells are seeded onto Bioflex wells, before being treated with mitomycin C to inhibit cell proliferation.

1. The medium is removed from the feeder cells and replaced with 2 ml of fresh medium, per 25 cm^2, containing 0.004 mg/ml mitomycin C and cells are incubated for 2–3 h at 37°C, 5% CO_2.

2. After 2–3 h, the mitomycin C-containing medium is removed and the cells are rinsed twice with PBS before KSC aggregates are added.

3.4 KSC Aggregate Formation in Aggrewell Plates

For generation of KSC aggregates in large numbers, Aggrewell 400Ex plates are used. These culture plates consist of six wells, each of which contains a plastic insert composed of 4,700 microwells, each with a width of 400 μm.

1. 2 ml of Aggrewell rinsing solution is added to each well of the Aggrewell plate, and the plates are then centrifuged at a maximum speed (minimum of $200 \times g$) for 10 min to remove any air bubbles (see Note 5).

2. The Aggrewell rinsing solution is then removed and the wells are rinsed with 1 ml complete FAD, before each well has 2 ml of 1.2 mM calcium complete FAD medium added.

3. Keratinocytes are harvested as per routine cell culture procedure (Subheading 3.1), and after centrifugation, the cells are resuspended in 1 ml of 1.2 mM calcium complete FAD medium per confluent 25 cm^2 surface area of culture vessel.

4. 10 μl of the cell suspension is mixed 1:1 with trypan blue and then counted using a hemocytometer to determine the number of viable cells.

5. Between 50 and 2,000 viable cells can be seeded per microwell of the Aggrewell plate (2.35×10^5 to 94×10^5 cells per six-well) and the volume made up to 3 ml with 1.2 mM calcium complete FAD.

6. The medium in each well is resuspended gently to evenly distribute the cells between the microwells and the plate is centrifuged at $100 \times g$ for 3 min to gently ensure all of the cells settle into the microwells (see Note 6).

7. The cells are incubated for 24 h at 32°C, 5% CO_2 to allow the formation of aggregates (Fig. 1).

3.5 KSC Aggregate Harvesting

1. The medium in each well of the Aggrewell plate is pipetted firmly up and down 2–3 times using a 1 ml micropipette tip to dislodge the aggregates.

2. The medium in the well is then passed through an inverted 40 μm cell strainer, placed over a 50 ml reaction tube to collect aggregates and remove single cells.

3. Using prewarmed 1.2 mM calcium complete FAD medium, the Aggrewell plate wells are washed 3–5 times with 1 ml medium to remove any remaining aggregates, and this is also passed over the inverted cell strainer.

4. The cell strainer is then turned over and placed into a fresh 50 ml tube and the aggregates are recovered by rinsing the cell strainer with 3–5 ml of 1.2 mM calcium complete FAD medium.

5. The aggregates are left for 5 min to settle down to the bottom of the tube before removing the excess medium and continuing with the protocol of choice (see Note 7).

3.6 Cryosectioning of KSC Aggregates

For immunofluorescence analysis, aggregates are frozen down in OCT for cryosectioning. Cryosectioning of aggregates in this manner should be performed on aggregates which have not been adhered to a membrane or been disrupted in any way, e.g., by trypsinizing or scraping from a surface.

1. After allowing them to settle in a 50-ml tube, the harvested aggregates are transferred in approximately 1 ml of medium into a 1.5-ml reaction tube (see Note 8).

2. The aggregates are left for 5 min to allow them to settle to the bottom of the tube and then as much medium as possible is removed without disturbing the aggregates (see Note 9).

3. The aggregates are then resuspended in 40 µl of 1.2% alginate solution, and this is dripped from a micropipette tip in one drop into 102 mM calcium chloride solution (see Note 10)

4. The alginate drop is left in the calcium chloride solution for 10 min before being transferred into a drop of OCT on parafilm, stained with 1 µl of 1% Neutral Red solution to help distinguish the alginate during sectioning (see Note 11).

5. The OCT drop containing the aggregates is placed, still on the parafilm, into precooled isopentane (cooled over liquid nitrogen) and left to set.

6. Once the drop of OCT has set it can be transferred into a precooled tube and stored at −80°C.

3.7 KSC Aggregate Stretching

1. Before seeding of KSC aggregates onto Bioflex plates, the Bioflex plates are collagen coated (see Note 1), and feeder cells are seeded onto the coated plates and left to adhere overnight before being inactivated with mitomycin C (see Subheadings 3.2 and 3.3).

2. After harvesting the aggregates from Aggrewell plates, they are added to the prepared Bioflex plate wells containing the inactivated feeder cells and left for 2–3 h at room temperature to

aid their adherence to the surface, before being placed in a 32°C, 5% CO_2 incubator for 24 h (see Note 12).

3. The next morning, the Flexcell system is set up.

4. Before the Bioflex plate is put into place, the feeder cells are removed by trypsinizing the wells for a short time (1–2 min).

5. To prevent any mechanical stresses from the set up affecting the results, the Bioflex plate is left in the Flexcell set-up for 2 h.

6. The aggregates can then be stretched using the Flexcell system, which allows precise control over the type and strength of the stretch applied.

7. After stretching, the aggregates can be processed for analysis, for example, by lysing or fixing onto the plates (Fig. 3).

3.8 Immuno-fluorescence of Stretched Aggregates

The silicone membrane of the Bioflex plates can be cut and is also transparent enough to be used with a fluorescent microscope. Immunofluorescence of stretched aggregates can be performed after fixing the aggregates to the Bioflex well surface. The silicone membrane can be cut out from the Bioflex plate (see Note 13), used for immunostaining and mounted upside down onto glass slides (Fig. 3).

4   Notes

1. Collagen I solution can be reused up to ten times, though the coating efficiency reduces with each subsequent use. This is more important when using the Bioflex plates as cell adherence to these is generally reduced compared to plastic culture vessels.

2. PBS must not contain Ca^{2+} and Mg^{2+} as these may induce keratinocytes to differentiate.

3. The alginate is difficult to dissolve. It is made by adding alginate salt to ddH_2O to make a 0.6% solution, then dissolved by heating in a microwave on full power (with occasional shaking), evaporating off half of the solution to give 1.2% alginate solution.

4. If after 10 min some cells are detached but others are still attached to the surface, the rest can usually be removed by forcefully tapping the side of the flask two or three times.

5. After centrifugation, the microwells should be checked under the microscope to ensure all air bubbles have been removed as any air bubbles will prevent aggregates forming.

6. After centrifuging, the microwells should be checked under the microscope to ensure even distribution of cells. If the cells

are not evenly distributed, they can be resuspended and then centrifuged again.

7. If the aggregates are to be used for subsequent cell culture experiments, this step can be omitted as the medium does not need to be removed, and the cells can be directly seeded in new culture vessels.

8. For analysis of some proteins, e.g., mechanical stress-induced proteins, the aggregates may have to be fixed at this stage to prevent activation of signaling pathways during the process of mounting them in the OCT. This can be done by transferring the aggregates to the 1.5-ml tube in the fixative then leaving them for the time required before continuing with the rest of this step.

9. Medium is easiest to remove by using a P200 pipette and tip to gently aspirate off the excess.

10. It is best to pipette the drop out slowly as pipetting too fast will result in the alginate breaking up into droplets before reaching the calcium chloride solution. The calcium chloride solution is kept at ambient temperature and can be reused for at least 10 times.

11. It is best to place the OCT drop onto a piece of parafilm which has been stretched in the middle into a concave shape using the base of a 1.5-ml conical tube to prevent the OCT from spreading out. The Neutral Red solution can be added by dipping a 10-μl pipette tip into the solution then mixing it into the OCT drop. The alginate should be removed from the calcium chloride gently with tweezers.

12. The amount of stretch applied over the surface of a Bioflex plate well is not evenly distributed, increasing from the center of the well to the edges (16). This distribution of stretch has to be taken into account when studying individual aggregates using immunofluorescence analysis and also for the analysis of lysates.

13. Membranes can be cut out carefully using a scalpel blade from underneath the plate.

Acknowledgments

We would like to thank Matthew D. Alexander for creating Fig. 4. We are grateful to the German Research Foundation for financial support (DFG; RE2673/1-1) and the Institute of Cellular Medicine and the Medical Faculty at Newcastle University for funding a studentship to Lee Wallace.

References

1. Vollmers A, Wallace L, Fullard N et al (2012) Two- and three-dimensional culture of keratinocyte stem cell and precursor cells derived from primary murine epidermal cultures. Stem Cell Rev 8:402–413
2. Ito M, Liu Y, Nguyen J et al (1995) Stem cells in the hair follicle bulge contribute to wound repair but not to homeostasis of the epidermis. Nat Med 11:1351–1354
3. Ehrlich PJ, Lanyon LE (2002) Mechanical strain and bone cell function: a review. Osteoporos Int 13:688–700
4. Liu L, Yuan W, Wang J (2010) Mechanisms for osteogenic differentiation of human mesenchymal stem cells induces by fluid shear stress. Biomech Model Mechanobiol 9:659–670
5. Reichelt J (2007) Mechanotransduction of keratinocytes in culture and in the epidermis. Eur J Cell Biol 86:807–816
6. Yano S, Komine M, Fujimoto M et al (2004) Mechanical stretching in vitro regulates signal transduction pathways and cellular proliferation in human epidermal keratinocytes. J Invest Dermatol 122:783–798
7. Vogel V, Sheetz M (2006) Local force and geometry sensing regulate cell functions. Nat Rev Mol Cell Biol 7:265–275
8. Ingber DE (2008) Tensegrity-based mechanosensing from macro to micro. Prog Biophys Mol Biol 97:163–179
9. O'Keefe EJ, Briggaman RA, Herman B (1987) Calcium-induced assembly of adherens junctions in keratinocytes. J Cell Biol 105:807–817
10. Duden R, Franke WW (1987) Organisation of desmosomal plaque proteins in cells growing at low calcium concentrations. J Cell Biol 107:1049–1063
11. Janmey PA (1998) The cytoskeleton and cell signalling: component localization and mechanical coupling. Physiol Rev 78:763–781
12. Dahl KN, Kalinowski A (2011) Nucleoskeleton mechanics at a glance. J Cell Sci 124:675–678
13. Keller GM (1995) In vitro differentiation of embryonic stem cells. Curr Opin Cell Biol 7:862–869
14. Watt FM, Mattey DL, Garrod DR (1984) Calcium-induced reorganisation of desmosomal components in cultured human keratinocytes. J Cell Biol 99:2211–2215
15. Zamansky GB, Nguyen U, Chou I (1991) An immunofluorescent study of the calcium-induced coordinated reorganisation of microfilaments, keratin intermediated filaments and microtubules in cultured human epidermal keratinocytes. J Invest Dermatol 97:985–994
16. Vande Geest JP, Di Martino ES, Vorp DA (2004) An analysis of the complete strain field within Flexercell membranes. J Biomech 27:1923–1928
17. Pellegrini G, Dellambra E, Golisano O et al (2001) p63 identifies keratinocyte stem cells. Proc Natl Acad Sci 98:3156–3161

Chapter 14

In Vivo Transplantation Assay at Limiting Dilution to Identify the Intrinsic Tissue Reconstitutive Capacity of Keratinocyte Stem Cells and Their Progeny

Holger Schlüter and Pritinder Kaur

Abstract

This protocol describes an in vivo grafting approach to investigate the intrinsic long-term tissue reconstitutive capabilities of interfollicular keratinocyte stem cells and their committed progeny—the committed progenitors or transit amplifying and early differentiating cells. This approach utilizes the previously described skin reconstitution rat trachea assay, which has been adapted to investigate differences between stem cells and their more committed progeny. Limiting dilutions of each cell fraction reveal that both stem cells and their progeny are capable of skin tissue reconstitution, but at a limiting dilution of 100 cells per rat trachea only the keratinocyte stem cells maintain the reconstituted skin for as long as 10 weeks.

The thorough analysis of reconstituted tissues using skin specific proliferation and stage-specific differentiation markers is also described because it provides qualitative distinction of the epithelial sheets reconstituted by stem cells and their progeny.

Key words Keratinocyte stem cells, Human skin regeneration, Keratinocytes, Rat trachea, Xenograft

---

1   Introduction

The epidermis of the skin is a constantly renewing tissue. The regenerative capacity of keratinocytes has been proven for many years, which led to the development of autologous grafts to regenerate an epidermis over full-thickness wounds (1–3). Partially due to the lack of knowledge about the roles of the specific keratinocyte populations involved in epidermal regeneration and wound healing, skin grafts can still fail. The rat trachea system has been developed as a long-term tissue reconstitution assay to study the growth and differentiation properties of freshly isolated keratinocytes from human and mouse skin (4–7) including the epidermal stem cell population (keratinocyte stem cells—KSC) and their progeny (committed progenitors—CP previously

Kursad Turksen (ed.), *Skin Stem Cells: Methods and Protocols*, Methods in Molecular Biology, vol. 989,
DOI 10.1007/978-1-62703-330-5_14, © Springer Science+Business Media New York 2013

termed transit amplifying—TA cells by us, and early differentiating—ED keratinocytes). The advantage of this in vivo assay is that the regenerated epithelium is maintained long-term and can easily be studied. Furthermore the assay allows the study of small numbers of rare cells.

Initially we transplanted relatively high numbers of cells in this assay, which did not reveal any differences in the tissue regenerative capacity between KSC and their progeny.

Here we describe how we modified the rat trachea model to study the different tissue regenerative properties of KSC, CP/TA, and ED cells, transplanting them at limiting cell dilutions. Furthermore we describe how to assess the regenerated tissue to reveal the differences of the tissue regenerative capacity of KSC and their progeny as reported by us recently (8).

2 Materials

2.1 Skin Sample Dissection

The dissection of keratinocytes from healthy neonatal human foreskin donors is performed in a laminar flow hood. Prepare listed materials:

1. 70% Ethanol.
2. Sterile PBS$^+$ (containing Mg^{2+} and Ca^{2+}) containing penicillin (final concentration 1.2 mg/mL), gentamycin (final concentration 16 mg/mL), and fluconazole (final concentration 3 mg/mL) and PBS^{3+} (containing Mg^{2+} and Ca^{2+}) with penicillin (final concentration 12 mg/mL), gentamicin (final concentration 160 mg/mL), and fluconazole (final concentration 6 mg/mL).
3. 10 cm Petri dishes.
4. Sterile dissecting curved forceps and scalpel holder with sterile blades (ProSciTech, Australia).
5. Sterile dispase solution: Neutral dispase II (Roche Diagnostics) at 4 mg/mL in PBS$^+$.
6. Sterile 5 mL screw-capped tubes (Sarstedt).

2.2 Keratinocyte Isolation

Keratinocyte isolation should be performed under sterile conditions in a laminar flow hood on ice. Set up the following:

1. Polystyrene box with ice.
2. Gilson pipettes P1000, P200, P20 and P10.
3. Sterile plastic transfer pipettes (Livingston).
4. 50 mL Falcon screw cap tubes.
5. 5 and 10 cm Petri dishes containing sterile PBS^{3+} on ice.
6. Sterile dissecting curved forceps and scalpel holder including a sterile blade.

7. Trypsin-EDTA 0.05% (GibCo).

8. Trypsin inhibitor: 0.1 g soybean trypsin inhibitor (Sigma), 0.5 g of BSA (tissue culture grade) in 500 mL of DMEM (GibCo).

9. Trypan blue solution: 0.4% (Sigma).

10. 70 µm cell strainer (BD Falcon).

11. Neubauer chamber.

12. EpiLife medium with 60 µM Ca^{2+} supplemented with HKGS supplement (GibCo).

13. 96-Well plate (Corning).

14. Tissue culture centrifuge at 4°C (e.g.).

15. Tissue culture flasks T75 or T125 (Corning).

2.3 Flow Cytometry

To obtain phenotypically different basal keratinocyte populations the isolated bulk keratinocytes need to be stained and single cells sorted using sterile conditions and a flow cytometer. Set up the following:

1. Sterile 5 mL round bottom FACS tubes (BD).

2. Blocking buffer (BB): 2% BSA + 2% FCS in PBS^+.

3. Staining buffer (SB): 2% BSA in PBS^+.

4. Antibodies:

(a) Mouse anti-human CD49f IgG1 1:100 in SB (BD).

(b) Anti CD45-PE conjugated 1:100 in SB (BD).

(c) Anti CD71-biotinylated 1:50 in SB (BD).

(d) Streptavidin-APC 1:100 in SB (BD).

(e) Anti rat IgG2a 1:100 in SB (Jackson Immunotech).

(f) Rat IgG2a isotype control 1:100 (BD).

(g) Biotinylated mouse isotype control 1:100 (BD).

(h) Mouse IgG1-PE conjugated isotype control 1:100 (BD).

5. DAPI 1:100 in SB (stock solution: 1 mg/mL, Roche, Germany).

6. EpiLife medium with 60 µM Ca^{2+} supplemented with HKGS supplement (GibCo).

7. Rotating wheel at 4°C.

2.4 Rat Tracheas

The rat tracheas were purchased from the Institute of Medical and Veterinary Science, Adelaide, South Australia, Australia. These can also be purchased from Charles River Laboratories, USA. For the preparation set up:

1. Sterile PBS^{3+}.

2. DMEM (GibCo).

3. 10 cm Petri dishes.

4. Sterile instruments: one pair of small dissecting scissors (ProSciTech), hemostats (forceps with clamping arm and serrated tips), dissection forceps (ProScitech), 18 G 1$^{1/2}$" drawing-up needle (Terumo), surgical McKenzie 6" clip applying forceps (Surgipro, USA), medium titanium-ligating clips (LT200, Ethicon Endo-Surgery, USA), sterile 10 mL syringe (Terumo), 1.5 cm long microbore polytetrafluoroethylene (PFTE) tubing (Cole-Parmer Instrument Company, USA) all prepared and autoclaved in advance.

5. Sterile 1.5 mL Eppendorf tubes.

2.5 Filler Keratinocytes

5×10^5 irradiated support passage (p) 0 or p1 neonatal human keratinocytes per rat trachea. To irradiate these cells in tissue culture flasks or in suspension, this equipment is required:

1. Caesium gamma irradiator (Nordion Gammacell 40 Irradiator).

2.6 Inoculation of the Keratinocytes into the Rat Tracheas

The following needs to be set up in a sterile laminar flow hood:

1. Transplant medium: 1:1 mix of EpiLife and DMEM (both GibCo) supplemented with 10% bovine serum, 20 ng/mL murine epidermal growth factor (EGF) (Sigma), 10 ng/mL cholera toxin (Sigma), 6 mg/mL penicillin, 80 mg/mL gentamycin (Life Technologies, USA) and 6 mg/mL diflucan (Pfizer). This complex medium is stable for 6 months if stored at 4°C. 30 μL of the transplantation medium is required per rat trachea.

2. Sterile 10 mL syringe with a drawing-up needle (as previously described) filled with cold PBS^{3+}.

3. Gilson 200 μL pipette with gel loader tips (Prot/Elec Tps-Bulk Bio-Rad).

4. Sterile instruments: hemostats, dissection forceps (ProSciTech), surgical McKenzie 6 clip applying forceps (Surgipro, USA), medium titanium-ligating clips (LT200, Ethicon Endo-Surgery, USA).

5. Sterile 15 mL yellow cap tubes.

2.7 In Vivo Grafting

Rat tracheas will be grafted under the skin of about 6–8 week old female SCID mice.

1. 50 mL Falcon tube containing sterile gauze.

2. Isoflurane (200 μL per mouse).

3. Anesthetic solution containing 2 mg/mL xylazine (Bayer Australia, Pymble, Australia) and 10 mg/mL ketamine (Parnell Laboratories, Sydney, NSW, Australia) prepared in PBS.

4. Cotton pads soaked with 70% Ethanol.

5. Sterile instruments: medium operating size scissors blunt/blunt (ProSciTech), two dissecting forceps (ProSciTech), one autoclip wound clip applier (BD), one box of 9 mm sterile autoclips (one clip per mouse) (BD), one autoclip wound clip remover (BD).

2.8 Harvesting the Grafts

To harvest the grafts from the murine recipients, the following materials are required:

1. CO_2 euthanasia chamber.

2. Operating scissors, dissecting forceps and scalpel holder with sterile blades (ProSciTech, Australia).

3. 10 cm Petri dishes.

4. Sterile PBS.

5. Decalcification solution: 5 mM EDTA in PBS, pH8. Dissolve 6.26 g of NaCl + 80 g EDTA, adjust the pH to 8 and the volume to 250 mL with water.

6. Fixative 4% (w/v) PFA in sodium cacodylate buffer.

7. Tissue-Tek O.C.T. Compound (Sakura, USA).

8. Eppendorf tubes.

9. Cryomold standard (Tissue Tek).

10. Rotating wheel at 4°C.

11. Three staining troughs with lid (Grale Scientific, Australia).

12. Tissue processor (Tissue Tek, Sakura, USA).

13. Histotec pastilles, embedding wax material (Merck).

14. Leica cryostat.

15. Slides (Menzel Glaeser, Germany).

2.9 Analysis of Harvested Grafts

Cryo- or wax sections can be stained using specific markers for proliferation and early versus late differentiation. To perform this staining set up the following items:

1. Xylene.

2. Ethanol series (100%, 70%).

3. Double distilled water.

4. Scotts tap water: 3.5 g of sodium bicarbonate + 10 g $MgSO_4$ in 1 L of tap water.

5. Haematoxylin solution.

6. Eosin solution.

7. Pressure cooker (Dako Cytomation, USA).

8. Slide holder for pressure cooker and staining chambers.

9. Citrate buffer pH6.0 for antigen retrieval of epitopes to Keratin 10 and Loricrin.

10. Tris–HCl buffer pH9.0 for antigen retrieval of epitopes to Ki67, Keratin 14 and 15 and Involucrin.

11. Kimwipe tissues.

12. Light-resistant humidified chamber.

13. Blocking solution: Tris buffered saline (TBS)+2% BSA+10% normal goat serum.

14. Primary antibodies.

    (a) Guinea pig anti Keratin15 1:2,000 (Progen, Heidelberg, Germany).

    (b) Mouse anti Ki67 1:100 (Dako, USA).

    (c) Mouse anti Keratin 10, neat LHP2 hybridoma supernatant (kindly provided by Dr. Irene Leigh, Royal London Hospital, London, UK).

    (d) Guinea pig anti Keratin 14 1:2,000 (Progen, Heidelberg, Germany).

    (e) Rabbit anti Involucrin 1:3,000 (a gift from Dr. Rob Rice, University of California, San Francisco, USA).

    (f) Rabbit anti Loricrin 1:500 (Covance, Princeton, NJ, USA).

15. Secondary antibodies.

    (a) Goat anti-mouse Alexa555 (Life Technologies).

    (b) Goat anti-mouse Alexa488 (Life Technologies).

    (c) Goat anti-guinea pig Alexa555 (Life Technologies).

    (d) Goat anti-guinea pig Alexa488 (Life Technologies).

    (e) Goat anti-rabbit Alexa555 (Life Technologies).

    (f) Goat anti-rabbit Alexa488 (Life Technologies).

16. DAPI 1:10,000 in SB (stock solution: 1 mg/mL, Roche, Germany).

17. ProLong® Gold antifade reagent (Life Technologies).

18. Coverslips (Menzel Glaeser, Germany).

19. Olympus BX51 fluorescent microscope.

20. Morphometric software, i.e., Metamorph software.

3 Methods

3.1 Keratinocyte Extraction

The isolation of the keratinocyte populations has been described in great detail previously (9) and has to be performed using auto-claved instruments under sterile conditions in a laminar flow hood.

After surgical excision the skin samples have to be processed rapidly—within 24 h.

1. Prepare Dispase solution (4 mg/mL in PBS$^+$), filter sterilize through a 0.22-μm filter, aliquot into 5 mL sterile tubes with screw cap and store on ice (3 mL of solution per sample per tube need to be prepared).

2. Wash the excised skin sample in 70% ethanol for 2 min by placing the samples using forceps into a 10-cm Petri dish filled with 70% ethanol.

3. Place the skin into a lid of a 10 cm Petri dish and trim excess fat and connective tissue from the dermal side using a scalpel with a fresh blade and forceps.

4. Cut the skin into small 4 mm^2 pieces.

5. Place the skin pieces into a 10 cm Petri dish filled with PBS^{3+} for washing using forceps.

6. Wash all trimmed skin pieces three times in PBS^{3+} in total.

7. Transfer small skin pieces into dispase solution in 5 mL tubes and rotate on rotating wheel at 4°C overnight.

8. Prewarm 10 mL trypsin-EDTA (in a 50 mL Falcon tube) and trypsin inhibitor to 37°C in water bath.

9. Proceed with steps 1–7 with all excised skin samples.

10. Hold the dermis with curved forceps and carefully peel off the epidermal sheets using a second forceps and put into a 5-mL Petri dish with PBS$^+$.

11. After peeling off all epidermal sheets, transfer them onto a Petri dish and cut into very small pieces using a scalpel blade, then transfer into 50 mL Falcon screw cap tubes containing 5 mL of prewarmed trypsin (see Note 1).

12. Trypsinize epidermal sheets by pipetting up and down vigor-ously using a plastic transfer pipette for 5 min.

13. Quench the reaction by adding an equal volume of trypsin inhibitor and mix gently.

From this step onwards cells should remain on ice or spun down in a centrifuge at 4°C to maintain optimal viability.

14. Filter the cell suspension through a 70-μm cell strainer into a 50 mL Falcon tube.

15. Wash cell strainer with additional 10–20 mL PBS$^+$ to collect any additional cells.

16. Centrifuge the keratinocyte cell suspension at $420 \times g$ for 5 min at 4°C and resuspend cell pellet in 10 mL.

    (a) Cold EpiLife medium for immediate culture for filler cells; or

    (b) Cold blocking buffer for FACS.

17. Count the keratinocytes and assess their viability using trypan blue (a viability of greater than 80% is needed for successful grafting).

The keratinocytes used for FACS to sort into keratinocyte stem cells and their progeny CP/TA and ED cells will now be treated as described in the FACS protocol below.

3.2 FACS Protocol

All procedures should be performed at 4°C.

1. Block keratinocytes for 15 min in blocking buffer and meanwhile count cells using a Neubauer chamber.

2. Aliquot 50,000 cells for unstained, single color and negative controls into 5 mL round bottom FACS tubes filled with 3 mL of blocking buffer.

   (a) Tube 1: unstained control in staining buffer.

   (b) Tube 2: will be the viability test tube. For now use staining buffer.

   (c) Tube 3: all three isotype control antibodies in staining buffer.

   (d) Tube 4: anti-α6 integrin in staining buffer.

   (e) Tube 5: anti-CD71 biotinylated in staining buffer.

   (f) Tube 6: anti-CD45-PE conjugate in staining buffer.

   (g) Tube 7: anti-α6 integrin + anti-CD71 + anti-CD45.

3. Pipette remaining cell suspension into a 5 mL round bottom FACS tube.

4. Centrifuge at $420 \times g$ for 5 min at 4°C and resuspend cell pellets in pre-made primary antibody solutions according to the pipetting scheme below and incubate for 45 min to 1 h.

5. Fill tube with 3 mL staining buffer to wash and centrifuge for 5 min at $420 \times g$ and 4°C.

6. Carefully aspirate buffer and repeat wash with 3 mL staining buffer and repeat centrifugation to pellet cells.

7. Resuspend cell pellets in secondary antibody solution according to antibody pipetting scheme below and incubate for 30–45 min:

(a) Tube 1: unstained control in staining buffer.

(b) Tube 2: Viability test tube, in staining buffer.

(c) Tube 3: anti-rat FITC conjugated + streptavidin-APC conjugated in staining buffer.

(d) Tube 4: anti-rat FITC conjugate in staining buffer.

(e) Tube 5: Streptavidin-APC conjugate in staining buffer.

(f) Tube 6: staining buffer.

(g) Tube 7: anti-rat FITC conjugate + streptavidin-APC conjugate in staining buffer.

8. Fill tubes with 3 mL staining buffer to wash and centrifuge for 5 min at $420 \times g$ and 4°C; repeat washing step.

9. Resuspend control cell pellets in 500 µL and test cell pellet (Tube 7) in 1,000 µL of staining buffer for FACS analysis.

10. Add DAPI to viability control tube number 2 and sorting tube number 7 and perform FACS.

11. Set up FACS sorter using unstained and single color controls using proper compensation between channels.

12. Sort keratinocyte populations based on gates as shown in Fig. 1d. Therefore the first gate is single cells (FSC-A versus FSC-H; Fig. 1b) followed by viability (DAPI versus FSC-A; Fig. 1a) and morphology (SSC-A versus FSC-A; Fig. 1c). For further details see Li and Kaur (10).

13. Gate out $CD45^+$ cells, selecting $CD45^-$ cells for α_6 versus CD71 analysis (Fig. 1d).

14. Reanalyze sorted KSC, CP/TA, and ED samples to confirm purity of cell sorting procedure.

Sorted keratinocyte populations should be kept on ice, combined with irradiated filler keratinocytes and inoculated into rat tracheas as quickly as possible on the same day.

3.3 Filler Keratinocytes

500,000 irradiated cultured filler passage 1 or passage 2 keratinocytes need to be mixed with sorted KSC, CP/TA and ED cells to provide soluble factors and the cellular scaffold necessary to sustain the process of skin regeneration from small numbers of the test keratinocyte subpopulations.

1. Prior to harvesting, feed the filler keratinocytes with fresh EpiLife medium twice weekly until they reach ~80% confluency.

2. Irradiate filler keratinocytes at a dose of 15 Gy gamma rays on the day of grafting.

3. Trypsinize, pool and resuspend irradiated keratinocytes in cold EpiLife medium. This represents the stock of filler cells used in the individual grafts.

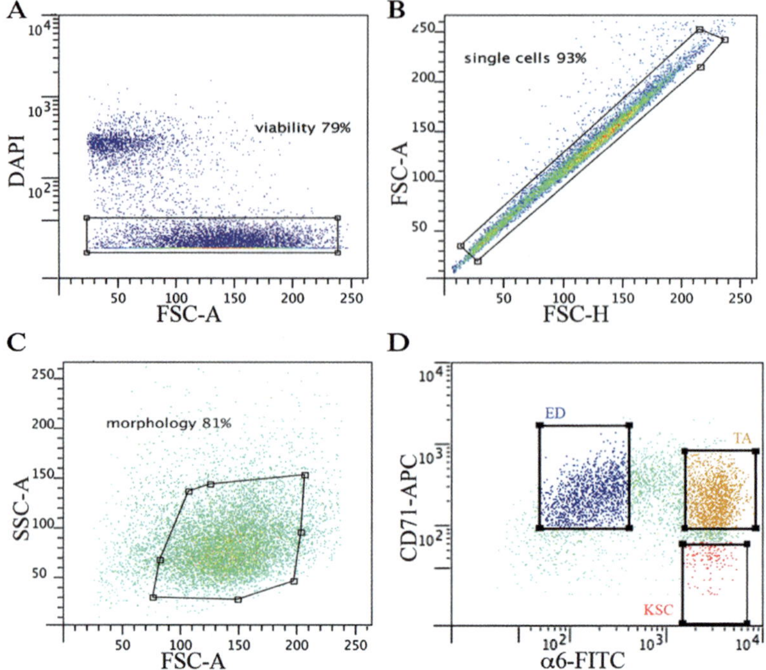

Fig. 1 Images of sorting gates for the setup of a KSC, CP/TA, and ED keratinocyte sort. (**a**) Viability gate, (**b**) single cell gate, (**c**) morphology gate, and (**d**) KSC, CP/TA, and ED sorting gates

3.4 Trachea Preparation

The cleaning and preparation of the rat tracheas and the grafting into the recipient mice is performed (6) under sterile conditions using a laminar flow hood.

1. Thaw rat tracheas and trim excess connective tissue in a sterile Petri dish using scissors and dissecting forceps.

2. Clamp together 1.5-cm long PTFE tubing to a slightly stretched rat trachea (~1 cm) at both ends using two hemostats to keep the tracheal lumen open.

3. Use LT400 ligaclips and clip, using Mackenzie clip applying forceps to secure trachea and tubing together.

4. Trim off the larynx and the bronchi (top and bottom parts of the trachea) using a scalpel.

5. Devitalize rat tracheas by freeze-thawing (−70°C and 37°C) three times which detaches the rat tracheal epithelium.

6. Cut tracheas (around 2 mm in size just below the top clip) with a pair of scissors and flush tracheas multiple times with cold PBS^{3+} using a 10 mL syringe with a needle affixed.

7. Aspirate all remaining fluid from the lumen of the trachea.

8. Immerse trachea in DMEM for ~30 s to test for leaks or infiltration into the wall of the trachea.

9. Aspirate off all remaining DMEM.

10. Non-leaky rat tracheas can be stored at –80°C for up to 1 year at –20°C for 6 months and for 1 week in the fridge.

3.5 Inoculation of Keratinocyte Cell Populations into Rat Tracheas

KSC, CP/TA and ED cell populations are inoculated at 100 and 1,000 cells per rat trachea combined with 5×10^5 irradiated filler keratinocytes. Cells are resuspended in 30 µL of transplantation medium per rat trachea in 1.5 mL prechilled sterile Eppendorf tubes on ice.

1. Thaw rat tracheas on ice.

2. Make an incision at one end of the trachea just under the sealing clip.

3. Wash the inside of the rat trachea with several flushes of PBS3+ through the incision using blunt ended needles, holding the rat trachea with a hemostat.

4. Aspirate remaining PBS from the lumen.

5. Resuspend the prepared cell populations (KSC, CP/TA, and ED plus irradiated filler keratinocytes) carefully creating a uniform cell suspension by pipetting up and down. Use gel loader tips and avoid generating air bubbles (see Note 2)!

6. Carefully pipet 30 mL of the cell suspension into the lumen of the rat trachea. Avoid air bubbles (see Note 3)!

7. Seal the rat trachea using titanium ligating clips just above the meniscus of the cell suspension and trim off excess rat trachea tissue.

3.6 Engrafting Rat Tracheas onto Mice

1. Anesthetize mice by injecting 250 mL of ketamine/xylazine cocktail intraperitoneally per 20 g mouse body weight and keep mice sedated by using a slow and steady flow of isoflurane.

2. Sterilize the back of the anesthetized mouse with 70% ethanol.

3. Use large forceps to pinch part of the back skin and pull up to make a small incision (~2 cm) with sterile scissors.

4. Insert scissor blades into the incision underneath the skin and enlarge space between skin and underlying fascia by opening and closing scissor blades multiple times until the space created reaches 3 cm × 3 cm square.

5. Insert one rat trachea into the space created towards the right flank and a second trachea towards the left flank of the mouse (see Note 4).

6. Pull skin back into place and close the cavity using two ligation wound clips.

7. Wound clips are removed 1 week after the procedure for the well being of the mouse.

3.7 Harvesting the Grafts

We have established that two time-points at 6 and 10 weeks post-transplant are sufficient in this skin reconstitution assay to obtain differences between the basal keratinocyte populations investigated.

1. Sacrifice mice carrying grafts by asphyxiation with CO_2, since cervical dislocation could damage grafts.

2. Cut mice open and recover rat tracheas.

3. Clean mouse tissue from rat tracheas using scissors and place tracheas in a 10 cm Petri dish filled with cold PBS.

4. Fix the cleaned rat tracheas in 4% (w/v) buffered formalin for 2 h at room temperature.

5. Cut the rat tracheas in half.

6. Decalcify one half of each harvested rat trachea in decalcifying solution for 24 h at 4°C on a rotating wheel (for cryosections), and at the same time fix the second half of each trachea for 18 h at 4°C on a rotating wheel (for paraffin embedding).

7. Place decalcified tracheas into OCT, incubate for 5 h at 4°C to allow OCT penetration and freeze for cryosections and immunostaining.

8. Decalcify second half of tracheas for 24 h at 4°C on a rotating wheel and embed into wax using tissue processor.

3.8 Analyzing the Grafts

Specific immunostaining for proliferation and differentiation markers and histological stains are used to investigate the success of the engraftment and the state of the reconstituted skin.

3.8.1 Histology H&E Stain

1. Cut wax sections of each embedded rat trachea.

2. Dry sections onto slides at 37°C overnight.

3. Bake sections onto slides for 1 h at 60°C.

4. De-wax and stain sections with H&E by submerging the slides in:

(a)	Histolene	3:00 min
(b)	Histolene	3:00 min
(c)	Histolene	3:00 min
(d)	Alcohol abs	1:00 min
(e)	Alcohol abs	1:00 min
(f)	Alcohol abs	1:00 min
(g)	Alcohol 70%	1:30 min
(h)	Water	1:00 min
(i)	Haematoxylin	4:00 min

(continued)

(continued)

(j)	Water	1:00 min
(k)	Water	1:00 min
(l)	Scotts tap water	0:45 min
(m)	Water	2:00 min
(n)	Eosin	4:00 min
(o)	Water	0:15 min
(p)	Alcohol abs	0:45 min
(q)	Alcohol abs	0:45 min
(r)	Alcohol abs	0:45 min
(s)	Alcohol abs	1:00 min
(t)	Histolene	2:00 min
(u)	Histolene	2:00 min
(v)	Histolene	2:00 min

5. Coverslip slides.

6. Use the microscope to analyze the morphology of the reconstituted skin (see Note 5).

3.8.2 Immunofluorescence Staining for Specific Skin Proliferation and Differentiation Markers

1. Cut wax sections.

2. Dry sections onto slides at 37°C overnight.

3. Bake sections onto slides for 1 h at 60°C.

4. De-wax sections by submerging the slides in:

(a)	Histolene	4 min
(b)	Histolene	4 min
(c)	Alcohol abs	1 min
(d)	Alcohol abs	1 min
(e)	Alcohol abs	1 min
(f)	Alcohol abs	1 min
(g)	Alcohol 70%	1 min
(h)	Demineralized water	

5. Perform antigen-retrieval by submerging slides in antigen-retrieval buffer and place into pressure cooker.

(a)	Antigen-retrieval buffer:	Pressure cooker protocol	Antibodies
1.	Tris–HCl pH 9.0	125°C for 3′	Ki67, K14 and 15, involucrin
2.	Citrate buffer pH 6.0	98°C for 2′	Loricrin, K10

6. Cool slides down to room temperature in retrieval buffer.

7. Wash slides three times for 5 min each in PBS.

   Slides must be incubated in light resistant humidified chambers from now on (see Note 6).

8. Tap and wipe off excess PBS with Kimwipes and apply blocking solution to sections at room temperature for 1 h to prevent unspecific antibody binding.

9. Tap and wipe off excess blocking solution and apply primary antibody dilutions.

10. Incubate primary antibody dilutions overnight at 4°C.

11. Wash slides in excess PBS three times for 10 min per wash.

12. Tap and wipe off excess PBS.

13. Apply and incubate sections with secondary antibody dilutions/combinations for 2 h at room temperature.

14. Wash slides submerged in PBS three times for 10 min per wash.

15. Counterstain with DAPI at 1:10,000 for 5 min and rinse with PBS.

16. Rinse slides with water to wash off excess salt and dip into absolute Ethanol.

17. Mount slides in ProLong Gold antifade mounting solution.

The proliferative index of the epithelial sheets can be obtained by determining the number of Ki67 positive basal cells per tissue section using a fluorescence microscope and Metamorph software (see Note 7). At least three independent transplant experiments with replicates for each experimental condition (e.g., cell type transplanted or time point analyzed) are performed to obtain robust data.

3.9 Conclusions

Careful histological analysis of the H&E stained slides of this in vivo tissue reconstitution assay reveals differences in the overall amount of epithelia generated measurable as differences in the thickness of the epithelium, extent of luminal coverage of the rat tracheal surface, and in the morphology of the basal layer (Fig. 2). More importantly, immunofluorescent staining with a combination of skin specific proliferation and differentiation marker antibodies reveals more profound differences in the quality of the epithelium reconstituted within the rat trachea.

While keratinocyte stem cells demonstrate a delay in the onset of proliferation and tissue reconstitution (8), their direct progeny CP/TA and ED cells regenerate skin at a much higher rate initially, contributing to the amount of epithelial tissue regenerated at earlier time points (Fig. 2). The initially slow rate of tissue reconstitution obtained

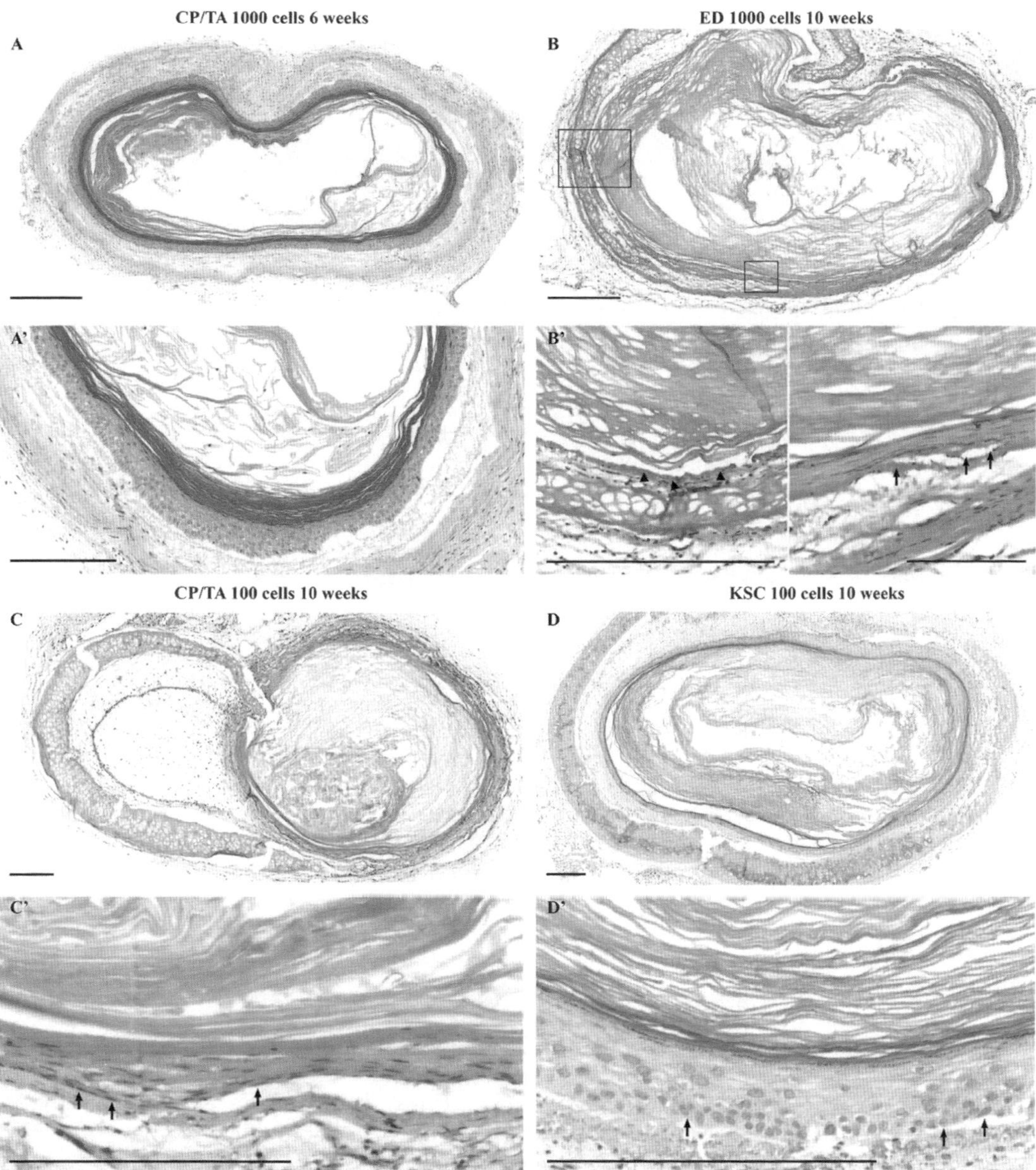

Fig. 2 H&E stained sections of skin regenerated by 1,000 CP/TA cells for 6 weeks (**a–a′**), 1,000 ED cells for 10 weeks (**b–b′**), 100 CP/TA cells for 10 weeks (**c–c′**), and 100 KSC for 10 weeks (**d–d′**). CP/TA-derived epidermal sheets produced good luminal coverage at 6 weeks after grafting independent of cell number inoculated (**a–a′**, **c–c′**). Whereas epidermal sheets derived from 100 CP/TA cells or 1,000 ED cells almost completely cover the rat tracheal lumen at 6 weeks, after 10 weeks of grafting these demonstrate signs of terminal differentiation and absence of a polarized basal layer (*arrows* in **b′**, **c′**) indicating reduced intrinsic long-term capacity to regenerate skin. However sheets derived from 100 KSC (**d, d′**) maintain a polarized basal layer (*arrows* in **d′**) and remain morphologically normal 10 weeks post-transplant demonstrating long-term tissue regenerative capacity at limit dilution—a hallmark of stem cells. Scale bars: 200 μm

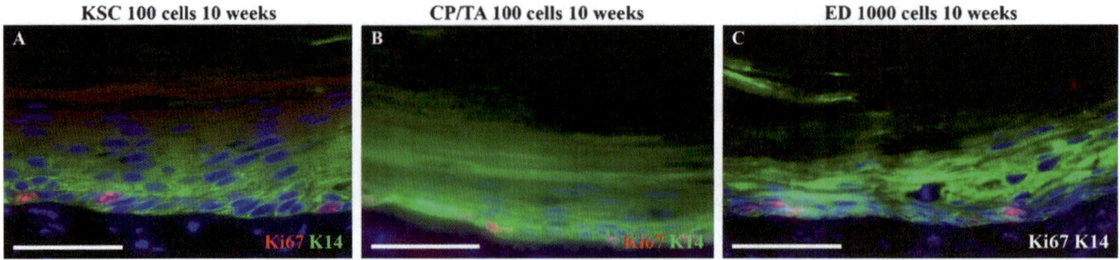

Fig. 3 Ki67 (*red*) and Keratin 14 (K14-*green*) immunofluorescence staining of skin regenerated in rat trachea grafts by KSC (**a**), CP/TA (**b**) or ED cells (**c**) showing maintenance of homeostatic cellular proliferation levels by KSCs (**a**) and the loss of a morphologically distinct basal layer in CP/TA and ED cell (**b–c**) skins

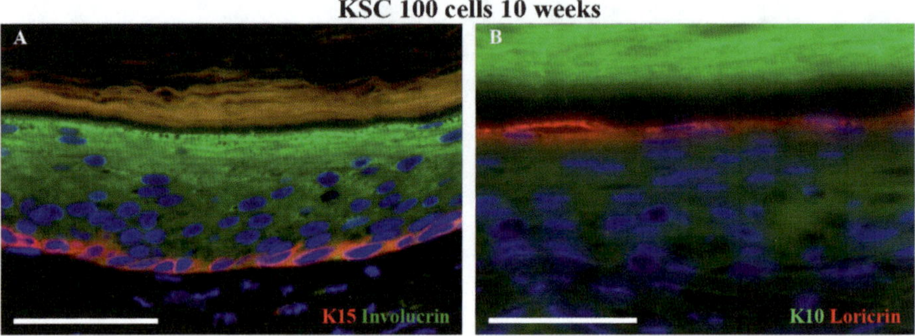

Fig. 4 Rat tracheas inoculated with 100 KSC xenografted for 10 weeks and immunostained with skin-specific homeostatic and differentiation marker antibodies to K15 (*red*) + Involucrin (*green*) (**a**), K10 (*green*) + Loricrin (*red*) (**b**). KSC-derived epithelial sheets contain an intact K15 positive basal layer and a completely normal differentiation pattern of expression of the differentiation markers K10, Loricrin (**b**) and Involucrin (**a**). Scale bar: 50 μm

from KSCs is probably due to the quiescent state of the stem cells at the time of isolation from the tissue (4). However, staining of the epithelial sheets with antibodies to the proliferation marker Ki67, reveal that at 6 weeks there are no differences in the amount of proliferating cells between the skin tissues derived from KSC and their progeny in the 1,000 and 100 cell grafts (Fig. 3). Importantly the stem cell population is the only one capable of maintaining a homeostatic level of proliferating cells long term at limiting dilution of 100 cells after 10 weeks of engraftment, whereas the CP/TA and ED populations exhibit clear signs of burnout over the long term with the absence of a morphologically distinct and mitotic basal layer (8).

Detailed analysis of the reconstituted skin tissue using antibodies to Keratin 15 (see Note 8), a marker for basal keratinocytes, and differentiation marker antibodies such as Keratin 10, Involucrin, and Loricrin demonstrate the differences of the intrinsic tissue reconstitutive capabilities between the stem cell population and their progeny. At 10 weeks after engraftment the skin reconstituted by 100 keratinocyte stem cells is the only reconstituted tissue with an intact Keratin 15 positive layer and a normal distribution of Keratin 10 (see Note 9), Involucrin and Loricrin, which is consistent with homeostasis (Fig. 4).

CP/TA 100 cells 10 weeks

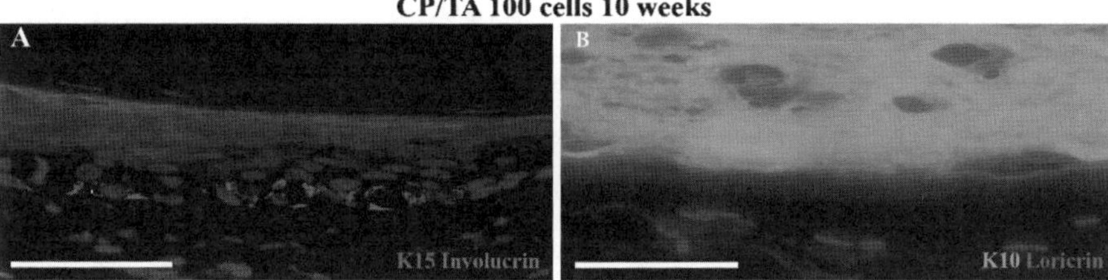

Fig. 5 Skin regenerated by 100 CP/TA cells xenografted for 10 weeks and immunostained with skin specific homeostatic and differentiation marker antibodies K15 + Involucrin (**a**), K10 + Loricrin (**b**). CP/TA cell-derived epithelial tissue shows significantly decreased K15 expression in the basal layer and clear signs of terminal differentiation proven by a loss of Loricrin expression. Scale bar: 50 μm

In contrast, 100 early differentiating cells fail to reconstitute skin most of the time; similarly tissue derived from 100 CP/TA cells exhibit only a thin epithelium with loss of an intact Keratin 15-positive basal layer and signs of terminal differentiation, e.g., expression of differentiation marker Keratin 10 down at the most basal layer and earlier onset of Involucrin and Loricrin expression (Fig. 5).

Therefore we conclude that transplantation of 100 keratinocytes reveals that the keratinocyte stem cells are the most potent long-term tissue reconstitutive fraction of the interfollicular epidermis.

4   Notes

1. During keratinocyte cell isolation remaining PBS containing Mg^{2+} and Ca^{2+} should be aspirated carefully before the trypsinization—this will give a better digest with increased cell numbers.

2. A cell viability of greater than 60% is necessary to achieve successful engraftment.

3. Air bubbles should be avoided at all times during the extraction of the keratinocytes to get a higher cell yield and also during inoculation into rat tracheas to obtain better engraftment.

4. Grafting of more than two tracheas per mouse leads to much poorer engraftment rates probably due to a lack of sufficient nutrients provided by the mouse and increased stress levels for the mouse.

5. When inoculating low cell numbers, epithelialization of rat trachea may be restricted to small regions of the rat trachea lumen and therefore it is recommended that serial sections at several

points along the length of each rat trachea are analyzed histologically.

6. During the immunofluorescence staining procedure, the drying out of tissue sections needs to be prevented in order to minimize unspecific background staining.

7. Decrease of Ki67-positive cells indicates downregulation of proliferation.

8. Loss of K15 indicates loss of basal layer.

9. Loss of K10 indicates loss of differentiating cell layers.

Acknowledgment

This work was supported by NIH grant RO1 AR050013-01A2.

References

1. Banks-Schlegel S (1981) Grafting of burns with cultured epithelium prepared from autologous epidermal cells. Lancet 1:75–78

2. Gallico GG 3rd, O'Connor NE, Compton CC, Kehinde O, Green H (1984) Permanent coverage of large burn wounds with autologous cultured human epithelium. N Engl J Med 311:448–451

3. Green H, Kehinde O, Thomas J (1979) Growth of cultured human epidermal cells into multiple epithelia suitable for grafting. Proc Natl Acad Sci U S A 76:5665–5668

4. Li A, Pouliot N, Redvers R, Kaur P (2004) Extensive tissue-regenerative capacity of neonatal human keratinocyte stem cells and their progeny. J Clin Invest 113:390–400

5. Li A, Simmons PJ, Kaur P (1998) Identification and isolation of candidate human keratinocyte stem cells based on cell surface phenotype. Proc Natl Acad Sci U S A 95:3902–3907

6. Paquet-Fifield S, Redvers RP, Pouliot N, Kaur P (2010) A transplant model for human epidermal skin regeneration. Methods Mol Biol 585:369–382

7. Redvers RP, Li A, Kaur P (2006) Side population in adult murine epidermis exhibits phenotypic and functional characteristics of keratinocyte stem cells. Proc Natl Acad Sci U S A 103:13168–13173

8. Schluter H, Paquet-Fifield S, Gangatirkar P, Li J, Kaur P (2011) Functional characterization of quiescent keratinocyte stem cells and their progeny reveals a hierarchical organization in human skin epidermis. Stem Cells 29:1256–1268

9. Gangatirkar P, Paquet-Fifield S, Li A, Rossi R, Kaur P (2007) Establishment of 3D organotypic cultures using human neonatal epidermal cells. Nat Protoc 2:178–186

10. Li A, Kaur P (2005) FACS enrichment of human keratinocyte stem cells. Methods Mol Biol 289:87–96

Chapter 15

Melanoblasts as Multipotent Cells in Murine Skin

Tsutomu Motohashi and Takahiro Kunisada

Abstract

Melanoblasts are melanocyte precursors that are derived from neural crest cells (NCCs). Recently we showed that melanoblasts differentiate into not only pigmented melanocytes but also into other NCCs derivatives. Here, we describe methods for the isolation of melanoblasts from mouse skin by flow-cytometry. Methods for culturing the isolated melanoblasts, allowing them to express their multipotentiality, are also described.

Key words Melanoblasts, Multipotency, Flow-cytometer, Kit, CD45

1 Introduction

Melanoblasts are melanocyte precursors that are derived from the neural crest cells (NCCs) that emerge from the dorsal region of the fusing neural tube. Streams of NCCs travel dorsolaterally at the dermamyotome of the embryo toward the ventrum. During this migration, NCCs are thought to become committed progenitors, melanoblasts, prior to invading the ectoderm to finally colonize various tissues or organs including skin and hair follicles, where they differentiate into mature melanocytes (1). At around embryonic day 8.5 (E8.5), melanoblasts emerge from the dorsal neural tube and migrate ventrally through the developing dermis starting from E10.5. At E14.5, they begin to invade the overlying epidermis and then migrate into the developing hair follicles, where they continue to proliferate and differentiate, before starting to synthesize pigment at around post natal day 4 (2, 3).

The immediate descendants of NCCs whose differentiation potential has become restricted only to melanocytes have been identified in avian and mammalian embryos. In the quail embryo, the NCCs that express the Kit molecule after emerging from the neural tube represent these precursors that differentiate only into melanocytes (4, 5). In the mouse embryo, the Kit-positive NCCs

Kursad Turksen (ed.), *Skin Stem Cells: Methods and Protocols*, Methods in Molecular Biology, vol. 989, DOI 10.1007/978-1-62703-330-5_15, © Springer Science+Business Media New York 2013

that emerge at the dorsal midline of the trunk neural tube at E9, precisely in the premigratory neural crest region, migrate only into the ectoderm and develop into melanocytes (6). Therefore, melanoblasts that have already migrated out from the NCC population, express the Kit molecule, and have moved into the embryonic skin, are thought to be fate-restricted precursors that differentiate only into melanocytes.

Recently, multipotent precursor cells were identified in the skin or the appendages of fetal and adult animals, and these cells have a differentiation capability similar to that of neural crest stem cells (NCSCs) (7–9). Other studies have shown that cells in adult hair follicles are also capable of differentiating into derivatives of NCCs, and these cells are thought to be derived from NCCs according to the results of cell lineage analysis (10–12). Furthermore, some reports have indicated that cells differentiated from NCCs also have a multipotential cell fate (13–17). These reports suggest that NCC-derived cells show unstable NCC phenotypes and can dedifferentiate and then redifferentiate into other NCC derivatives (17). In this context, we reinvestigated the differentiation of melanoblasts and identified that they can differentiate into not only pigmented melanocytes but also into neurons, glial cells, and smooth muscle cells, which are cells not previously thought to be generated from melanoblasts (18). This finding suggests that the melanoblasts, which are thought to be restricted in their fate to melanocytes, might actually have a multipotential cell-fate, even when they have already migrated toward the target sites in the skin (18).

In this chapter we describe methods for the flow-cytometric isolation of melanoblasts from embryonic and neonatal skin and hair follicles as well as methods for culturing melanoblasts such that they show their multipotency. Melanoblasts isolated as Kit-positive cells from the skin and cultured on monolayers of ST2 stromal cells differentiate into not only pigmented melanocytes but also into neurons, glial cells, and smooth muscle cells.

2   Materials

2.1 Preparation of Cell Suspensions from Mouse Skin

1. Dispase II solution: Dilute 10× dispase II (Sanko-Junyaku CO., LTD.) with Ca, Mg-free Phosphate buffer saline (PBS) and sterilize by passage through a ϕ 0.20-μm membrane filter.

2. SM (Staining medium): PBS containing 3% fetal calf serum (FCS).

3. 0.25% Collagenase type I solution: Dilute Collagenase type I (Wako-Junyaku CO., LTD.) with Ca, Mg-free PBS and sterilize by passage through a ϕ 0.20-μm membrane filter.

4. Hanks' balanced salt solution containing 0.005% DNaseI (Roche) and 20% FCS (JRH Biosciences, Inc.).

5. Cell dissociation buffer (Invitrogen).

6. 0.05% trypsin solution: 0.05% trypsin (Invitrogen)/0.5 mM EDTA (Invitrogen) in PBS.

7. Binocular microscope (such as Carl Zeiss, Stereomicroscope DV4).

2.2 Immunostaining of Skin Cell Suspension for Flow-Cytometric Analysis

1. Rat anti-mouse Fc gamma receptor (2.4-G2; BD Bioscience).

2. FITC-conjugated rat anti-mouse CD45 (30-F11; BD Bioscience).

3. Allophycocyanin-conjugated rat-anti mouse Kit (2B8; BD Bioscience).

4. PI solution: SM containing 3 μg/mL propidium iodide (Calbiochem).

5. Flow-cytometer (such as Becton-Dickinson, FACS Vantage).

2.3 Culture of Isolated Melanoblasts

1. RPMI-1640 (Invitrogen).

2. αMEM (Invitrogen).

3. Medium for maintenance of ST2 stromal cells: RPMI-1640 (Invitrogen) supplemented with 5% FCS, 50 μM 2-mercapto-ethanol, 50 μg/mL streptomycin, and 50 U/mL penicillin. Store at 4°C and use within 2 months.

4. Medium for differentiation of melanoblasts: αMEM (Invitrogen) supplemented with 10% FCS, 10^{-7} M dexamethasone (Dex; Sigma), 20 pM fibroblast growth factor-2 (bFGF; R&D Systems), 10 pM cholera toxin (CT; Sigma), 100 ng/mL human recombinant endothelin-3 (EDN3; Peptide Institute, Inc.), 50 μg/mL streptomycin, and 50 U/mL penicillin. Store at 4°C and use within 2 months.

5. Dexamethasone (Dex; Sigma): Store original stock solution at 10^{-2} M in ethanol at –70°C. Prepare the working stock solution at 10^{-3} M in ethanol at 4°C and dilute from this stock solution for each use. Stable for at least 1 year (stock at –70°C) or 3 months (stock at 4°C).

6. Human recombinant fibroblast growth factor-2 (bFGF; R&D Systems): Store the stock solution at 200 nM in 0.1% bovine serum albumin (BSA)/PBS at –70°C. Store at 4°C after thawing. Stable for at least for 6 months (stock at –70°C) or 1 month (stock at 4°C).

7. Cholera toxin (CT; Sigma): Store the stock solution at 50 nM in distilled water at –70°C. Store at 4°C after thawing. Stable for at least 1 year (stock at –70°C) or 1 month (stock at 4°C).

8. Human recombinant endothelin-3 (EDN3; Peptide Institute, Inc.): Store the stock solution at 100 μg/mL in 0.1% acetic acid bovine serum albumin (BSA)/PBS at –70°C. Store at 4°C after thawing. Stable for at least 1 year (stock at –70°C) or 1 month (stock at 4°C).

2.4 Immunohisto-chemical Analysis of Cultured Melanoblasts

1. 4% PFA (Paraformaldehyde) in PBS: pH=7.0–7.5 (see Note 1).

2. 0.1% Triton X-100 in 0.5% BSA PBS.

3. Blocking solution: 3% goat serum or 5% BSA in PBS.

4. Primary antibodies: mouse anti-neuronal class III β-tubulin (TuJ-1; BABCO), rabbit anti-mouse glial fibrillary acidic protein (GFAP; Z0334, Dakocytomation).

5. Secondary antibodies: Texas-Red-conjugated anti-mouse IgG (Molecular Probes), Alexa-Fluor 488-conjugated anti-rabbit IgG (Molecular Probes).

6. Fluorescence microscope (such as Olympus, IX-71).

3   Methods

3.1 Maintenance of ST2 Stromal Cells and Preparation for Use

ST2 cells are maintained in RPMI 1640 media supplemented with 5% FCS and 50 μM 2-ME (see Note 2).

1. Obtain confluent ST2 cells in a dish or flask and trypsinize them with 0.05% trypsin/0.5 mM EDTA at 37°C for 4 min.

2. Add medium, dissociate the cells by pipetting, centrifuge them at 180–200×g for 4 min, and split them 1:4 into 100-mm dishes.

3. Maintain the cells by regularly passing them every 3 or 4 days.

4. To prepare feeder layers, seed one-fourth of the confluent cells from a 100-mm dish equally into one plate type, i.e., 6-well or 96-well. 2 days later, the cells reach confluence and are ready to use for the differentiation of melanoblasts. Neither irradiation nor treatment with mitomycin C is needed.

3.2 Preparation of Cell Suspensions from Mouse Embryonic Skin, Neonatal Skin, and Hair Follicles

1. Put male mice and female mice in the same cage in the evening and allow them to mate overnight (see Note 3).

2. Check the vaginal plug in female mice the next morning. Noon of the day that the plug was detected is designated day 0.5 of gestation (E0.5, see Note 4).

3.2.1 Preparation of Pregnant Mice

3.2.2 Skin of E12.5–E16.5 Embryos

1. The skin is removed from the dorsal lateral trunk region with fine forceps.

2. Skin samples are incubated for 6 min at 37°C in Dispase II solution (see Note 5).

3. Gently dissociate the cells by passing them through an 18- to 21-gauge needle.

4. Add 2 volumes of SM for quenching the digestion, and then centrifuge the cells at 180–200×g for 4 min (see Note 6).

3.2.3 Skin of E17.5–E19.5 Embryos and P0.5–P6 Neonates

1. The back skin is removed with scissors and incubated for 40 min at 37°C in 0.25% Collagenase type I solution in PBS (see Note 7).

2. Epidermal sheets are peeled mechanically with fine forceps from the digested dermal tissues during observation through a binocular microscope.

3. The peeled epidermal sheet and dermal tissues are each washed with 5 mL of Hanks' balanced salt solution.

4. The samples are cut finely and then incubated in cell-dissociation buffer at 37°C for 10–15 min (see Note 8).

5. Dissociate the cells by passage through an 18- to 21-gauge needle (see Note 9).

6. Add 2 volumes of SM for quenching the digestion, and then centrifuge the dissociated cells at 180–200×g for 4 min (see Note 10).

3.2.4 Vibrissa Follicles

1. The upper lip containing the vibrissa pad of neonates is cut to expose its inner surface.

2. The vibrissa follicles are gently dissected from the vibrissa pad under a microscope.

3. The isolate follicles are then washed and incubated for 30 min at 37°C in 0.05% trypsin solution (see Note 11).

4. The samples are gently dissociated by passage through 21-gauge needles.

5. Add 2 volumes of SM for quenching the digestion, and then centrifuge the dissociated cells at 180–200×g for 4 min (see Note 12).

3.3 Immunostaining of Skin Cell Suspensions for Flow-Cytometry

Dissociated cells are immunostained with anti-Kit antibodies and anti-CD45 antibodies. The Kit molecule is a melanoblast marker, and CD45 is a hematopoietic cell-specific marker. Although Kit molecules are good melanoblast marker, they are also expressed on the hematopoietic cells in the skin (19). Therefore, to eliminate the hematopoietic cells, use the anti-CD45 antibodies and isolate the Kit-positive and CD45-negative cells as melanoblasts from fetal skin by flow-cytometry (Fig. 1).

1. The dissociated cells are washed with SM twice.

2. Add the rat anti-mouse Fc gamma receptor and incubate the cells on ice for 30–40 min (see Note 13).

3. Wash two times with SM.

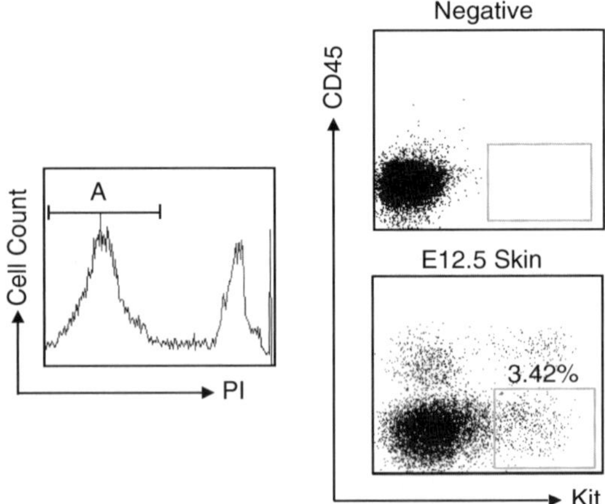

Fig. 1 Analysis and isolation of skin Kit+/CD45-cells by flow-cytometry. Flow-cytometric analysis of Kit+/CD45-cells in skin from E12.5 embryos. The skin cells were stained with anti-Kit-APC, anti-CD45-FITC antibodies, and PI. The analysis is performed by using PI-negative cells ("A" *line gate*). "Negative" indicates no incubation with antibodies

4. Add the FITC-conjugated rat anti-mouse CD45 and allophyco-cyanin-conjugated rat-anti mouse Kit (see Notes 14 and 15).

5. Incubate on ice for 30–40 min.

6. Wash the cells two times with SM.

7. Resuspend the cells in PI solution (see Note 16).

3.4 Isolation of Melanoblasts by Flow-Cytometry and Culture of the Cells

The immunostained cell suspensions prepared in Subheading 3.3 are analyzed and sorted with a flow-cytometer (see Note 17).

1. Isolate the Kit-positive and CD45-negative cells (Kit+/CD45-cells) from the skin cell suspension by performing flow-cytometry (see Note 18).

2. For culturing the sorted cells, directly inoculate the sorted cells (100–200 or single ones) into the wells of culture plates by using the flow-cytometry system (see Notes 19 and 20). The wells should have been previously seeded with ST2 stromal cells and contain medium for differentiation of melanoblasts (see Note 21).

3. Incubate the sorted cells in a 5% CO_2 incubator at 37°C. The medium is changed every 2 days. The day when the sorted cells are seeded onto the ST2 monolayers is defined as day 0.

3.5 Culture and Immunohistochemical Analysis of Isolated Melanoblasts

After 10–14 days of culture, the sorted cells form colonies that contain pigmented melanocytes (M, Fig. 2). After 21 days, some colonies contain TuJ-1-positive neurons (N) and GFAP-positive glial cells (G) together with M (Fig. 2). The Kit+/CD45- cells also form colonies composed of two cell types (M/N, M/G, and N/G) and of one cell type (M, N, and G). It is thus conceivable that the

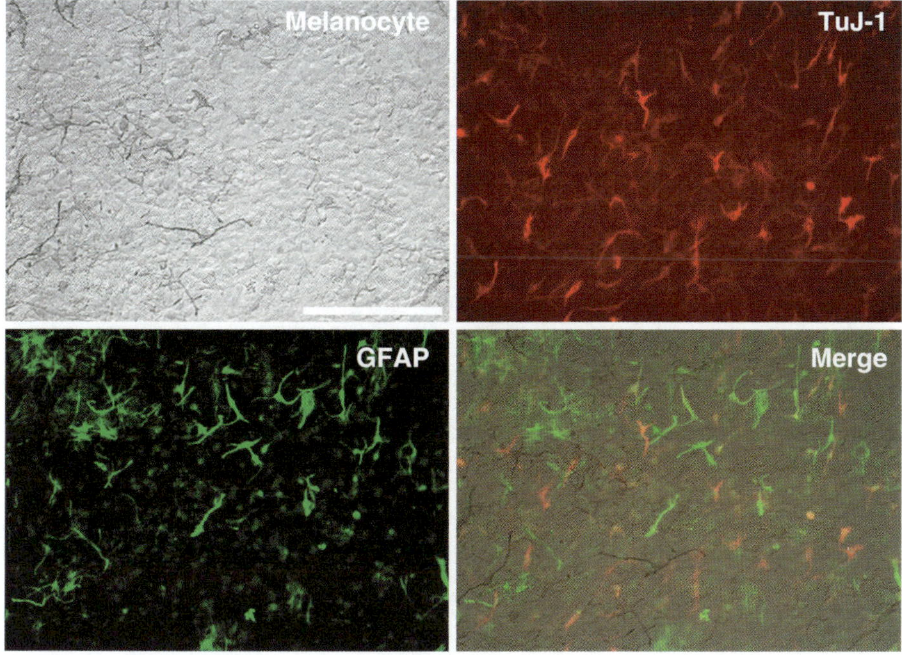

Fig. 2 Kit+/CD45-cells differentiate into melanocytes, neurons, and glial cells. Kit+/CD45-cells were sorted by flow-cytometry (Fig. 1, *gray line gate*) from E12.5 skin and inoculated at 200 cells/well onto ST2 monolayers in a 6-well dish. After 21 days in culture, the colonies were immunostained for neuronal marker β-tubulin (TuJ-1) and glial marker (GFAP). Melanocytes were detected as pigmented cells. Merge indicate the merged image. The same visual field is shown in each photo. Scale bar = 200 μm

formerly designated melanoblasts have the potential to differentiate into multilineage cells.

3.5.1 Immunohistochemical Analysis of Isolated Melanoblasts

1. Aspirate the culture medium from the cultures and wash them three times with PBS.

2. Add 4% PFA in PBS to fix the cells (see Note 22).

3. Incubate at room temperature for 15 min.

4. Wash three times with PBS.

5. Add 500 μL of 0.1% Triton X-100 in 0.5% BSA PBS to render the cells permeable (see Note 23).

6. Incubate at room temperature for 30 min.

7. Wash three times with PBS.

8. Add a mixture of 3% goat serum or 5% BSA in PBS to block the nonspecific binding of antibodies.

9. Incubate at room temperature for at least 30 min.

10. Wash three times with PBS.

11. Dilute the primary antibody (TuJ-1) in 0.5% BSA PBS (1:500, see Note 15), and then add it to the cells at room temperature (see Note 24).

12. Wash three times with PBS.

13. Dilute the secondary antibody (Texas-Red-conjugated anti-mouse IgG) in 0.5% BSA PBS (1:1,000, see Note 15), and then add it to react with the primary antibody.

14. Wash the cells three times with PBS, react them with the anti-GFAP (1:500, see Note 15), and then with the Alexa-Fluor 488-conjugated-anti-rabbit IgG (1:1,000, see Note 15) in the same manner.

15. Wash the stained cells three times with PBS and examine the colonies under the fluorescence microscope (see Note 25).

---

4 Notes

1. Do not make a stock solution. Prepare the fresh solution when needed.

2. ST2 cultures, when cultured for long periods, sometimes change in appearance (i.e., change to a more dendritic shape or senescent appearance). Discard such cultures and use freshly thawed ST2 cells from the frozen stock.

3. We mainly use C57BL/6 mice in our experiments.

4. The developmental stages of embryos are judged by their morphological appearance as described in "The Mouse" (20). All animal experiments should be performed in accordance with the appropriate regulations for animal experiments.

5. We use 1 mL of dispase II solution for five embryonic skins and incubate them for up to 15 min for dissociation.

6. Approximately 5×10^6 cells are collected from eight embryonic skins.

7. We use 10 mL of collagenase solution for three neonatal skins and incubate them for up to 60 min for dissociation.

8. We use 5 mL of cell-dissociation buffer for three neonatal skins, with incubation for up to 15 min for dissociation.

9. Passing the sheet several times through a blue-tip Pipetman with its tip cut back approx. 2–5 mm before using the needle helps the dissociation process using the needle.

10. Approximately 2×10^7 cells are collected from three neonatal skins.

11. We use 500 μL of trypsin solution for each neonatal follicle.

12. Approximately 3×10^6 cells are collected from vibrissa follicles of one neonate.

13. The antibodies against the Fc gamma receptor block the nonspecific cell-surface binding of antibodies. Add an appropriate volume of the diluted antibody solution according to the supplier's recommendation.

14. Keep aside a small amount of sample without addition of antibodies. This sample is used as a negative control in the flow-cytometric analysis.

15. Add an appropriate volume of the diluted antibody solution according to the supplier's recommendation.

16. Only dead cells are stained with propidium iodide solution (Fig. 1).

17. We use a FACS Vantage from Becton-Dickinson.

18. Exclude the PI-positive dead cells (Fig. 1). About 3.5% of the total E12.5 embryonic skin cells are Kit+/CD45-cells according to flow-cytometric analysis (Fig. 1).

19. We use the CloneCyt Plus v3.1 System (Beckton-Dickinson).

20. One hundred to two hundred and single sorted cells are introduced into 6-well and 96-well plates, respectively.

21. The volume of medium is 2 mL/well and 200 μL/well in 6-well and 96-well plates, respectively.

22. The volume of fixative is 1 mL/well and 200 μL/well of 4% PFA for 6-well and 96-well plates, respectively.

23. The volume of Triton-X is 500 μL/well and 200 μL/well for 6-well and 96-well plates, respectively.

24. Incubate at room temperature for more than 30 min or at 4°C overnight.

25. We use an Olympus IX-71 fluorescence microscope.

Acknowledgments

This study was supported by a grant from the program Grants-in-Aid for Scientific Research (C) from the Japan Society for Promotion of Science.

References

1. Nicole M, Le Douarin CK (1999) The neural crest. Cambridge University Press, Cambridge

2. Mayer TC (1973) The migratory pathway of neural crest cells into the skin of mouse embryos. Dev Biol 34:39–46

3. Jordan SA, Jackson IJ (2000) MGF (KIT ligand) is a chemokinetic factor for melanoblast migration into hair follicles. Dev Biol 225:424–436

4. Henion PD, Weston JA (1997) Timing and pattern of cell fate restrictions in the neural crest lineage. Development 124:4351–4359

5. Luo R, Gao J, Wehrle-Haller B et al (2003) Molecular identification of distinct neurogenic and melanogenic neural crest sublineages. Development 130:321–330

6. Wilson YM, Richards KL, Ford-Perriss ML et al (2004) Neural crest cell lineage segregation in the mouse neural tube. Development 131:6153–6162

7. Fernandes KJ, McKenzie IA, Mill P et al (2004) A dermal niche for multipotent adult skin-derived precursor cells. Nat Cell Biol 6:1082–1093

8. Toma JG, McKenzie IA, Bagli D et al (2005) Isolation and characterization of multipotent skin-derived precursors from human skin. Stem Cells 23:727–737

9. Wong CE, Paratore C, Dours-Zimmermann MT et al (2006) Neural crest-derived cells with stem cell features can be traced back to multiple lineages in the adult skin. J Cell Biol 175:1005–1015

10. Sieber-Blum M, Grim M, Hu YF et al (2004) Pluripotent neural crest stem cells in the adult hair follicle. Dev Dyn 231:258–269

11. Amoh Y, Li L, Katsuoka K et al (2005) Multipotent nestin-positive, keratin-negative hair-follicle bulge stem cells can form neurons. Proc Natl Acad Sci U S A 102:5530–5534

12. Yu H, Fang D, Kumar SM et al (2006) Isolation of a novel population of multipotent adult stem cells from human hair follicles. Am J Pathol 168:1879–1888

13. Dupin E, Glavieux C, Vaigot P et al (2000) Endothelin 3 induces the reversion of melanocytes to glia through a neural crest-derived glial-melanocytic progenitor. Proc Natl Acad Sci U S A 97:7882–7887

14. Trentin A, Glavieux-Pardanaud C, Le Douarin NM et al (2004) Self-renewal capacity is a widespread property of various types of neural crest precursor cells. Proc Natl Acad Sci U S A 101:4495–4500

15. Real C, Glavieux-Pardanaud C, Le Douarin NM et al (2006) Clonally cultured differentiated pigment cells can dedifferentiate and generate multipotent progenitors with self-renewing potential. Dev Biol 300:656–669

16. Dupin E, Real C, Glavieux-Pardanaud C et al (2003) Reversal of developmental restrictions in neural crest lineages: transition from Schwann cells to glial-melanocytic precursors in vitro. Proc Natl Acad Sci U S A 100: 5229–5233

17. Real C, Glavieux-Pardanaud C, Vaigot P et al (2005) The instability of the neural crest phenotypes: Schwann cells can differentiate into myofibroblasts. Int J Dev Biol 49: 151–159

18. Motohashi T, Yamanaka K, Chiba K et al (2009) Unexpected multipotency of melanoblasts isolated from murine skin. Stem Cells 27:888–897

19. Lagasse E, Connors H, Al-Dhalimy M et al (2000) Purified hematopoietic stem cells can differentiate into hepatocytes in vivo. Nat Med 6:1229–1234

20. Rugh R (1990) The mouse. Oxford University Press, Oxford

Chapter 16

Generation of Human Melanocytes from Induced Pluripotent Stem Cells

Shigeki Ohta, Yoichi Imaizumi, Wado Akamatsu, Hideyuki Okano, and Yutaka Kawakami

Abstract

The discovery of human induced pluripotent stem cells (iPSCs) has provided a model system for studying early events during human development. Developmentally melanocytes originate from migratory neural crest cells that emerge from the neural plate during embryogenesis after a complex process of differentiation, proliferation, and migration out of the neural tube along defined pathways. In the adult, human melanocytes are located in the basal layer of the epidermis, hair follicles, uvea, inner ear, and meninges. In the epidermis, melanocytes produce melanin pigment that gives color to the skin as well as providing protection from ultraviolet light damage. In addition, melanocytes transfer melanin pigment to hair matrix keratinocytes during each hair cycle to maintain hair pigmentation. Characterization of mouse melanocyte stem cells (MELSCs) is more complete than for humans. MELSCs are located in the bulge region of hair follicles, where hair follicle stem cells (HFSCs) also reside. Recently, it has been demonstrated that HFSCs provide a functional nice for MELSCs. According to current cancer stem cell theory, melanomas are considered to evolve from MELSCs, although the exact mechanism remains to be elucidated fully. In humans, importantly, the lack of more specific markers of MELSCs, current understanding of the molecular regulations of melanocyte development remains incomplete. Recently, the generation of melanocytes from iPSCs has lead to some clarification of human melanocyte development in vitro. Utilization of iPSC-derived melanocytes may prove invaluable in further study of human melanocytic development and novel therapies for patients suffering with pigmentation disorders and melanoma.

Key words Induced pluripotent stem cells, Neural crest cells, Melanocytes, Melanoma

1 Introduction

Pigmented cells including epidermal melanocytes play an important physiological role in providing protection from harmful ultraviolet rays and are also implicated in various pigmented cell disorders. Either defects in or a lack of melanocytes and/or MELSCs can lead to pigment disorders including vitiligo. Vitiligo is a common disease affecting approximately 0.1–2.0% of the world population, although the pathogenesis has not been completely understood (1).

Kursad Turksen (ed.), *Skin Stem Cells: Methods and Protocols*, Methods in Molecular Biology, vol. 989,
DOI 10.1007/978-1-62703-330-5_16, © Springer Science+Business Media New York 2013

Autologous cultured melanocytes may be useful for the treatment of vitiligo (2). In contrast to foreskin melanocytes, expansion of adult epidermal melanocytes is challenging. Thus, development of methods to generate large numbers of autologous melanocytes is required, considering the therapeutic implications. Melanocyte generation from embryonic stem cells (ESCs) has been previously reported in mice and human (3, 4). Alternatively, iPSCs have specific advantages compared to ESCs. Besides avoiding ethical issues, iPSCs can be propagated as autologous cells meaning that melanocytes from autologous iPSCs are not likely to be immunologically rejected if transplanted for the treatment of pigment cell disorders. To date, many genetic disorders of hypopigmentation including Waardenburg syndrome, piebaldism, albinism, Chediak–Higashi syndrome, and Griscelli syndrome, which are developed by either melanoblast or melanocyte's functional defects, are known (5). The importance of generation of patient-specific iPSCs has been shown previously in amyotrophic lateral sclerosis (ALS), familial dysautonomia, Parkinson's disease, and spinal muscular atrophy (SMA) for the development of therapeutic strategies (6–9). The generation of melanocytes from iPSCs of the pigmented disorder patients can faithfully recapitulate the disease phenotype and contribute significantly to revealing disease mechanisms and developing drug screening systems for disease. With these considerations, human iPSCs are a superior starting cell source to generate melanocytes.

Melanocytes are specialized cells derived from the neural crest cells. During embryonic development, the precursors of melanocytes are nonpigmented melanoblasts derived from the neural crest cells. A number of studies have shown that cell factors such as Wnt, c-Kit, and endothelian-3 (ET-3) are important for melanocyte development (10). Indeed, these factors were included in melanocyte differentiation medium in this study. A transcription factor, microphthalmia-associated transcription factor (MITF), can regulate the melanocyte lineage in part by regulating several pigmentation enzymes including dopachrome tautomerase (DCT), tyrosine-related protein 1 (TYRP1), and tyrosinase and has many gene variants (11). In particular, MITF-M is abundantly expressed in epidermal melanocytes not but in retinal pigmented epithelia (12). Thus, MITF-M is a good marker to examine the generation of epidermal melanocytes. It has been reported that mouse MELSCs locate in the bulge region of hair follicles, where HFSC support the MELSC niche by producing collagen17a, TGF-β, and Wnt (13, 14); however, localization of MELSCs in human skin is yet to be defined due to lack of definitive markers. It is also difficult to analyze the developmental cell lineages of skin melanocytes in all stages in humans. Thus, it is desirable to develop a new in vitro system for generating human melanocytes through MELSCs, which mimics in vivo differentiation processes to better understand human melanocyte development. This developed in vitro culture system will accelerate the study of human MELSCs.

Melanoma is one of the most aggressive types of human cancers and is suspected to arise from MELSCs. The recent progress of cancer stem cell studies supports the hypothesis that melanoma stem cells (MMSCs) which are resistant to chemotherapy may exist and are thus important therapeutic targets (15, 16). It is proposed that MMSCs are generated from MELSCs through accumulation of genetic changes and may have some similar phenotypes to MELSCs (17). Thus, understanding the biology of human MELSCs and MMSCs is critically important. In addition, to permit the investigation of MELSCs and MMSCs, their generation and purification is required. Although definition of MELSCs and MMSCs markers is controversial, collecting knowledge using this method will permit the development of therapeutic treatments against MMSCs, including immunotherapy.

In our in vitro development system, we observed the appearance of some neural crest cell marker-positive cells during EBs differentiation, supporting the notion that iPSCs-derived melanocytes may be sequentially generated from neural crest cells, mimicking in vivo development. One of the methods to increase the productivity of melanocytes from iPSCs is to enrich neural crest populations by developing a new culture method or molecular biological technique. Of note, it may be important to find technology to select iPSC clones that can generate melanocytes with high yield, as individual iPSC clone may have different differentiation potency. In the future, we predict we aim to achieve much higher efficiencies of melanocyte derivation by enrichment of either neural crest cells or MELSCs through chemical and/or physical means, and these approaches may contribute to the cell therapy using melanocytes in the future.

In the present study, we used retrovirally expressed Yamanaka factors (18). Many new methods to avoid transgenes integration including episomal vectors, micro RNA (miR), and protein transfection have been developed (19–21). In future studies, it may be important to choose a "transgene-free" method of iPSCs generation, which will not accompany the potential of risk of tumorigenicity. In the future, it may also be possible to directly differentiation of melanocytes from other cell sources such as fibroblasts, as shown for other cell types (22, 23). In addition, in our study, we used dermal fibroblasts for the generation of iPSCs. Recently, it has been reported in the early iPSC passage, iPSCs continue to keep some extent of epigenetic memory derived from cell sources (24). Thus, it will be important to examine the best cell sources considering the origin of cell type, although it is also important to choose cell sources that have easy accessibility such as blood.

In this study, we have established an in vitro system for generating human melanocytes from iPSCs, apparently through a neural crest cell intermediate (25). This system may contribute to the cell therapy using melanocytes and understanding of human melanocyte development, various pigment cell disorders, and melanoma leading to the development of new therapeutic strategies. Accordingly, in this chapter we show a method of how to generate melanocytes from human iPSCs and the methods to validate the process and generated melanocytes using molecular biological and immunochemical methods.

2   Materials

2.1   Cell Culture and Derivation

1. Human iPSCs (WA29; 4 Yamanaka factor-iPSC and WDT2; 3 Yamanaka factor-iPSC) (25). Established iPSCs are available upon request to H. Okano (Keio University). 201 B7 is distributed from RIKEN Cell Bank (RIKEN Bioresource Center, Cell Bank, Tsukuba, Ibaraki, Japan, Cat. No. HPS0001; http://www.brc.riken.jp/lab/cell).

2. Mouse embryo fibroblasts (SNL 76/7) (DS Pharma Biomedical, Osaka, Japan).

3. Cultured normal human foreskin-derived epidermal melanocyte (NHEM)(Kurabo, Osaka, Japan, Cat. No. KM-4009).

4. Dullbecco's phosphate-buffered saline (PBS) (Invitrogen, Carlsbad, CA, USA, Cat. No. 14190).

5. Fetal bovine serum (FBS) (JRH Biosciences, Lenexa, KS).

6. 0.25% trypsin (Invitrogen, Cat. No. 15050).

7. Mytomycin-C (Sigma-Aldrich, St. Louis, MO, USA, Cat. No. M4287): working stock solution (400 µg/ml) is prepared by dissolving mytomycin-C powder in tissue culture grade water (Sigma-Aldrich, W3500). Store at −20°C, after filtration (0.22 µm).

8. Y-27632 (Wako Pure Chemical, Osaka, Japan, Cat. No. 253–00513). 10 mM stock. Reconstitute 5 mg in 15 ml of cooled sterilized distilled water to prepare 100× stock. Store at −20°C in 100–500 µl aliquots. The final concentration should be 10 µM and protect from light.

9. Human fibronectin (BD Bioscience, Bedford, MA, Cat. No. 354008).

10. Gelatin (Sigma-Aldrich, Cat. No. G9391). Dissolve 1 g gelatin powder in 1 L of distilled water, autoclave, and store at 4°C.

11. Cryovials (Nalge Nunc International, Rochester, NY, USA, Cat. No. 5000–1012).

12. Cell scraper (Sumitomo Bakelite, Tokyo, Japan, Cat. No. MS-93100).

13. Cell strainer 40 µm (BD Falcon, Cat. No. 352340).

14. Culture slides: four chamber polystyrene vessel tissue culture-treated glass slide (BD Falcon, Cat. No. 354114).

15. Tissue culture plastic: Tissue culture plate 6 well (Nalge Nunc International, Cat. No. 140675), tissue culture dish 100-mm, (BD Falcon, Cat. No. 353003), Petri dish 100-mm (Kord-Valmark, Ontario, Canada, Cat. No. 2910).

16. CO_2 incubator (Sanyo, Tokyo, Japan, model MCO-20A1C).

17. Liquid nitrogen.

18. Liquid nitrogen tank.

2.2 Gene Expression Analysis (PCR Technology and Gene Chip)

1. Trizol reagent (Invitrogen, Cat. No. 15596).

2. QIAShredder homogenizer (Qiagen, Hilden, Germany, Cat, No. 79654).

3. PureLink RNA Mini Kit (Invitrogen, Cat. No. 12183).

4. DNase I (Takara Bio, Otsu, Japan, Cat. No. 2270).

5. High Capacity RNA-to-cDNA Master Mix (Applied Biosystems, Forster City, CA, Cat. No. 4390777).

6. THUNDERBIRD SYBR qPCR Mix (Toyobo, Osaka, Japan, Cat. No. QPS-101).

7. Diethyl pyrocarbonate (DPEC)-treated water (Nacalai Tesque, Kyoto, Japan, Cat. No. 36420).

8. Nanodrop ND 1000 spectrophotometer (NanoDrop Technologies, Wilmington, DE, USA).

9. PCR thermocycler Applied Biosystems 7900 HT Real-time PCR System (Applied Biosystems).

10. RNA 6000 Nano Complete kit (Agilent Technologies, Palo Alto, CA , Cat. No. 5067–1511).

11. Quick Amp Labeling Kit, One-Color (Agilent Technologies, Cat. No. 5190–0442).

12. Whole Human Genome Microarray 4 X44K (Agilent technologies, Cat. No. G4110F).

13. Gene expression Hybridization Kit (Agilent Technologies, Cat. No. 5188–5242).

14. Gene Expression Wash Pack (Agilent Technologies, Cat. No. 5188–5327).

15. Agilent 2100 Bioanalyzer (Agilent Technologies).

16. Hybridization Oven (Agilent Technologies).

17. DNA Microarray Scanner system high resolution (Agilent Technologies).

18. Feature Extraction Software (Agilent Technologies).

19. GeneSpring software 7.3.1 (Agilent Technologies).

2.3 Immunocyto-chemistry

1. 4% Paraformaldehyde (PFA). For 100 ml, add 4 g paraformaldehyde (Sigma-Aldrich, Cat. No. P6148) to 80 ml preheated dH_2O (60°C) while carefully stirring in a fume hood. Add 1 N NaOH one drop at a time until clear solution and cool to room temperature. After addition of 10 ml 10× PBS (Invitrogen, Cat. No. 70011), adjust the pH (pH 7.4) with 1 N HCl and bring total volume to 100 ml. After filtration, this solution can be stored at 4°C in the dark for 1 week.

2. DAPI (4′, 6-diamidino-2-phenylindole, dihydrochloride) (Invitrogen, Cat. No. D1306). Stock solution (5mg/ml in deionized water) stored at −20°C at dark site is diluted to 300 nM in PBS as a working solution.

3. PermaFluor Aqueous Mounting Medium Liquid (Beckman Coulter, Brea, CA, Cat. No.IM0752).

4. Micro cover glass 24×60 mm (Matsunami, Tokyo, Cat. No. C024601).

5. Normal goat serum (Vector laboratories, Burlingame, CA, Cat. No. S-1000).

6. O.C.T. Compound (Sakura Finetek Tissue-Tek, Tokyo, Japan, Cat. No. 4583).

2.4 Electron Microscopy

1. Glutaraldehayde (TAAB Laboratories, Aldermaston, Berkshire, UK, Cat. No.G004).

2. Sodium cacodylate trihydrate (Wako Pure Chemical, Cat. No. 194–04852).

3. OsO4 (1%, TAAB Laboratories, Cat. No. O008).

4. Acetone: *n*-butyl glycidyl ether (QY-1, Oken Shoji, Tokyo, Japan).

5. Epok 812 (Oken Shoji).

6. Uranyl acetate (TAAB Laboratories, Cat. No. U008).

7. Lead citrate (TAAB Laboratories, Cat. No. L018).

2.5 Photography

1. Zeiss LMS confocal laser scanning microscope (Carl Zeiss, Gottingen, Germany).

2. Olympus IX71 microscope (Olympus, Tokyo, Japan) equipped with digital camera (Axiocam MR5, Carl Zeiss).

3. Electron microscope JEOL-1230 (Japanese Electronic Optical Laboratories, Tokyo, Japan).

4. Olympus inverted microscope CKX31.

2.6 Flow Cytometry

1. FACS buffer: 0.5% BSA and 2 mM EDTA in PBS (pH 7.2).

2. TrypLE Select Animal-Origin-Free (Invitrogen, Cat. No. 12563).

3. PE-anti-human p75 NGFR (CD271, BioLegend, Brea, CA, Cat. No. 345105). FITC-anti-human HNK-1 (CD57, Beckman coulter, Cat. No. IM0466).

4. Mouse IgG1-PE (BD Bioscience, Cat. No. 555749). Mouse-IgM-FITC (BioLegend, Cat. No. 401605).

5. Propidium iodide (Sigma, Cat. No. P4170).

3   Methods

The schematic for generating melanocytes from human iPSCs is shown in Fig. 1, and typical cell morphology of each steps are shown in Fig. 2. It is very important to start culture with healthy iPSC colonies without differentiation and one can expect to see pigmented cells at around 10–11 weeks from starting culture.

3.1 Mytomycin-C Inactivation of Feeder Cells (iPSCs Culture Procedure)

1. Culture SNL feeder cells in 100-mm dish using SNL medium (Table 1) to confluence, changing the medium on alternative days.

2. Inactivate feeder cells by adding 310 µl mytomycin-C (400 µg/ml) for a 100-mm dish at 37°C, 5% CO_2 for 2–2.5h.

3. Suction mytomycin-C containing medium and wash the cells with PBS twice.

4. Add 1 ml of 0.25% trypsin for a 100-mm dish and incubate using rocking motion at room temperature for 3–5 min.

5. Add 5 ml SNL medium to dishes and transfer cell suspension to 15-ml tube and centrifuge at $200 \times g$ for 5 min at room temperature.

6. Resuspend cell pellet with appropriate volume of culture medium and collect all cell suspension in a 50-ml Falcon tube.

7. Count cells and seed 1.5×10^6 cells per gelatin-coated 100-mm dish (see Note 1).

8. Return the plate into incubator at 37°C in a humidified atmosphere of 5% CO_2 (see Note 2).

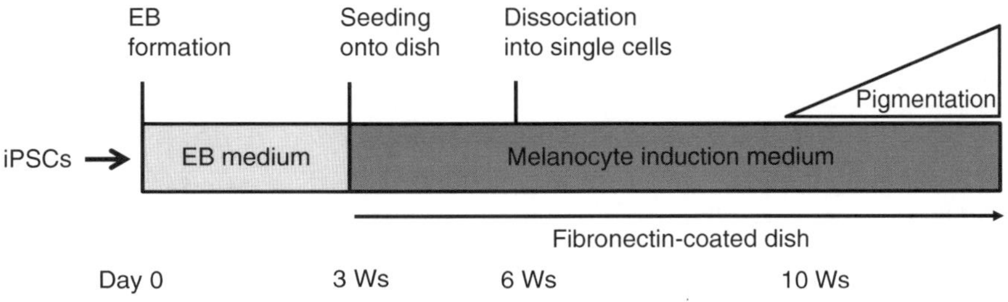

Fig. 1 Schematic of protocol for generating melanocytes from human iPSCs

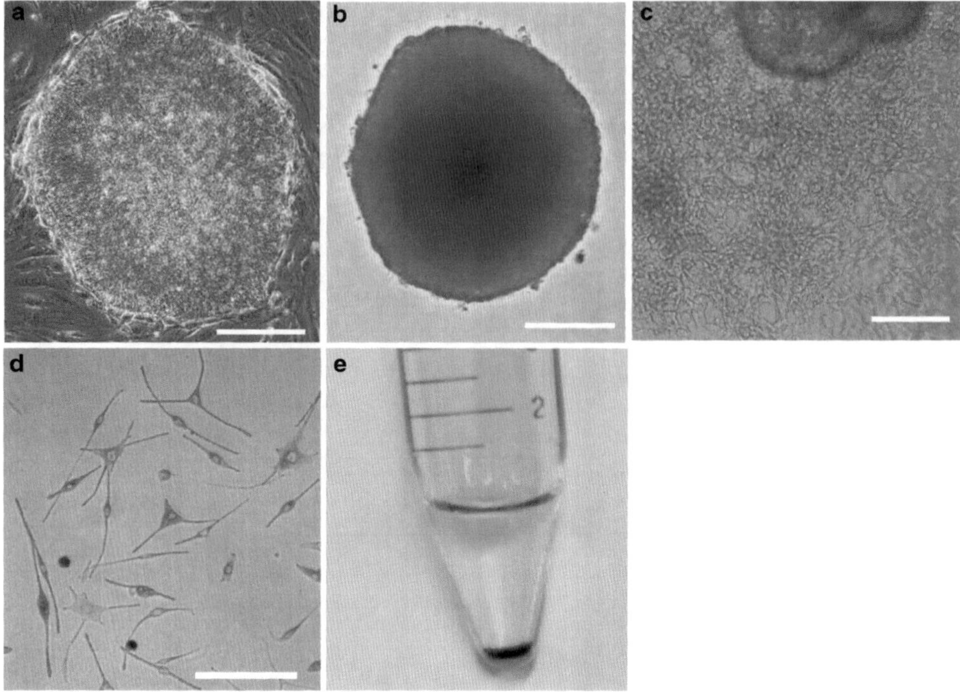

Fig. 2 Generation of melanocytes from iPSCs. Efficient differentiation of human iPSCs into pigmented melano-cytes in defined conditions. (**a**) A representative image of a 3F-iPSC colony grown on a mitomycin-C-treated STO feeder layer. (**b**) Embryoid (EBs) formed in suspension under feeder-free conditions. (**c**) After 1 week in differentiation medium, cells were observed migrating out from EBs on a fibronectin-coated plate. (**d**) Morphology of putative melanocytes differentiated from human EBs 7 weeks after differentiation. Cells display melanocytic morphology and pigmentation. (**e**) Pigmented cell pellet of iPSC-derived melanocytes 9 weeks after differentiation. Scale bar, 100 μm (**a–c**), 40 μm (**d**) (reproduced from ref. 25)

Table 1
SNL medium

Component	Company	Catalogue number	Final concentration
D-MEM	Sigma	D5796	1×
FBS	JRH Biosciences	#12603C	7%
L-Glutamine	Invitrogen	25030	2 mM
Penicillin	Invitrogen	15140	50 U/ml
Streptomycin	Invitrogen	15140	50 μg/ml

3.2 iPSC Culture (Cell Passage Procedure)

1. Discard iPSC medium (Table 2) and wash 100-mm dish of confluent iPSC with PBS once.

2. Add 1–2 ml CTK (Table 3) solution in a 100-mm dish and incubate at room temperature for 3–5 min.

Table 2
Human iPSCs medium

Component	Company	Catalogue number	Final concentration
D-MEM/F-12	Sigma	D6421	1×
Knockout serum replacement (KSR)	Invitrogen	10828	20%
L-Glutamine	Invitrogen	25030	2 mM
Nonessential amino acids	Sigma	M7145	0.8×
2-Mercaptoethanol	Sigma	M3148	0.1 mM
FGF-2	Peprotech	100-18B	4 ng/ml
Penicillin	Invitrogen	15140	50 U/ml
Streptomycin	Invitrogen	15140	50 µg/ml

Table 3
CTK solution

Component	Company	Catalogue number	Final concentration
Trypsin	Invitrogen	15090-046	0.25%
Knockout serum replacement (KSR)	Invitrogen	10828-028	20%
Collagenase IV	Invitrogen	17104-019	0.1 mg/ml
$CaCl_2$	Sigma	C7902	1 mM
PBS	Invitrogen	14190	1×

3. After checking for signs of curling up at the edges of colonies under microscope, suction excess CTK solutions and rinse the cells with 10 ml of PBS to remove feeder cells (see Note 3).

4. Add 10 ml of iPSC medium and break colonies by using cell scraper or pipetting and then make colonies into small cell clumps (few hundreds cells/clump) by pipetting up and down (see Note 4).

5. Transfer the cell suspension to the 15-ml tube and centrifuge at $200 \times g$ for 5 min at room temperature.

6. Dilute cell suspension at 1:5–1:8 with iPSC medium depending on colony confluency prior to passage and transfer to a new 100-mm dish seeded with mytomycin-C-treated feeder cells filled with iPSC medium.

7. Return plate to incubator at 37°C in a humidified atmosphere of 3% CO_2. Change iPSC medium every day (see Note 5).

8. Enzymatic passaging was performed routinely every 5–8 days depending cell growth, prior to iPSC colony differentiation.

3.3 Storage of iPSCs

1. Discard iPSC medium and wash the cells with PBS once (see Note 6).

2. Add 1–2 ml CTK solution in a 100-mm dish and incubate at room temperature for 3–5 min.

3. After checking for signs of curling up at the edges of colonies under microscope, discard CTK solutions, and rinse the cells with 10 ml of PBS to remove feeder cells.

4. Discard PBS and add 10 ml of iPSC medium.

5. Detach the colonies from the dish by gentle pipetting or cell scraper, avoiding breaking of colonies.

6. Transfer the cells to 15 ml tube and centrifuge at $200 \times g$ for 5 min at 4°C.

7. Remove the supernatant, finger tap the tube gently to disperse the pellet, and resuspend the cell pellet in 0.2–0.3 ml of pre-chilled DAP213 solution (Table 4).

8. Transfer the cell suspension to prechilled cryovials and put the vials quickly into liquid nitrogen (see Note 7).

3.4 Thawing iPSCs

1. Prepare a 15-ml tube with 10 ml of iPSC medium prewarmed to 37°C.

2. Add 0.5–1 ml of prewarmed iPSC medium to the vial drop by drop and thaw the frozen cells by pipetting gently and quickly.

3. Transfer the cell suspension to the tube prepared in step1 quickly and centrifuge at $200 \times g$ for 5 min at room temperature.

4. Discard the supernatant, and add 10 ml of iPSC medium containing Y-27632 (final concentration 10 μM) and transfer the cell suspension to a 100-mm dish seeded with mytomycin C-treated feeder cells.

5. Return the plate into incubator at 37°C in a humidified atmosphere of 3% CO_2.

6. A day after culture, change medium using iPSC medium without Y-27632.

3.5 Embryoid Bodies Formation

Embryoid body formation is a procedure to differentiate ESCs and iPSCs into three types of germ layer cells (ectodermal, mesodermal, endodermal) losing the attachment with feeder cells. Our protocol for the generation of melanocytes from iPSC includes this step.

1. Select iPSC colonies approximately 70–80% confluency for the embryoid body formation.

2. Discard iPSC medium and wash cells with PBS once.

Table 4
DAP213 solution

Component	Company	Catalogue number	Final concentration
iPSC medium			1×
DMSO	Sigma	D5879	14.3%
Acetamide	Sigma	A0500	1 M
Propylene glycol	Sigma	398039	22%

Store up to 1 month at −80°C.

3. Add 1–2 ml CTK solution in a 100-mm dish and incubate at room temperature for 3–5 min.

4. After checking for signs of curling up at the edges of colonies under microscope, suction excess CTK solutions, and rinse the cells with 10 ml of PBS to remove feeder cells.

5. Add 10 ml of iPSC medium and detach iPSC colonies from the dish by gentle pipetting.

6. Transfer the cells to 15 ml tube, pipetting gently the cells with up and down, and centrifuge at $200 \times g$ for 5 min at room temperature.

7. Resuspend the cells in 10 ml of iPSC medium and seed onto gelatin-coated dish and incubate for 2 h at 37°C in a humidified atmosphere of 3% CO_2.

8. Collect cell suspension and transfer to bacterial culture dishes and incubate in iPSC medium for a day at 37°C in a humidified atmosphere of 3% CO_2.

9. Change medium to EB medium (Table 5) after 24 h of culture in the bacterial culture dishes, and incubate EBs at 37°C in a humidified atmosphere of 3% CO_2.

10. Change medium using EB medium every 3–4 day for three weeks without supporting feeder cells. In every medium change, let cell clumps settle by gravity to the bottom of the 15-ml tube, and discard the supernatant, add new EB medium, and collect EBs carefully without damaging spherical shape (see Note 8).

3.6 Melanocyte Induction

1. Prepare human fibronectin-coated 6-well plate. Incubate 2 ml of fibronectin (1 µg/cm^2) per well of 6-well plate in the CO_2 incubator at least 12 h.

2. Suction excess fibronectin and rinse wells with PBS once.

Table 5
Embryoid bodies (EBs) medium

Component	Company	Catalogue number	Final concentration
D-MEM/F-12	Sigma	D6421	1×
Knockout serum replacement (KSR)	Invitrogen	10828	5%
L-Glutamine	Invitrogen	25030	2 mM
2-Mercaptoethanol	Sigma	M3148	0.1 mM
Nonessential amino acids	Sigma	M7145	1×

3. Collect EBs aggregates in a 15-ml tube and allow settle down for 5–10 min. Discard the supernatant and resuspend in appropriate volume of fresh melanocyte induction medium (Table 6).

4. Place approximately 20–40 EBs aggregates in melanocyte induction medium to each well of fibronectin-coated 6-well plates and place plates in CO_2 incubator overnight at 37°C in a humidified atmosphere of 5% CO_2. A day after culture, gently change the melanocyte induction medium to discard floating unattached cell aggregates. At this point, cells migration from the EB body can be observed (see Note 9).

5. Change medium every 4 days for 3 weeks (see Note 10). A typical image of 7 days after differentiation is shown in Fig. 2c.

6. After 3 weeks of starting of melanocyte differentiation, dissociate attached cells using 1ml of TrypLE Select per well of a 6-well plate. Incubate plates in the 5% CO_2 incubator at 37°C for 5 min and dissociate gently by pipetting. Add 5 ml of DMEM containing 1% FBS and pass the cells through 40 μm cell strainer.

7. Transfer flow-through suspension cells to 15-ml tube and centrifuge 5 min at $200 \times g$ at room temperature.

8. Resuspend cell pellet in desired amount of melanocyte induction medium and seed cells onto fibronectin-coated 6-well plates at a split ratio of 1:3.

9. Before reaching confluent condition, passage cells using TrypLE Select every 4–6 days until pigmented cells appear (usually takes 4–5 weeks after cell dissociation, Fig. 2d). Majority of bipolar and/or dendritic pigmented cells are observed in wells, and the pigmentation of cells usually develop in a time-dependent manner. In the passage procedure, cell pellet is highly pigmented. It is possible to maintain pigmented cells with the proliferative cell ability for few months.

Table 6
Melanocyte induction medium

Component	Company	Catalogue number	Final concentration
D-MEM	Invitrogen	12696	79%
MCDB201	Sigma	M6770	20%
FBS	JRH Biosciences	#12603C	1%
END3	American Peptide Company	88-5-10	100 nM
FGF-2	Peprotech	100-18B	4 ng/ml
SCF	R&D systems	255-SC	50 ng/ml
Wnt3a	R&D systems	5036-WN	50 ng/ml
L-Ascorbic acid	Sigma	A4403	0.1 mM
Cholera toxin	Wako	030-20621	20 pM
Dexamethasone	Sigma	D4902	50 nM
Linoleic acid Albumin	Sigma	L8384	1 mg/ml
ITS	Sigma	I1884	1×
TPA	Sigma	P1585	50 nM
Penicillin	Invitrogen	15140	50 U/ml
Streptomycin	Invitrogen	15140	50 µg/ml

10. To characterize pigmented cells perform immunocytochemical analysis for the melanocytic markers, MITF, TYR, TYRP1, SILV, and S100 (Fig. 3a–o) and assess gene expression of melanocytes marker, MITF-M and c-KIT (Fig. 4). Comparison of global gene expression profiles between primary melanocytes and iPS-derived melanocytes using DNA microarray is a useful method (Fig. 5) (see Note 11).

3.7 Immunocyto-chemistry

The presence of melanocytes can be confirmed by immunocytochemical analysis using melanocyte specific markers (Fig. 3a–o). In addition, we also identified the generation of neural crest cells in the differentiation procedure by immunocytochemical analysis (Fig. 7) (25). The source of primary and secondary antibodies and dilutions used are summarized in Tables 7 and 8.

1. Aspirate culture medium and fix the cells with 4% PFA in PBS for 20 min at room temperature (see Note 12).

2. 4% PFA is discarded and cells are washed three times with PBS for 10 min each at room temperature.

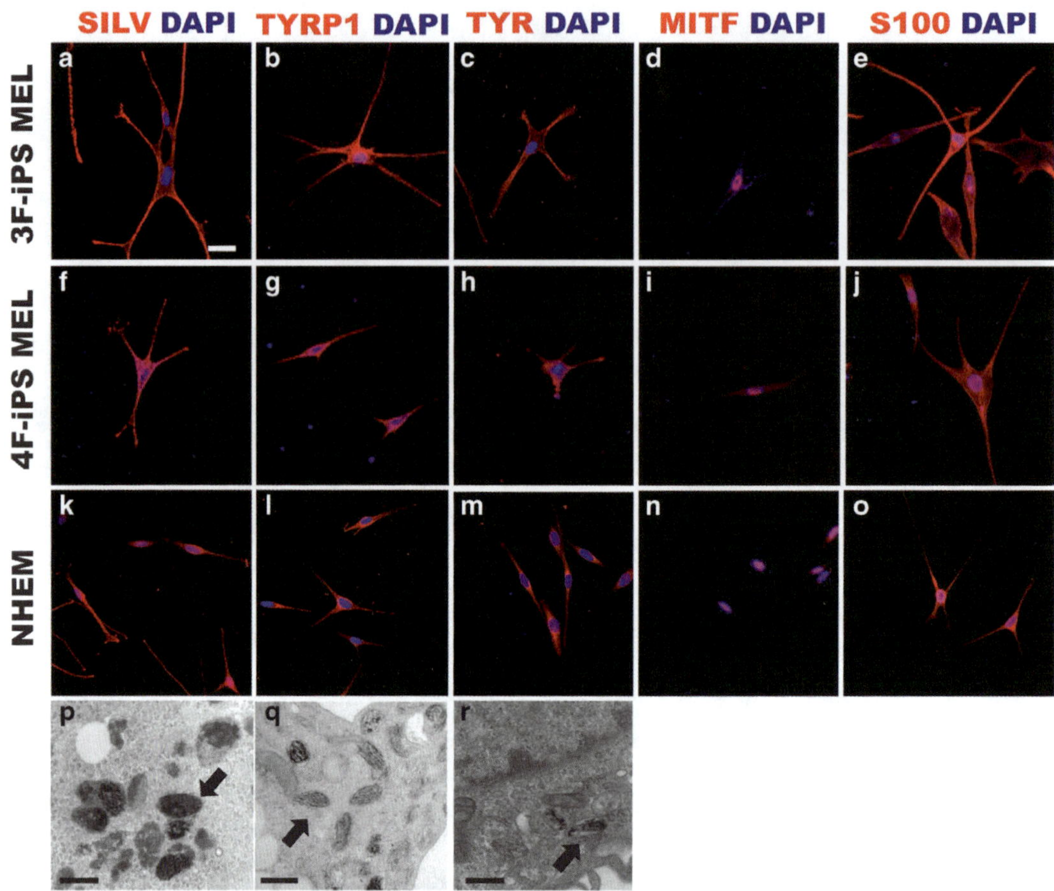

Fig. 3 Marker expression characteristic of the melanocyte lineage in human iPSC-derived melanocytes observed 8 weeks after differentiation. Either 3F- (3F-iPSC MEL, **a–e**) or 4F- (4F-iPSC MEL, **f–j**) iPSC-derived melanocytes were positive for an array of melanocyte markers: SILV (silver protein), TYRP1 (tyrosinase-related protein-1), TYR (tyrosinase), MITF (microphthalmia-associated transcription factor), and S100. Normal human epidermal melanocytes (NHEM) were also stained with the same antibodies (**k–o**). Transmission electron microscopy images of either 3F (**p**), 4F (**q**) iPSC-derived melanocytes, or NHEM (**r**) showed that those melanocytes have many melanosomes in the cytoplasm. Scale bar, 20 μm (**a–o**), 0.5 μm (**p–r**) (reproduced from ref. 25)

3. Incubate cells with blocking solution containing 10% NGS in PBST (PBS containing 0.3% Triton-X100) or PBS for 1 h (for cell surface marker staining, use PBS) at room temperature.

4. Aspirate blocking buffer and add cells with primary antibody solution containing appropriate amount of antibody in 5% NGS in PBS for 2 h at 37°C or overnight at 4°C.

5. Aspirate primary antibody solution, and wash cells three times with PBS for 10 min each at room temperature.

6. Incubate the cells with secondary antibody solution under dark conditions for 1 h at room temperature. Steps following incubation with secondary antibody should be performed in dimmed light.

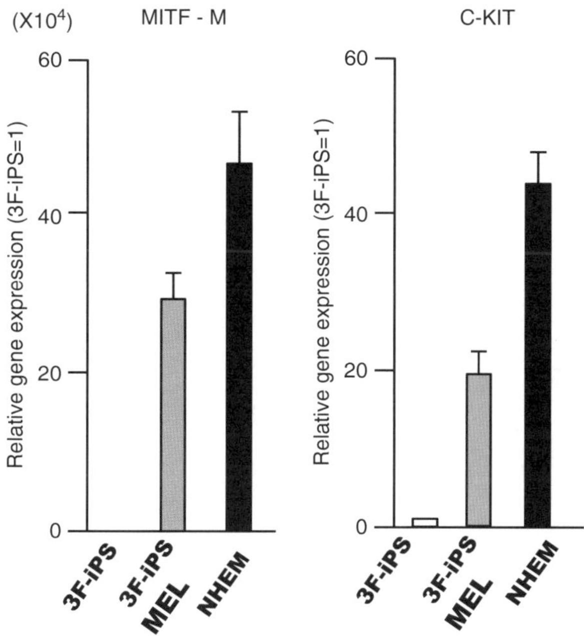

Fig. 4 Characterization of melanocytes. Quantitative real-time PCR analysis of MITF-M and c-Kit in 3F-iPSC, 3F-iPSC MEL, and NHEM. The data are normalized to GAPDH and represented as fold change relative to RNA levels in 3F-iPSCs. The graphs show the average of two independent experiments. Error bar indicates mean S.E.M (reproduced from ref. 25)

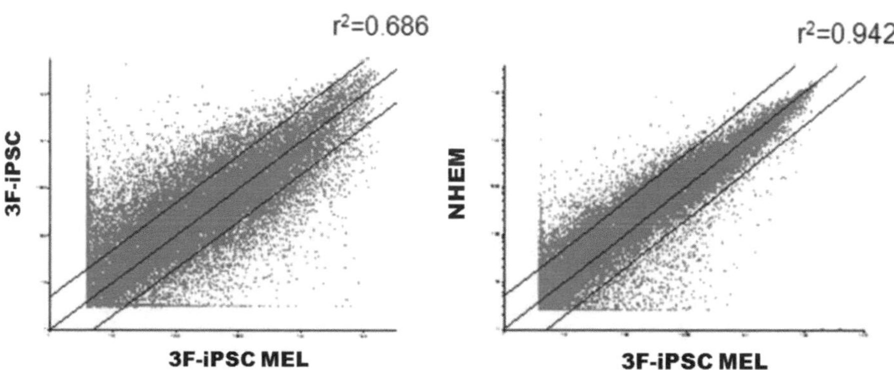

Fig. 5 DNA Microarray analysis of iPSC-derived melanocytes. The global gene-expression patterns were compared between human 3F-iPSCs and 3F-iPSCs derived-MEL (3F-iPSC MEL), and between NHEM and 3F-iPSC MEL. The *lines* indicate the linear equivalent and fivefold differences on either side in gene expression levels between the two samples (reproduced from ref. 25)

7. Secondary antibody solution is discarded and cells are washed three times with PBS 10 min each.

8. Counterstain cells with DAPI for 10 min to visualize nuclei and rinse in PBS once.

9. After rinse in distilled water, mounting solution is added and the cells are covered with a glass coverslip without introduction of air.

Table 7
List of primary antibodies

Primary antibodies	Species	Company	Catalogue number	Dilution antibodies
AP2a	Rabbit	Cell Signaling Technology	#3215	1:50
HNK-1	Mouse	Sigma	C6680	1:200
MITF	Mouse	Thermo Scientific	#MS-772-P	1:50
P75	Mouse	Advanced Targeting system	AB-N07	1:100
PAX3	Mouse	R&D Systems	MAB2457	1:100
S100	Rabbit	DAKO	Z0311	1:500
SILV	Mouse	DAKO	M0634	1:50
SOX10	Rabbit	Abcam	ab25978	1:100
Tyrosinase	Goat	Santa Cruz Biotechnology	sc-7833	1:100
TYRP1	Mouse	COVANCE	SIG-38150	1:20

Table 8
List of secondary antibodies

Secondary antibodies	Company	Catalogue number	Dilution antibodies
Alexa Fluor 488 goat anti-rabbit IgG	Invitrogen	A11034	1:400
Alexa Fluo 568 goat anti-rabbit IgG	Invitrogen	A11036	1:400
Alexa Fluo 568 goat anti mouse IgG	Invitrogen	A11031	1:400
Alexa Fluo 568 donkey anti goat IgG	Invitrogen	A11057	1:400
Alexa Fluo 568 goat anti mouse IgM	Invitrogen	A21043	1:400

3.8 Preparation Samples for Electron Microscopy

Analysis of melanosome in generated melanocytes is the most reliable method to examine melanocyte generation (Fig. 3p–r).

1. Samples are washed three times with PBS, and then fixed with 2.5% glutaraldehyde diluted from ultrapure glutaraldehyde, 2–4% PFA in 0.1 M cacodylate buffer (pH 7.4) for 2 h at room temperature.

2. Wash samples three times with PBS, and add 1% OsO_4 in 0.1 M PB buffer (pH 7.4) for 2 h at 4°C.

3. Dehydration procedure: Wash samples with 50% EtOH, 70% EtOH, 80% EtOH, 90% EtOH, 100% EtOH (Each step, 20 min).

4. After *n*-butyl glycidyl ether (QY-1) treatment at room temperature, the samples are finally embedded in EpoK 812 resin for 3 days at 60°C, and then ultramicrotomed. Sections are contrasted with uranyl acetate and lead citrate.

3.9 Preparation of Samples for Quantitative RT-PCR Analysis

To assess lineage of the neural crest and melanocyte differentiation from iPSCs, RT-PCR is used to examine the marker genes of melanocytes (MITF-M, c-KIT) (Fig. 4) and neural crest (SLUG, PAX3) (25). PCR primer sets are listed in Table 9.

3.9.1 Total RNA Extraction

1. Add 1 ml Trizol and mix by pipetting.

2. Transfer the lysate into a QIAshredder spin column and centrifuge for 2 min at $17,000 \times g$ to obtain the homogenate. The homogenate can be stored at –80°C.

3. Transfer the supernatant of the flow-through to a new microcentrifuge tube and add 0.2 ml of chloroform and shake tubes vigorously by hand for 15 s and incubate them at room temperature for 2–3 min. Centrifuge the samples at 12,000 x*g* for 15 min at 4°C.

4. Add one volume of 70% EtOH to upper aqueous phase in step 3 and mix gently by pipetting and pipette the solution to the Spin Cartridge provided from PureLink RNA Mini Kit.

5. Follow Handbook for the instructions on how to use the PureLink RNA Mini Kit as to on column DNaseI treatment.

6. Eluted RNA samples after washing procedure are subjected to RNA yield and quality analyses using spectrophotometer and can be stored at –80°C.

Table 9
Primer sequences

Target	Primer name	Sequence	Size (bp)
c-KIT	c-KIT Fw	GTTCTGCTCCTACTGCTTCGC	135
	c-KIT Rev	TAACAGCCTAATCTCGTCGCC	
MITF-M	MITF-M Fw	ACCTTCTCTTTGCCAGTCCATCT	262
	MITF-M Rev	GGACATGGCAAGCTCAGGACT	
PAX3	PAX3 Fw	AGCCGCATCCTGAGAAGTAA	148
	PAX3 Rev	CTTCATCTGATTGGGGTGCT	
SLUG	SLUG Fw	CATACAGCCCCATCACTGTG	137
	SLUG Rev	CTTGGAGGAGGTGTCAGATG	
GAPDH	GAPDH Fw	TGAACGGGAAGCTCACTGG	307
	GAPDH Rev	TCCACCACCCTGTTGCTGTA	

Table 10
cDNA synthesis reaction mixture

Reagents	Volume	Final concentration
Master Mix	4 µl	1×
RNA	1.0 µg	
DPEC-H$_2$O	X µl	
Total	20 µl	

Table 11
Quantitative PCR mixture

Reagents	Volume (µl)	Final concentration
THUNDERBIRD™ SYBR® qPCR Mix	25	1×
Fw primer (15 pmol/µl)	1.0	0.3 µM
Rev primer (15 pmol/µl)	1.0	0.3 µM
Template cDNA	1.0	
50× ROX reference dye	1.0	1×
Distilled water	X	
Total	50	

3.9.2 cDNA Synthesis

1. Gently mix reaction mixtures as shown in Table 10 and briefly centrifuge.

2. In a thermal cycler, incubate samples at 25°C for 5 min followed by 42°C for 30 min and 85°C for 5 min for the inactivation of reverse transcriptase.

3. Finally, cool down synthesized cDNA to 4°C and stored at −20°C.

3.9.3 Real-Time Quantitative PCR

1. Use 0.5–1 µl of the reverse transcribed cDNA solutions in a single real-time PCR reaction.

2. Primer sets used in the study are listed in Table 9. Endogenous housekeeping gene, glyceraldehyde-3-phosphate dehydrogenase (GAPDH) is used for the normalization.

3. For Real-Time PCR, Thunderbird SYBR qPCR Mix is used following to manufacturer's protocol (Table 11). The reaction cycles are at 95°C for 5 min, followed by 40 cycles of 95°C for 30 s, 60°C for 60 s, and 72°C for 60 s.

4. Reported threshold cycle (Ct) values are analyzed using ΔΔCt calculation method, normalized against GAPDH gene expression levels, and relative expression levels are identified (Note 13).

3.10 DNA Microarray

For characterization of melanocytes generated from iPSCs, the global gene expression profile can be compared between melanocytes-generated iPSCs and normal human foreskin-derived epidermal melanocyte using DNA microarray (Fig. 5) or RNA sequencing based on next-generation sequencing technologies.

1. Prepare total RNA from cells for the DNA microarray analysis using commercialized RNA separation kit (e.g., Invitrogen, PureLink RNA Mini Kit).

2. Check the RNA quality and quantity using Agilent RNA 6000 Nano kit and Agilent 2100 Bioanalyzer (Agilent Technologies) (see Note 14).

3. The RNA from each sample (1 μg) is reverse transcribed using RT-PCR to cDNA and labeled with fluorescent dyes such as cyanine 3-CTP and cyanine 5-CTP using labeling kit (Agilent Technologies, Quick Amp Labeling Kit).

4. The cRNA probes are hybridized to the DNA microarray using hybridization kit (Agilent Technologies) in a hybridization oven. Then, wash DNA array with wash solutions (see Note 15).

5. Fluorescent hybridization signals are scanned and detected using microarray scanner. Data is the filtered and processed using Feature Extraction Software (Agilent Technologies).

6. Data analysis is conducted using DNA Microarray bioinformatic software such as GeneSpring software 7.3.1 (Agilent Technologies).

3.11 Flow Cytometric Analysis

Human neural crest precursor cells are enriched in $p75^+HNK^+$ population. In the present study, we performed the flow cytometric analysis to examine the generation of neural crest cells in the differentiation process (Fig. 6).

1. Trypsinize rosetta cells with TrypLE Select for 5 min at room temperature.

2. After pipetting several times, add cells to 5 ml FACS buffer and pass cells through a 40-μm pore cell strainer (BD Falcon), spin the cells, and resuspend in 100 μl FACS buffer.

3. Add appropriate amount of antibodies and incubate for 30 min on ice in a dark place.

4. Add 5 ml FACS buffer, spin $200 \times g$ for 5 min at 4°C, and aspirate buffer and resuspend the cell pellet in 200 μl FACS buffer.

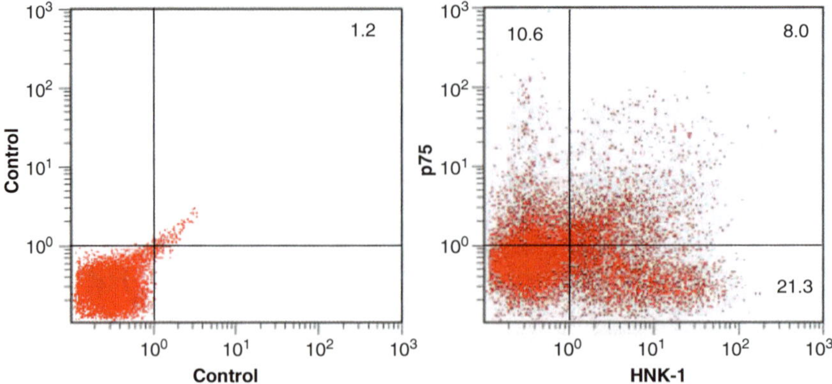

Fig. 6 Flow cytometry analysis to identify neural crest precursor cell population. Representative flow cytometry results for HNK-1 and p75 staining in 3F-iPSC EB-derived cells cultured in melanocyte differentiation medium for 1 week. Dead cells were excluded using PI staining (reproduced from ref. 25)

5. To remove dead cells, add PI solution (final concentration, 1 μg/ml).

6. For cell analysis, set the combination of the FACS gates and count 10,000 cells.

4 Notes

1. To prepare gelatin-coated dish incubate approximately 5 ml of gelatin solution a 100-mm dish in CO_2 incubator at 37°C in a humidified atmosphere of 5% CO_2 for at least 1 h. Suction excess gelatin prior to use.

2. MEF, which are prepared from embryonic mouse body, are also useful for iPSC culture. Commercialized MEF are also available from many companies. Mitomycin-C-treated STO should be used within 5 days.

3. Remove the differentiated colonies by using the pipette tip (100–200 μl) under microscope. Usually iPSCs of passage number between 20 and 35 are used for the generation of melanocytes.

4. Checking cell conditions during dissociation procedure under microscope is recommended. Do not overdissociate iPSC colonies to single cells.

5. The day after passage, medium change is not necessary, and in the case that many living colonies are floating in medium, transfer medium into a 15-ml tube and let the cells settle to the bottom of the tube. Discard the medium, add 10 ml of iPSC medium, and transfer cells back to the culture dish to increase

the recovery of iPSC colonies. Daily medium change is recommended thereafter. To date, commercialized nonfeeder iPSC culture systems using matrigel (BD bioscience, Cat. No. 354277) and mTeSR™1 (Stem cell technologies, Vancouver, Cat. No. 05850) are available and avoid daily iPSC medium changing.

6. Addition of Y-27632 (final concentration 10 μM) for confluent iPSCs at 37°C for 1 h before detaching the cells from dishes can be useful to decrease cell damage.

7. Cryovials should be precooled on ice and the freezing procedure after addition of DAP213 should be completed within 15 s to keep cell viability.

8. We use EB medium within 2 weeks after preparation. During EB culture, some amount of EBs usually touch down to the surface of bacterial culture dishes. Consideration of the cell loss is important to design following experiments. In the EBs handling procedure, we usually use 5-ml pipette (Greiner bio one, Kremsmünster, Austria, Cat. No. 606180) to prevent cell clump break down.

9. After overnight culture, healthy EBs should attach to the fibronectin-coated plates. If only a small proportion of EBs have attached overnight, the experiment should be discontinued at this point.

10. Over crowding in cell culture starts to change medium color to yellow, which is not favorable conditions. In some cases, migrated cells overall reach confluency between 2 and 3 weeks after differentiation. If this occurs, process to the next step. One week after differentiation, migrating out cells contain neural crest cells. Generation of neural crest cells can be analyzed by the gene expression (25), flow cytometry (Fig. 6), and immunocytochemistry (Fig. 7) (25).

11. One of the reliable methods to examine the successful generation of melanocytes from iPSC is to examine expression of MITF-M, which is abundant and is expressed in neural crest-derived melanocytes. In addition, three-dimensional reconstruction studies using human keratinocytes in vitro may be supportive to examine function of iPS-derived melanocytes. Usually, over 60% of pigmented SILV-positive cells are observed 7–8 weeks after differentiation under melanocyte induction medium. After culture for a few months, pigmented cells gradually loose the color and decrease the ability of proliferation.

12. For immunocytochemical analysis, fibronectin-coated cell culture slides can be used.

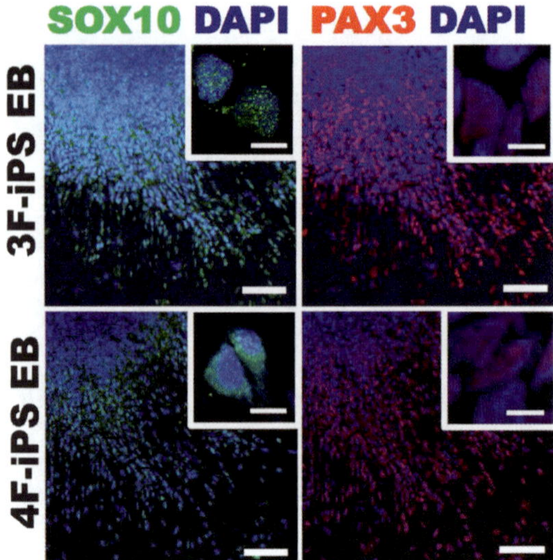

Fig. 7 3F- and 4F-iPSCs can be differentiated into neural crest cells and/or MELSCs. 3F- and 4F-iPSCs can be differentiated into neural crest cells and/or MELSCs. Images show cells positive for neural crest cell marker SOX10, as well as neural crest cell and/or possible MELSC marker, PAX3 in both 3F- and 4F-iPSCs-derived EB-differentiated cells 7 days after differentiation. Scale bar, 50 μm and 10 μm (*inset images*) (reproduced from ref. 25)

13. PCR cycle conditions can be changed depending on PCR primers, qRT-PCR kit, and thermal cycler. The concentration of ROX dye should be determined depending on qPCR devices following to manufacturer's instruction.

14. It is important to examine for RNA degradation.

15. In our study, microarray, Whole Human Genome Microarray $4 \times 44K$ (Agilent technologies), which contains over 44K probe sets for analyzing the expression level of over 19K transcripts was used.

Acknowledgments

The authors would like to acknowledge the following collaborators; Dr. Yohei Okada, Dr. Yumi Matsuzaki, Ms. Reiko Kuwahara, Dr. Manabu Ohyama, Dr. Masayuki Amagai (Keio University), and Dr. Shinya Yamanaka (Kyoto University). For technical assistance we thank Dr. Toshihiro. Nagai (Keio University) for electron microscopy studies, Ms. Mari. Fujiwara (Keio University) for Microarray data analysis and the Core Instrumentation Facility (Keio University). For critical reading of the manuscript we thank Dr. Ophelia Veraitch (Keio University) and Dr. Knut Woltjen

(Kyoto University). The study was supported by a grant from MEXT (Ministry of Education, Culture, Sports, Science and Technology), the project for realization of regenerative medicine and support for the core institutes for iPSC research from MEXT, and a Grant-in-aid for the Global COE program to Keio University from MEXT.

References

1. Alkhateeb A, Fain PR, Thody A, Bennett DC, Spritz RA (2003) Epidemiology of vitiligo and associated autoimmune diseases in Caucasian probands and their families. Pigment Cell Res 16(3):208–214

2. Czajkowski R et al (2007) Autologous cultured melanocytes in vitiligo treatment. Dermatol Surg 33(9):1027–1036, discussion 1035–1026

3. Yamane T, Hayashi S, Mizoguchi M, Yamazaki H, Kunisada T (1999) Derivation of melanocytes from embryonic stem cells in culture. Dev Dyn 216(4–5):450–458

4. Fang D et al (2006) Defining the conditions for the generation of melanocytes from human embryonic stem cells. Stem Cells 24(7):1668–1677

5. Dessinioti C, Stratigos AJ, Rigopoulos D, Katsambas AD (2009) A review of genetic disorders of hypopigmentation: lessons learned from the biology of melanocytes. Exp Dermatol 18(9):741–749

6. Ebert AD et al (2009) Induced pluripotent stem cells from a spinal muscular atrophy patient. Nature 457(7227):277–280

7. Dimos JT et al (2008) Induced pluripotent stem cells generated from patients with ALS can be differentiated into motor neurons. Science 321(5893):1218–1221

8. Soldner F et al (2009) Parkinson's disease patient-derived induced pluripotent stem cells free of viral reprogramming factors. Cell 136(5):964–977

9. Lee G et al (2009) Modelling pathogenesis and treatment of familial dysautonomia using patient-specific iPSCs. Nature 461(7262):402–406

10. White RM, Zon LI (2008) Melanocytes in development, regeneration, and cancer. Cell Stem Cell 3(3):242–252

11. Levy C, Khaled M, Fisher DE (2006) MITF: master regulator of melanocyte development and melanoma oncogene. Trends Mol Med 12(9):406–414

12. Amae S et al (1998) Identification of a novel isoform of microphthalmia-associated transcription factor that is enriched in retinal pigment epithelium. Biochem Biophys Res Commun 247(3):710–715

13. Nishimura EK (2011) Melanocyte stem cells: a melanocyte reservoir in hair follicles for hair and skin pigmentation. Pigment Cell Melanoma Res 24(3):401–410

14. Jahoda CA, Christiano AM (2011) Nich crosstalk: intercellular signals at the hair follicle. Cell 146(5):678–681

15. Grichnik JM (2008) Melanoma, nevogenesis, and stem cell biology. J Invest Dermatol 128(10):2365–2380

16. Schatton T, Frank MH (2008) Cancer stem cells and human malignant melanoma. Pigment Cell Melanoma Res 21(1):39–55

17. Sabatino M, Stroncek DF, Klein H, Marincola FM, Wang E (2009) Stem cells in melanoma development. Cancer Lett 279(2):119–125

18. Takahashi K et al (2007) Induction of pluripotent stem cells from adult human fibroblasts by defined factors. Cell 131(5):861–872

19. Okita K et al (2011) A more efficient method to generate integration-free human iPS cells. Nat Methods 8(5):409–412

20. Sridharan R, Plath K (2011) Small RNAs loom large during reprogramming. Cell Stem Cell 8:599–601

21. Zhou H, Wu S, Joo JY, Zhu S, Han DW et al (2009) Generation of induced pluripotent stem cells using recombinant proteins. Cell Stem Cell 4:381–384

22. Szabo E et al (2010) Direct conversion of human fibroblasts to multilineage blood progenitors. Nature 468(7323):521–526

23. Vierbuchen T et al (2010) Direct conversion of fibroblasts to functional neurons by defined factors. Nature 463(7284):1035–1041

24. Kim K et al (2010) Epigenetic memory in induced pluripotent stem cells. Nature 467(7313):285–290

25. Ohta S et al (2011) Generation of human melanocytes from induced pluripotent stem cells. PLoS One 6(1):e16182

Chapter 17

Cellular Populations Isolated from Newborn Mouse Skin Including Mesenchymal Stem Cells

Lauren Kimlin and Victoria Virador

Abstract

We developed protocols for isolation and characterization of mesenchymal progenitors from murine dermis. Our protocols are part of a more general isolation procedure starting with neonatal murine skin, which has been described in detail by U. Lichti and coauthors (Nat Protoc 3(5):799–810, 2008). We list Lichti's procedures in an abbreviated form as part of this methods section. Our methods to isolate mesenchymal stem cells are presented as a continuous workflow of isolation and characterization, including flow cytometry, cell survival assays, colony formation assays, immunoblotting, immunostaining, multipotential differentiation assays, and in vivo engraftment. In most cases, the protocols are standard; in others, they were adapted to our particular purpose. We made special emphasis on the use of in vitro three-dimensional cultures to cue mesenchymal progenitors into epidermal cells.

Key words Epidermis, Dermis, Flow cytometry, Organotypic, Adipocyte, Chondrocyte, Osteocyte, Murine skin, In vitro 3D culture

1 Introduction

The skin provides a protective barrier for the body from the outside world. Human epidermis can be subdivided into five layers (from bottom to top): stratum basale, stratum spinosum, stratum granulosum, stratum lucidum, and stratum corneum. The bottom layer, stratum basale, is composed of columnar-shaped cells known as keratinocytes (Kc). Stratum basale Kc divide and push already formed cells into higher layers where they gradually flatten out, lose nuclear integrity and form the top layer of flat, dead cells known as the stratum corneum. In addition to Kc, there are other specialized cell types, which include melanocytes, Langerhans cells, and Merkel's cells. Melanocytes give the skin its pigment while Langerhans cells are the front line of immune defense in the skin. Merkel's cell function is not well understood yet. Fibrous components of the dermis include collagen, reticular, and elastic fibers

Kursad Turksen (ed.), *Skin Stem Cells: Methods and Protocols*, Methods in Molecular Biology, vol. 989, DOI 10.1007/978-1-62703-330-5_17, © Springer Science+Business Media New York 2013

arranged to form the papillary (upper) and reticular (lower layers) of the dermis. In addition to fibroblasts (Fb), abundant in the dermis, there are specialized structures contained within the dermis including hair follicles (HFs), blood vessels, the erector pilus muscle, sebaceous, apocrine and eccrine glands, sensory nerves, Meissner's and Vater-Pacini corpuscles. Murine skin is relatively simpler than human skin, containing two main layers, the epidermis and the dermis with numerous interspersed hair follicles (Fig. 1).

The hair follicle is a structure that spans the epidermis and deep into the dermis. It is a very complex three-dimensional structure that contains at least seven epithelial cell types. Unique differentiation programs work together to form mature hair. Follicle morphogenesis during embryonic development is induced by interactions between epidermal cells and the underlying mesenchyme. The hair follicle in most mammals undergoes cyclic phases of growth, regression, and inactivity. Signals controlling the onset and duration of these phases have not yet been fully elucidated (1). Major compromise to skin is life threatening therefore it is important to have redundant reservoirs of adult stem cells for tissue regeneration. Such reservoirs exist in the hair follicle and in the underlying dermis. The main function of epidermal stem cells from bulge and dermal papilla is to restore epidermal integrity (1, 2), while those in the dermis restore mesenchymal cell types and cooperate to restore epidermis if needed (3). This redundancy in the stem-progenitor populations in the skin is apparent when there is a need to repair serious damage.

The Yuspa Laboratory has pioneered research of murine keratinocyte and hair follicle biology for various decades (4–8). Ulrike Lichti recently compiled their protocols for preparation of primary Kc, their short-term culture, immortalization into cell lines, and cellular transformation by exogenous DNA (9). Our contribution to this body of knowledge was a detailed understanding of cellular populations in neonate murine dermis (3). In order to understand progenitor potential in cells isolated from murine dermis, we developed protocols to isolate various subpopulations, we tested their long-term viability and their ability to differentiate with standard multilineage differentiation methods, which we modified to suit our purpose. We also tested the ability of the mesenchymal subpopulations to form model skin in three dimensions. Prior to our report, only stem cells from the hair follicles were thought to have this ability, at least in the mouse. Here we detail those protocols as a continuum process of isolation, characterization, and culture. Finally, we used two different grafting approaches to demonstrate differentiation ability of these populations in vivo.

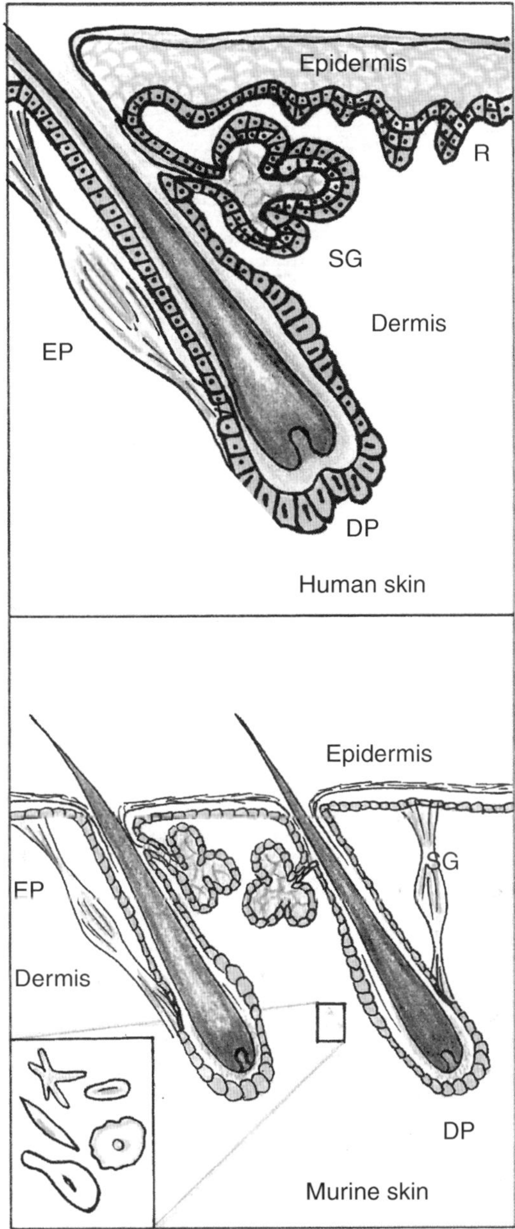

Fig. 1 Schematic of human and murine skin. Epidermis is more complex in human skin, yet human skin has fewer hair follicles. *R* rete ridges, *EP* erector pilus muscle, *SG* sebaceous gland, *DP* dermal papilla. Bulge region, where a subpopulation of stem cells resides, is indicated with an *arrowhead*. *Lower inset* lists various cell types resident in the dermis including Langerhans cells, endothelial cells, adipocytes, fibroblasts, and mesenchymal stem cells. Original art courtesy of Dr. Giovanna Casagrande

2   Materials

In vitro culture of Kc requires maintenance of a low known calcium concentration in all media (10) thus the emphasis on procedures that remove Ca from media and serum and on techniques to reliably assess this Ca concentration. Once mesenchymal populations have been separated from epithelial populations, maintaining low calcium concentration in culture media is no longer important for culture maintenance but is still a major requirement for trans-differentiation into the epidermal lineage.

Prepare all culture media and culture reagents with ultrapure water and analytical grade reagents by autoclaving or filter-sterilization (0.2 μm filter). Most of our chemicals are from Sigma and most of our media is from Gibco/Invitrogen unless indicated.

2.1   Cell Isolation and Purification Reagents and Solutions

1. Trypsin: 0.25%, without EDTA.

2. Betadine (Express Medical Supply).

3. Collagenase type I (Worthington): 0.35% (w/v) in Fibroblast (Fb) culture medium. Agitate on rocker for 10 min to dissolve, centrifuge at $2,800 \times g$ for 10 min, filter sterilize, store at 4°C.

4. DNAse I (Worthington): 20,000 U/ml in PBS. Filter sterilize, store frozen in small aliquots.

5. Chelex 100 (BioRAD): Used to remove calcium, see below.

6. Bovine collagen type I (PureCol, Inamed): 3 mg/ml in 0.01 N HCl (see Note 1).

7. Ficoll PM400. For preparations of hair follicles a ficoll gradient is best, we do not cover these protocols here but refer the reader to ref. 11.

2.2   Culture Media and Supplements

1. Fetal bovine serum (FBS), pretested lots for cell viability, treated with chelating resin to remove calcium (see Note 2).

2. Calcium chloride stock solution: 0.25–0.3 M. Filter sterilize, actual calcium concentration can be determined by atomic absorption (see Note 3).

3. Standard medium: To 500-ml bottle minimum essential medium (S-MEM) without calcium, add 5 ml penicillin–streptomycin (1% v/v).

4. Low calcium (LoCa) medium: 0.05 mM Ca^{2+}. Standard medium supplemented with 8% (v/v) chelexed serum.

5. High calcium (HiCa) medium: 1.3 mM Ca^{2+}. Add 0.25 M calcium stock solution to LoCa medium.

6. Fibroblast (Fb) culture medium: DMEM supplemented with 10% (v/v) newborn calf serum plus 5 ml GlutaMAX (1%), 5 ml

sodium pyruvate (1%), and 5 ml penicillin–streptomycin (1%) stock per 500 ml bottle.

7. Antibiotic–antimycotic (Gibco/Invitrogen) (see Note 4): 10%.

2.3 Other Reagents	1. Osteogenic medium: DMEM, 10% FBS, 1% penicillin-streptomycin, nonessential amino acids, 50 µg/ml ascorbic acid, 10 mM ß-glycerol phosphate, and 10 nM dexamethasone.

1. Osteogenic medium: DMEM, 10% FBS, 1% penicillin-streptomycin, nonessential amino acids, 50 µg/ml ascorbic acid, 10 mM ß-glycerol phosphate, and 10 nM dexamethasone.

2. Alizarin Red S: 2% (w/w) adjusted to pH 4.1 using ammonium hydroxide.

3. Adipogenic medium: DMEM, 10% FBS, 1% penicillin–streptomycin, nonessential amino acids, 10% rabbit serum, 20 µM ETYA, 10 nM dexamethasone, and 25 µg/ml insulin.

4. Oil Red O stain: 0.5% in isopropanol, Sigma.

5. Chondrogenic medium: HiCa medium, 50 µg/ml ascorbic acid, 100nM dexamethasone, 1 ng/ml TGF-ß.

6. Safranin O stain: 0.2% prepared in 0.2 M acetic acid.

7. Fast green counterstain: 0.0001%.

8. Myogenic differentiation: adenoviral infection to express GFP (10–30 MOI). See procedures.

9. Keratinocyte growth factor (KGF): 8.4 ng/ml.

10. Ascorbic Acid (40 µg/ml).

11. Paraformaldehyde: 4% in PBS. Electron microscopy Sciences 16% solution ampoules are a convenient source of high quality paraformaldehyde.

12. MTT staining kit (Roche Diagnostics, Indianapolis, IN, USA).

13. Crystal violet Solution: 0.5% in 25% methanol (500 mg Crystal violet, 25 ml methanol, 75 ml water).

2.4 Other Supplies and Equipment

1. Chromatography column packed with washed Chelex-100. Size of column depends on needs; for detailed procedures see ref. 11.

2. Sterile dissection/surgical instruments.

3. Cell strainers 50 µm, BD Falcon.

4. Sterile plastic domes for grafting onto back of nude mice (Cole equipment Co). Should have a 2.5-mm hole on the top for introducing cells with a pipette tip.

5. Sterile 50-ml tubes and 250-ml Erlenmeyer flasks.

6. Insert membranes: Millipore-uncoated (MU) membrane inserts or Costar-coated (CC) membrane inserts. These are 12 mm diameter inserts coated with equimolar mixture of

bovine placenta-derived Type I or Type III collagen. We used inserts of 0.4 or 3-μm pore sizes.

7. Histological punches, sponges, and cassettes (Daigger, Vernon Hills, IL).

8. Poly-L-lysine-coated slides (see Note 5).

9. Wide bore pipette tips.

10. Multi-well plates for tissue culture, various well sizes.

11. Tabletop centrifuge with swing buckets. We use Sorvall Legend RT.

12. Standard tissue culture inverted phase microscope.

13. Zeiss LSM 510 confocal microscope (Carl Zeiss Inc., Thornwood, NY, USA).

14. FACSCalibur flow cytometer (BD Biosciences, San Jose, CA, USA).

15. FACSVantage SE with DiVa option (BD Biosciences).

2.5 Antibodies and Conjugates

1. For one-step staining with fluorescently labeled antibodies: CD49f and CD90 (FITC, PharMingen, San Jose, CA, USA), CD34 and CD45 (PE, PharMingen), CD11b (PE) and CD117 (CyChrome, Caltag, CA, USA).

2. For immunofluorescence: K14-FITC, K10-FITC (Covance) at 1:200 dilution, anti-involucrin (a gift from Dr. Richard Eckhart) at 1:500 dilution, anti-K15 (BD Pharmingen) at 1:100 dilution, anti-MRP8 (a gift from Dr. Joannes Roth) at 1:200 dilution, and αpep8/DCT (a gift from Dr. Vincent Hearing) at 1:500 dilution. Anti-smooth muscle actin (Biocare Medical, Concord, CA, USA) at 1:300 dilution and anti-desmin (DAKO, Carpenteria, CA, USA) at 1:500 dilution.

3 Methods

3.1 Steps for Obtaining Dermis and Epidermis from Newborn Mice (11)

1. Place neonatal BalbC mice (see Note 6) into culture dish without overcrowding and transfer to a CO_2 euthanasia chamber for 20 min.

2. Immerse mice in betadine for 2 min two times and rinse with deionized water.

3. Wash mice twice in 70% ethanol for 2 min and then place dishes on ice. Keep dishes on ice for remaining steps.

4. Working on the lid of the culture dish, remove forelimbs and hind limbs above wrist and ankle joints, and remove the tail close to the body.

5. Cut skin along dorsal midline starting at tail incision and ending at the end of nose.

6. Using a pair of forceps loosen skin at both sides of dorsal midline cut.

7. Pull skin over hind limb stumps gathering skin on ventral side and pulling it towards head.

8. Flatten skin dorsal side down on culture surface of the culture dish. Stretching the skin on the dry surface of a culture dish is very important to make the floating of skins on the surface of a trypsin solution easier. Stretch eight to ten skins on one dish; the dish can then be refrigerated for up to 2 h before floating the skins on trypsin (see Note 7).

9. In a sterile hood, set up the following: sterile forceps, freshly thawed, cold trypsin (see Note 8) (0.25% without EDTA), and new culture dishes. Pour 50 ml trypsin into 150-mm dish. The stretched skins will now be set to float on the trypsin.

10. Grasp opposite ends of one side of the stretched skin with forceps, lift it, and transfer it gently, dermis side down, to the trypsin containing dish. Typically the skin will float perfectly flat. If not, use the rough end of the forceps to reach under the skin and gently uncurl edges while steadying the skin with the other forceps. All ten skins from the stretching dish can be floated in one 150-mm dish overnight at 4°C.

11. The following day transfer floating skins to the dry side of a tissue culture dish, epidermis side down.

12. Lift dermis, which should come off easily while holding epidermis down with sterile forceps and transfer dermises to a sterile tube containing HiCa medium to rinse.

13. Fold edges of epidermis towards the center to protect epidermal cells from becoming dislodged and transfer to a glass dish containing HiCa medium.

14. Mince epidermises finely with sterile scissors enough so that the resulting fragments can enter a 10-ml pipette tip (see Note 9).

15. Pipet up and down the suspension at least ten times, then transfer to a 50-ml conical tube (most of the stratum corneum sheets are left behind). Rinse dish with HiCa medium and transfer to the same 50-ml conical tube.

16. Centrifuge cell suspensions at $150 \times g$ for 5 min at 4°C. It is advisable to visually inspect cell or media aliquots through a phase contrast microscope at every step of the preparation to ensure desired cells are not discarded and estimate cell yields. We consider one mouse equivalent (ME) of a particular cell population the typical yield obtained from one skin.

17. Aspirate supernatant. This should include fragments of stratum corneum sheets that could reduce cell yields in the next filtration step.

18. Resuspend pellet in HiCa medium and filter through 100-μm cell strainer into a 50-ml conical tube. Rinse cell strainer with HiCa (see Note 10).

19. Centrifuge the cells again at $150 \times g$ for 5 min at 4°C.

20. Aspirate supernatant and resuspend pellet in HiCa medium to 1 ME/ml of primary keratinocytes.

21. Count an aliquot with a cell counter. Typical primary Kc yield will be about 8–10 million cells/ml (1 ME) (see Note 11). Trypan blue-excluding cells can be calculated using a hemocytometer. Cell suspensions can be kept at 4°C for several hours before plating.

22. Plate the desired number of cells in culture vessel. We typically use 1 ME per 100-mm dish, scale appropriately bearing in mind that plating efficiency depends on the age and strain of the mouse and should be experimentally determined (see Note 12). To plate, dilute suspension in HiCa medium with seven volumes of LoCa (0.05 mM Ca). The resulting medium concentration will be about 0.2 mM Ca-containing medium. Cells should be at about 1/8 ME/ml. This calcium concentration favors overnight adhesion of cells and is not high enough to promote early differentiation.

23. Move cells to culture incubator; next day medium should be changed to LoCa.

24. Feed cells every other day thereafter. Cultures will be confluent and ready for desired application by day 3 or 4. We do not recommend using primary keratinocyte cultures older than 7–10 days when most Kc should be already senescent.

3.2 Procedures for Isolation of Dermal Fractions (Fig. 2) (3)

1. Chop dermises coarsely and add 2 ml of 0.35% collagenase solution per dermis.

2. Transfer up to ten dermises into a 250-ml sterile erlenmeyer flask. Incubate with agitation in a water bath at 37°C for 30 min.

3. Add DNase I at 12.5 μl per dermis and incubate for 10 min (see Note 13).

4. Filter the suspension through a 100 μm Nytex filter.

5. Dilute filtered suspension with 100 ml Fb medium per five dermises.

6. Centrifuge at 800 rpm ($140 \times g$) for 5 min.

7. Collect the supernatant from 800 rpm ($140 \times g$) spin into new tubes and centrifuge at 1,400 rpm ($425 \times g$) for 5 min. Pool the pellets and designate as Fraction D.

8. Dissociate the 800 rpm ($140 \times g$) spin pellet and centrifuge at 300 rpm ($20 \times g$) for 5 min yielding a pellet of dermal hair follicles (DHF).

ISOLATION OF EPIDERMAL AND DERMAL FRACTIONS

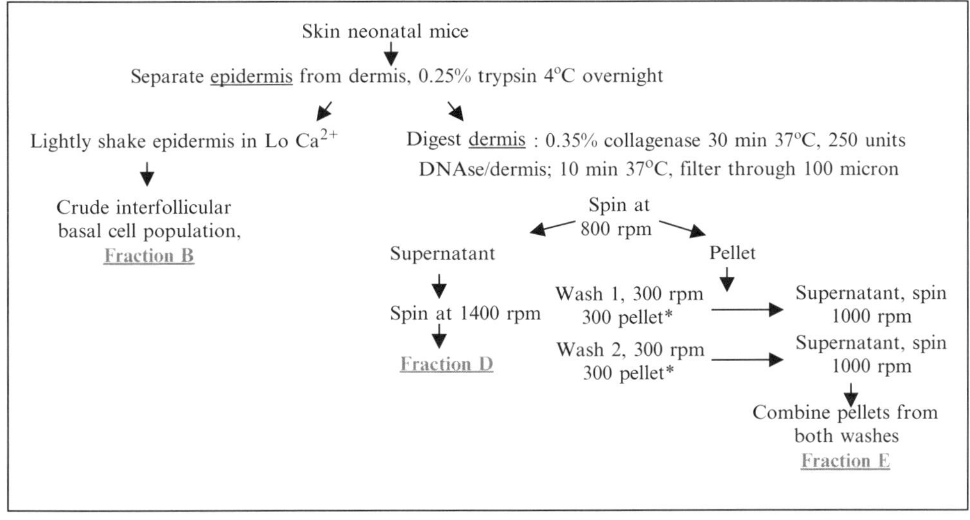

Fig. 2 Simplified schematic of fraction preparation (for more detailed schematic, see Supplementary Fig. 1 in ref. 3). *Asterisk* indicates hair follicle containing fractions

9. Centrifuge the 300 rpm $(20 \times g)$ supernatant twice at 1,000 rpm $(215 \times g)$. Pool both pellets and designate as Fraction E. This fraction contains the mesenchymal progenitor cells (3).

3.3 Organotypic (3D) Culture (see Note 14)

1. Cells adhere overnight to upper side of insert membrane while submerged in 1.4 mM calcium-containing media.

2. To allow expansion of cell population, change medium frequently without disturbing cell layer. For epidermal cell expansion, change the medium to 0.05 mM calcium-containing medium and keep submerged for 72 h. Keratinocyte growth factor (KGF) may be added (8.4 ng/ml).

3. For epidermal differentiation air-expose inserts containing the expanded subpopulations on the upper side of a membrane for 72 h while placed in contact with a submerged, serum-rinsed dermis on the underside of the membrane. This dermis is submerged in medium containing 0.25 mM calcium and 40 μg/ml ascorbic acid to facilitate epidermal differentiation.

4. Cultures are fixed in 3% formalin for 5 min (on day 6) and moved to 70% ethanol (see Note 15). For histological processing we use histological punches to cut desired size punches from membrane and place them between two histology sponges within a histological cassette. This keeps the fragile samples together before paraffin embedding. Sections are H&E stained to analyze their viability and morphology. Unstained sections can be immunostained with various markers.

3.4 Cell Viability Assays to Demonstrate Ability to Survive Minimal Media Conditions

1. Place fractionated cells are in monolayer culture using media containing different levels of calcium (1.4 mM, 0.25 mM or 0.05 mM calcium) and serum (8% or 2%) fetal bovine serum (see Note 16).

2. After approximately 4 weeks, perform MTT (see Note 17) analysis is to locate any living cells. The MTT assay stains active mitochondria thus measuring metabolic activity in viable cells. Cells are incubated with the MTT solution for approximately 4 h. After this incubation period, a water-insoluble formazan dye is formed. At this point, the cultures can be photographed. If wells larger than 96 are used, scale up the amounts of reagents. After solubilization, the formazan dye is quantitated using a microplate reader at a wavelength between 500 and 600 nm. Absorbance directly correlates to the number of viable cells. If the assay is done in a larger well, then triplicate aliquots of 100 μl of solubilized dye can be transferred to 96-well plate for quantitation.

3. Alternatively, nuclear crystal violet stain (0.5% crystal violet solution in 25% methanol) can be used to demonstrate viability. For this assay, fix cells for 10 min with ice-cold 100% methanol. Remove methanol and cover cells with 0.5% crystal violet solution in 25% methanol. Incubate for 10 min. Remove crystal violet. Wash the cells in water several times, until the dye stops coming off. Allow the cells to dry at room temperature (see Note 18). Dye can be eluted from stained cells by 5-min incubation with solubilization buffer (a 50/50 mixture of 0.1 M NaH_2PO_4, pH 4.5 and 50% EtOH). Absorbance of triplicate aliquots can be read at 540–570 nm on a microplate reader and correlates with viability as compared to control cells.

3.5 Immunofluorescence Staining in Suspension

We developed a method to preserve isolated cell populations for future characterization. This is a variation of the chondrogenic assay described in Subheading 3.9.

1. Coat slides with poly L-Lysine (Sigma, prepare solution per manufacturers instructions). These slides can be stored in a clean coplin jar at room temperature for a few weeks.

2. Make one or more circles with Pap-Pen in the places where cells will be deposited.

3. Pellet isolated cells and remove most of the medium.

4. Having the slide sitting on dry ice, use a wide bore pipette tip to carefully deposit the cell pellet in a bead within the Pap-Pen circle. The slides can then be stored at −80°C inside flat slide holders for future use.

5. When ready to perform immunofluorescence, thaw quickly. While removing liquid with a tissue tip add similar volume of the appropriate fixative in small microliter-size aliquots. Cells are now fixed and ready to be stained.

3.6 Culturing Cells on Coverslips for Immunofluorescence Staining

1. Place clean coverslips in 75 or 100-mm tissue culture dishes (2 or 3 coverslips per dish), soak in ethanol in the hood, rinse with PBS, and air-dry.

2. Plate fresh isolated cells on these dishes containing coverslips. Depending on cell type they may adhere to their own matrix to be kept in culture for 2–3 days (see Note 19).

3. To harvest cells, carefully remove coverslips using sterile forceps and transfer to 24-well dishes for staining. Cells remaining in the tissue culture dish can then be harvested for immunoblotting or other end points.

4. All immunofluorescence procedures are done using PBS containing Mg^{2+} and Ca^{2+}, denoted PBS+. Cells are washed three times with PBS+ and fixed with 4% paraformaldehyde for 10 min at 4°C (see Note 20). At this point cells can be stored at 4°C in PBS+ overnight.

5. Optional, permeabilize with 0.1% Triton-X or 0.05% saponin for 5 min followed by washing three times with PBS+ for 3 min.

6. Block cells using 5% BSA in PBS+ (see Note 21) for 1 h at 4°C and then wash three times in PBS+ for 3 min.

7. Add primary antibodies diluted with 5% BSA in PBS+. Negative controls can be normal serum from the same species as the primary antibody or no primary antibody. Coverslips are incubated overnight at 4°C in a humidified chamber.

8. Next day cells are washed thrice with PBS+ for 30 min. Secondary antibodies (for example Alexa-488 and Alexa-594 conjugates, molecular probes) are diluted generally between 1:300 and 1:500 in 5% NGS in PBS or in 5% BSA and incubated for 2 h at room temperature. Coverslips are then washed three times for 5 min.

9. To mount coverslips for fluorescent viewing, lift each coverslip with fine jewelers forceps, touch edge to kimwipe to remove extra fluid, which would increase autofluorescence. Add one drop of VectaShield medium for fluorescence (with DAPI) and place coverslip, cells side down onto slide. After coverslips dry, secure coverslip onto the slide with a few dots of nail polish. Slides may be stored in a humidified, dark chamber at 4°C for 3–4 days before imaging (see Note 22).

10. Fluorescent cells can be examined with suitable fluorescence microscope. We have used a Zeiss LSM 510 confocal microscope (Carl Zeiss Inc., Thornwood, NY, USA) with a 40×1.3 NA Plan Neofluar objective and collected images sequentially. FITC, TX-RED, and DAPI signals are collected with a BP 505–550 filter, LP560 filter, and BP 385–470 filter after excitation with 488 nm, 543 nm, and 364 nm laser lines, respectively.

3.7 One-Step Staining for Flow Cytometry with Fluorescent Antibodies (see Note 23)

1. Transfer cells to a conical tube add staining medium and centrifuge at $300 \times g$, 4°C for 5 min. Discard the supernatant.

2. Add 5 µL of diluted primary antibody conjugated to a fluorescent tag to the cell pellet.

3. Flick the tube to resuspend the cell pellet. Mix well and incubate on ice for 25–30 min.

4. Discard the supernatant and resuspend the cells with 0.5 ml of staining medium.

5. Filter the cell suspension into 12 × 75 mm tubes before analysis or sorting the cells by flow cytometry (see Note 24). We have used a FACSCalibur flow cytometer (BD Biosciences, San Jose, CA, USA)

3.8 Staining for Side Population Phenotype

1. Incubate cells (1×10^6 cells/ml in DMEM with 2% FBS) with 3 µg/ml Hoechst 33342 (Molecular probes, Eugene, OR, USA) for 90 min (37°C, 5% CO_2, 100% relative humidity).

2. Wash cells once in cold DMEM with 2% FBS and resuspended for sorting.

3. Sort cells. We have used a FACSVantage SE with DiVa option (BD Biosciences). Compare SP percents from the unfractionated dermis and individual dermal fractions.

3.9 Differentiation Assays

Cells are plated at or near confluence therefore adjust the culture surface to cell number (depending on the number of starting cells 24-, 12-, or 6-well plates can be used, except in the case of chondrogenesis as detailed below). Use HiCa or DMEM medium for plating.

1. Osteogenic differentiation. Plate cells in osteogenic differentiation medium and keep up to 3 weeks with weekly medium changes. The osteogenic assay verifies mineralization ability. Cells are stained with Alizarin Red S for 5 min at room temperature followed by a wash with deionized water. Osteoblasts are shown by the formation of calcium-rich hydroxyapatite in the extracellular space, which stains orange red with Alizarin Red S.

2. Chondrogenic differentiation. We have used one well in a round-bottom 96-well dish (see Note 25). Plate five million cells per well and treat with chondrogenic differentiation medium for 3 weeks. To change medium, carefully pipet away from pellet and replenish with fresh medium. Collect cell clumps with a wide bore pipette tip and spread onto poly-L-lysine-coated slides (see Note 26). Flash-freeze by placing the slide on dry ice. Cells may now be stored at −80°C if not used immediately. When needed, thaw partially then add very small volumes of 4% paraformaldehyde while carefully removing the

thawed fluid with a piece of blotting paper and replacing with more fixative. Carefully rinse once with PBS following the same procedure. At this point cell clumps are adhered and fixed onto the slide. Stain with 0.2% Safranin O for 1 h followed by a rinse with water for 10 min. Counterstain with 0.001% Fast Green solution for 10 min or alternatively with 0.5% Light Green (Polysciences). Dehydrate and clear with 95% ethanol, absolute ethanol, and xylene (two changes each), and mount with permount. With this stain, cartilage, mucin, and mast cell granules stain orange to red (see Note 27).

3. Adipocyte differentiation. Treat confluent cultures with complete adipogenic differentiation medium for up to 3 weeks. Subsequent feedings lack dexamethasone and have an increasing amount of rabbit serum concentration up to 15%. The adipocyte differentiation assay relies on visualization of fat droplets within adipocytes. Cells are fixed with 10% formalin, followed by 60% isopropanol, then stained with Oil Red O stain: (0.5% in isopropanol, Sigma).

4. Myocyte differentiation. Myocytes form on confluent cultures either on HiCa or DMEM media during a period of up to 3 weeks with limited medium changes. Myocytes can be appreciated by their elongated morphology. When plated onto membranes cells can be observed to "beat" under a low magnification scope. In order to demonstrate contractile behavior, these authors infected cultures with a GFP expressing adenovirus in serum-free medium containing 2.5 µg/ml polybrene at 10–30 multiplicity of infection for 30 min at room temperature (see Note 28). Following infection, add fresh complete medium. Observe cells and photograph. Myocytes are especially prominent when live cultures are plated on a membrane. A movie can be acquired using fluorescence microscopy.

5. Epidermal differentiation, see organotypic (3D) cultures, Subheading 3.3.

3.10 Engraftment of Isolated Cells

We have used two different ways of engraftment of the isolated cells onto immunodeficient mice: subcutaneous injection or depositing the cells under a silicon dome covering the graft bed, as typical epidermal grafts, as described in ref. 11.

1. Cells need to be counted and aliquoted in the different groups. Typically we use cells from two mouse equivalent (ME) for each experiment using four replicate mice for each cellular population tested (see Note 29). To inject subcutaneously, use a short hypodermic needle, pinch the back skin creating a "bubble" between shoulders and aim the needle at this space. Two weeks post injection, animals are euthanized according to institutional guidelines and the tissues processed for H & E staining.

2. For epidermal grafts, a skilled technician performs a circular wound on the back of the animal under anesthesia. Silicon domes are inserted under folds of the wound and cells from 4 ME are pipeted into the dome through an orifice on top using two replicate mice for each population tested. Five days post-grafting, the silicon domes are removed to facilitate wound healing. Animals are sacrificed and tissues taken at the desired times, after the wound is healed, typically at 2 weeks post-grafting.

4   Notes

1. Fibronectin or gelatin can be used as coating matrices for tissue culture plates where primary Kc are maintained. We successfully use a mixture of fibronectin/collagen prepared as follows: to 200 ml culture medium, add 2 mg fibronectin, 1 ml 2% BSA, 4 ml 1 M HEPES, and 2 ml of this collagen I solution. Filter-sterilize and keep at 4°C. To coat plates, add a few ml of this solution to each plate while in sterile hood. Assure even coating and remove extra liquid. Allow drying in the hood. Dishes are now ready for plating.

2. Serum chelexing to remove calcium has been described in minute detail (11). Chellex-100 resin should be well washed and packed in a radial flow column (we used 500-ml column). If FBS contains approximately 3 mM calcium, it should be efficiently decreased to about 0.01 mM when serum is put through this column at a flow rate of 1,500–2,000 ml/h. We mix five volumes of chellexed serum with one volume of unchellexed serum to adjust the calcium concentration of the serum so that other cations removed by chellexing are partially restored. LoCa serum made with 8% (v/v) of this serum mixture will have a Ca^{2+} concentration close to 0.05 mM. Finally, sterilize this serum mixture, aliquot, and store at –20°C.

3. If atomic absorption is not available, new batches of chellexed serum can be empirically tested by assuming Ca^{2+} concentration is 0.04 mM, then increasing it to 0.2 and 0.6 for example. Plate a small aliquot of primary Kc in those conditions and look at their morphology 1 and 2 days post-plating. For comparison, see pictures of culture morphology in ref. 11. It is important to prepare a large enough batch of chellexed serum to assure experimental reproducibility.

4. If cultures are older than a week, they become susceptible to fungal contamination. In some cases we use antifungal to prevent loss of precious cultures.

5. You can coat your own slides with a solution from Sigma, P-8920, preparing 10× stock solution according to manufacturer's

instructions. Make a 1× solution with water and place it in a coplin jar. Submerge slides for about 30 min, remove with forceps, and air dry. You may store these slides in a clean container for further use.

6. All procedures involving live rodents must conform to National and Institutional animal care and research regulations, in the case of the National Cancer institute at NIH, these guidelines are found at http://web.ncifcrf.gov/rtp/lasp/intra/acuc/default.asp.

7. This is done so epidermis and dermis separate easily the following day.

8. The shelf life of freshly thawed trypsin is limited. Pretesting new lots is advised.

9. In order to obtain most epidermal Kc containing differentiated and undifferentiated cells proceed through steps 15–17. If only undifferentiated Kc are sufficient, follow simplified process in Fig. 2 to obtain fraction B.

10. One filter can process approx 30 epidermises (ME) before it becomes clogged and should be discarded.

11. The cell suspensions will include single cells and small clusters of immature hair follicle buds.

12. Plating efficiencies compared to their parental strain can be reported as part of the initial characterization of new genetically engineered mouse models.

13. DNase I decreases cross-contamination of epidermal cells and HFs within the isolated dermal fraction. These contaminating cells are undetectable beyond passage 3 in high calcium-containing medium and beyond passage 2 if DNase I is included in this step.

14. The conditions for organotypic culture must be optimized to the purpose of each experiment. In our experience, growing cells in membranes allows them to differentiate more readily in response to external cues such as media component or differentiation stimuli. For example, some cells can be observed to "beat" when plated on the membranes. Epidermal differentiation could be induced on membranes by combining increased Ca with Ascorbic acid and air-exposure.

15. Samples can be stored at this point for future processing.

16. We prefer to do this assay in larger than 96 wells because we observe better cell viability in a larger culture surface.

17. We prefer the traditional MTT assay to the one-step XTT because, in the MTT assay, formazan precipitates can be visualized and photographed allowing cells with viable mitochondria to be clearly identified. This is especially important when one wants to demonstrate survival potential of a few remaining cells.

18. Dried cells can be stored for months at room temperature.

19. Alternatively coverslips can be first coated with various matrices and cells plated for 1 day.

20. In some cases a mixture of 50/50 methanol acetone is preferred; this depends on the primary antibody used and needs to be determined beforehand.

21. Other common blocking solution is 10% NGS in PBS+ but be mindful of the animal species of your antibodies.

22. We find we can keep slides inside their flat cardboard holders in a refrigerator for long time without drastic loss of fluorescence.

23. Typically we follow protocols for surface marker analysis, unless fixing and permeabilization are needed to study cytosolic or nuclear proteins.

24. For negative controls, prepare cells that have not been stained with antibody and cells stained with an isotype control.

25. Use a low binding plastic well such as those used for BCA or similar assays, not tissue culture treated, as cells should grow in bulk and not adhere to cultureware.

26. You may make a circle with pap-pen in the area where cells will be placed in order to find them better during staining.

27. For a positive control, we used mouse skin treated with TPA (12-0-tetradecanoylphorbol-13-acetate) to induce intraepidermal inflammation, fixed with ethanol and stained in the same way.

28. Alternative methods to introduce GFP in cultures may be used.

29. We have used cells isolated from C57BL/6J mice injected onto athymic mice. Dark melanocytes contained in the isolated cells are easily appreciated in the graft when stained with Fontana-Masson.

Acknowledgments

This work was supported by the Intramural Research Program, Center for Cancer Research, National Cancer Institute, National Institutes of Health.

The authors have no conflicting financial interests to declare.

References

1. Weinberg WC, Goodman LV, George C, Morgan DL, Ledbetter S, Yuspa SH, Lichti U (1993) Reconstitution of hair follicle development in vivo: determination of follicle formation, hair growth, and hair quality by dermal cells. J Invest Dermatol 100(3):229–236

2. Lichti U, Weinberg WC, Goodman L, Ledbetter S, Dooley T, Morgan D, Yuspa SH

(1993) In vivo regulation of murine hair growth: insights from grafting defined cell populations onto nude mice. J Invest Dermatol 101(1 Suppl):124S–129S

3. Crigler L, Kazhanie A, Yoon TJ, Zakhari J, Anders J, Taylor B, Virador VM (2007) Isolation of a mesenchymal cell population from murine dermis that contains progenitors of multiple cell lineages. FASEB J 21(9):2050–2063. doi:fj.06-5880com [pii], 10.1096/fj.06-5880com

4. Dlugosz AA, Cheng C, Williams EK, Dharia AG, Denning MF, Yuspa SH (1994) Alterations in murine keratinocyte differentiation induced by activated rasHa genes are mediated by protein kinase C-alpha. Cancer Res 54(24):6413–6420

5. Dlugosz AA, Yuspa SH (1994) Protein kinase C regulates keratinocyte transglutaminase (TGK) gene expression in cultured primary mouse epidermal keratinocytes induced to terminally differentiate by calcium. J Invest Dermatol 102(4):409–414

6. Hennings H, Lowry DT, Robinson VA, Morgan DL, Fujiki H, Yuspa SH (1992) Activity of diverse tumor promoters in a keratinocyte co-culture model of initiated epidermis. Carcinogenesis 13(11):2145–2151

7. Li L, Tennenbaum T, Yuspa SH (1996) Suspension-induced murine keratinocyte differentiation is mediated by calcium. J Invest Dermatol 106(2):254–260

8. Tennenbaum T, Li L, Belanger AJ, De Luca LM, Yuspa SH (1996) Selective changes in laminin adhesion and alpha 6 beta 4 integrin regulation are associated with the initial steps in keratinocyte maturation. Cell Growth Differ 7(5):615–628

9. Al-Batran SE, Astner ST, Supthut M, Gamarra F, Brueckner K, Welsch U, Knuechel R, Huber RM (1999) Three-dimensional in vitro cocultivation of lung carcinoma cells with human bronchial organ culture as a model for bronchial carcinoma. Am J Respir Cell Mol Biol 21(2):200–208

10. Fuchs E (1990) Epidermal differentiation: the bare essentials. J Cell Biol 111(6 Pt 2):2807–2814

11. Lichti U, Anders J, Yuspa SH (2008) Isolation and short-term culture of primary keratinocytes, hair follicle populations and dermal cells from newborn mice and keratinocytes from adult mice for in vitro analysis and for grafting to immunodeficient mice. Nat Protoc 3(5):799–810. doi:nprot.2008.50 [pii], 10.1038/nprot.2008.50

Chapter 18

Isolation, Characterization, and Differentiation of Human Multipotent Dermal Stem Cells

Ling Li, Mizuho Fukunaga-Kalabis, and Meenhard Herlyn

Abstract

Skin, as the body's largest organ, has been extensively used to study adult stem cells. Most previous skin-related studies have focused on stem cells isolated from hair follicles and from keratinocytes. Here we present a protocol to isolate multipotent neural crest stem-like dermis-derived stem cells (termed dermal stem cells or DSCs) from human neonatal foreskins. DSCs grow like neural spheres in human embryonic stem cell medium and gain the ability to self-renew and differentiate into several cell lineages including melanocytes, neuronal cells, Schwann cells, smooth muscle cells, adipocytes, and chondrocytes. These cells express neural crest stem cell markers (NGFRp75 and nestin) as well as an embryonic stem cell marker (OCT4).

Key words Skin, Stem cells, 3-D skin reconstruct, Cell differentiation

1   Introduction

Adult stem cells are well known for their potential therapeutic value and their accessibility from patient-derived samples to avoid ethical problems. Skin contains somatic stem cells that generate keratinocyte, melanocyte, and mesenchymal cell lineages. These somatic stem cells have traditionally been thought to be restricted in their differentiation and regeneration potential to the tissues in which they reside (1). The advantage of skin stem cell research is that skin is a readily accessible organ from which to obtain a biopsy. Hair follicle stem cells and keratinocyte stem cells from human epidermis have been well characterized. Human induced pluripotent stem (iPS) cells can be generated from human dermal fibroblasts following upregulation of different transcription factors, including Oct3/4, Sox2, c-Myc, Klf4, Lin28, and Nanog, using retroviral and/or lentiviral vectors (2–4). However, some issues of iPS cells have been recently observed such as low efficiency of the reprogramming process, high number of retroviral insertions, and tumorigenesis by reactivation of the c-Myc transgene (5).

Kursad Turksen (ed.), *Skin Stem Cells: Methods and Protocols*, Methods in Molecular Biology, vol. 989, DOI 10.1007/978-1-62703-330-5_18, © Springer Science+Business Media New York 2013

The dermis is a source of progenitor cells and stem cells for multiple lineages. The dermal papilla from whisker follicles has been reported to be of neural crest origin and to harbor skin-derived precursor cells (SKP) (6). We have developed a protocol to isolate dermal stem cells (DSCs) from human neonatal foreskins. These DSCs grow like three-dimensional (3-D) spheres when cultured in human embryonic stem cell medium HESCM4. DSCs differ from SKPs that are also derived from human foreskins (7). The neural crest marker NGFRp75, which has been used as a marker for isolating a pure or enriched population of neural crest stem cells (8), was expressed only at low or undetectable levels in SKPs, whereas it is highly expressed in DSCs. DSCs also express another neural crest marker nestin and stem cell marker OCT4 (9). NGFRp75 and OCT4 double-positive cells can be found in human foreskin dermis by immunofluorescence staining. DSCs show a self-renewal ability and can be differentiated into melanocytes, neuronal cells, smooth muscle cells, adipocytes, chondrocytes, and Schwann cells. We also demonstrated that DSCs can migrate to the epidermis, localize at the basement membrane, and become melanocytic marker positive cells when incorporated in 3-D skin reconstructs cultured together with dermal fibroblasts and epidermal keratinocytes. These data suggest that DSCs represent a potential reservoir for epidermal melanocytes in human skin (10, 11). We describe here the protocol for the isolation and differentiation of DSCs. This technique does not require the introduction of genetic materials, is robust and easily reproducible.

2 Materials

2.1 Medium and Solution Preparation

1. *Foreskin transporting medium*: Dulbecco's modification of Eagle's medium (DMEM; Cellgro #10-017-CM) supplemented with gentamycin (100 µg/mL; Cellgro #30-005-CR). After sterilization through a 0.2 µm filter, the medium is transferred into sterile containers in 20 mL aliquots and stored at 4°C for up to 1 month.

2. *Dispase solution (0.48%)*: Dispase (grade II, 0.5 U/mg; Boehringer Mannheim #165859) 0.48 g is dissolved in 100 mL phosphate-buffered saline (PBS) without Ca^{2+} and Mg^{2+} (Cellgro #MT21-031-CM). Sterilize the enzyme solution through a 0.2 µm filter, aliquot into 5 mL/tubes, and store at –20°C for up to 3 months.

3. *Collagenase solution (1 mg/mL)*: Collagenase type IV (Invitrogen #17104-019) 100 mg is dissolved in 100 mL DMEM to yield a final concentration of 1 mg/mL. Sterilize the enzyme solution through a 0.2 µm filter, aliquot into 5 mL/tubes, and store at –20°C for up to 3 months.

4. *Mouse embryonic fibroblast (MEF) derivation medium*: The medium contains 87% DMEM (Invitrogen #11965-092), 10% defined FBS (Invitrogen #16000-044; heat inactivate for 30 min at 57°C), 1% 200 mM L-glutamine (Invitrogen #21051-024), 1% nonessential amino acids 100× (Invitrogen #11140) and 1% penicillin-streptomycin 100×.

5. *MEF growth medium*: MEF derivation medium without penicillin–streptomycin.

6. *Human embryonic stem cell medium* (HES): 78% DMEM/F-12 (Invitrogen #11330-032), 20% Knockout-Serum Replacer (Invitrogen #10828-028), 1% 100 mM L-glutamine + β-mercaptoethanol (Invitrogen #21051-024—add 7 µL β-mercaptoethanol to 10 mL L-glutamine), 1% nonessential amino acids 100× (Invitrogen #11140), 4 ng/mL basic fibroblast growth factor (bFGF; Fitzgerald Industries #30R-AF015).

7. *Human embryonic stem cell medium 4* (HESCM4): 70% MEF-conditioned HES medium and 30% HES medium, sterilize through a 0.2 µm filter.

8. *L-Wnt3a cell growth medium and conditioning medium*: 90% DMEM (Cellgro #10-017-CM), add 10% FBS and 0.4 µg/mL G418 (Sigma #G-8168). Conditioning medium including 99% DMEM and 1% FBS.

9. *Differentiation medium*:

    (a) For melanocyte differentiation (100 mL): 50 mL Wnt3a conditioned medium, 30 mL DMEM-Low Glucose (Invitrogen #11885), 20 mL MCDB201 (Sigma #M6770), 1× ITS Liquid Medium Supplement (Sigma #I-3146), 1 mg/mL linoleic acid-BSA (LA-BSA; Sigma #L-9530), 10^{-4} M L-ascorbic acid (Sigma #A-4403), 100 ng/mL stem cell factor (SCF; Fitzgerald Industries, #RDI-307-255X), 0.05 µM dexamethasone (Sigma #D-2915), 20 pM Cholera toxin (Sigma #C-3012), 50 nM TPA (Sigma #P-1583), 4 ng/mL bFGF (Fitzgerald Industries #30R-AF015), 100 nM endothelin-3 (ET-3; American Peptide Co. #88-5-10).

    (b) For neuronal cell differentiation (100 mL): (1) 60 mL DMEM, 30 mL F12 (GIBCO #11765), 10 mL FBS, 40 ng/mL bFGF. (2) 60 mL DMEM, 30 mL F12, 10 mL FBS, 10 ng/mL nerve growth factor (Millipore #GF028), 10 ng/mL brain-derived neurotrophic factor (Peprotech #450-02-10), 10 ng/mL NT-3 (Stem Cell Technologies #02508).

    (c) For smooth muscle cell differentiation (100 mL): 90 mL DMEM-F12 (GIBCO m#11330), 10 mL FBS, 0.1 M nonessential amino acids solution and 60 pM transforming growth factor-β1 (TGF-β1; R&D Systems #240B).

(d) For adipocyte differentiation (100 mL): 90 mL low-glucose DMEM (Sigma #D6046), 10 mL horse serum (Invitrogen #26050-070), 1× ITS, 1 mg/mL LA-BSA, 1 µM hydrocortisone (Sigma #H4001), 60 µM indomethacin (Sigma #I7378), 0.5 mM isobutylmethylxanthine (Sigma #I5879).

(e) For chondrocyte differentiation (100 mL): 90 mL high-glucose DMEM, 10 mL FBS, 1× ITS, 1 mg/mL LA-BSA, 50 nM dexamethasone, 60 pM TGF-β1.

(f) For Schwann cell differentiation (100 mL): (1) 60 mL DMEM, 30 mL F12 (GIBCO #11765), 10 mL FBS. (2) Medium I plus 4 µM foskolin (Sigma #F3917).

(g) Skin reconstruct medium: Basic medium: 400 mL Keratinocyte-SFM (Invitrogen #10724), 2% dialyzed FCS (Gibco LTI #16440-034), 60 ng/mL bovine pituitary extract (BPE; Invitrogen #13028-014), 4.5 ng/mL bFGF, 100 nM ET3 (American Peptide Co. #88-5-10), 10 µg/mL SCF (Fitzgerald Industries #RDI-307-255X). (1) Add 1 ng/mL EGF (Invitrogen #10450-013) to 100 mL basic medium; (2) Add 0.2 ng/mL EGF (Invitrogen #10450-013) to 100 mL basic medium; (3) Add 2.4 mM $CaCl_2$ (Sigma; #C-7902) to 200 mL basic medium.

2.2 Additional Reagents and Equipment

0.25% trypsin/EDTA (Cellgro #25-053-CI).

Minimal essential medium with Eagle salts (10× EMEM) (Lonza #12-684F).

Fetal bovine serum (FBS) (Hyclone #SH-30071.02).

Sodium bicarbonate (Cambrex #17-613E).

Bovine tendon acid-extracted collagen I (Organogenesis #200/50).

10% buffered formalin (Surgipath #00600).

Forceps (Roboz, #RS5070).

Scissors (Roboz, #RS5840).

Iris scissors (Roboz, #RS5913).

Surgical blades (Feather, #2976).

Cell strainers 100, 70, and 40 µm (Becton Dickinson #08-771-19; #08-771-2; #08-771-1).

Shaker (Taitec, microincubator M-36).

CO_2 incubator (CO_2 at 5% (vol/vol); humidified, $T = 37°C$).

Cytospin 2 (Shandon).

Microscope slides (Fisher Scientific #12-550-15).

4-well chamber slides (Fisher #12-565-21).

Tissue culture 6-well trays with transwell (Organogenesis #9285).

Multi cassettes (Surgipath, #02293-BX).

TBS biopsy papers (Triangle Biomedical Sciences, #BP-B).

2.3 Animals, Cells, and Sources

1. CF-1 female mice (day 13–14 gestation) for MEF derivation (Jackson Laboratory).

2. L-Wnt-3A cells (ATCC, #CRL-2647).

3. Human fibroblasts (from neonatal human foreskins).

4. Human keratinocytes (from neonatal human foreskins).

3   Methods

3.1 Generation of Conditioned HES Medium from Mouse Embryonic Fibroblasts

1. Sacrifice a female CF-1 mouse (at day 13 or day 14 of pregnancy).

2. Sterilize the abdomen with 70% ethanol.

3. Pull up the skin using forceps, separate the hide from the peritoneum, and cut a nick in the skin with scissors.

4. Cut the peritoneum to expose the abdominal cavity. Separate the uterine horns from the abdominal cavity with a blunt-point forceps and scissors. Transfer the uterine horns to a 100 mm cell culture dish that contains 10 mL DPBS without Ca^{2+} and Mg^{2+}.

5. Wash the uterine horns with 10 mL DPBS without Ca^{2+} and Mg^{2+} three times.

6. Tease open the uterine walls using two fine-pointed forceps or a scissors to release the embryos into the culture dish.

7. Separate each embryo from its placenta and surrounding membranes.

8. Transfer the embryos to a new culture dish using forceps one by one and wash them three times using 10 mL DPBS without Ca^{2+} and Mg^{2+}.

9. Cut away brain and dark red organs from each embryo using a fine-tipped forceps, individually (see Note 1).

10. Wash the embryos three times with 10 mL DPBS without Ca^{2+} and Mg^{2+}, change to a new dish every time.

11. Remove the DPBS and add 2 mL 0.25% trypsin/EDTA solution to the washed embryos.

12. Mince each embryo finely using curved Iris scissors; add 5 mL 0.25% trypsin/EDTA solution.

13. Leave the dish on a shaker at 37°C for 20–30 min until individual cells are visible under the microscope.

14. Add 20 mL MEF derivation media to the plate after the incubation. Transfer the cells with media to a 50 mL tube.

15. Add 3 mL to rinse the remaining tissue in the plate. Transfer to the 50 mL tube. Mix by pipetting a few times and wait for 1 min to allow the debris in the suspension to settle to the bottom of the tube.

16. Transfer the top 18 mL of the suspension containing the individualized cells into a new tube and spin down, resuspend the pellet in MEF derivation medium and divide into T75 flasks (three embryos/flask).

17. Spin down the remaining 12 mL with the debris, resuspend with MEF derivation medium and add into one T75 flask.

18. Bring the final volume to 20 mL/flask by adding additional MEF derivation medium.

19. Incubate the flasks in a humidified incubator at 5% CO_2 and 37°C for 2–3 days, until 80–90% confluent. At this time, the MEF are ready to be harvested and passaged.

20. MEF cells are split at 1:4 into 0.1% gelatin (Sigma #G1890) coated T75 flasks with MEF growth medium and grow to reach 80% confluence.

21. Aspirate the MEF growth medium. Add 20 mL HES medium to each flask and incubate 24 h.

22. Collect the conditioned medium, add 20 mL fresh HES medium, and incubate another 24 h. Collect the conditioned medium and mix with the day 1 conditioned medium, store at –70°C. This is the HES conditioned medium.

3.2 Generation of L-Wnt3a-Conditioned Medium

1. L-Wnt3a cells are cultured in L-Wnt3a growth medium.

2. Split 1:10 into 100 mm culture dishes when cells reach confluence, add 10 mL L-Wnt3a conditioning medium in each dish. Incubate cells 4 days at 37°C and 5% CO_2.

3. Collect the medium after 4 days and filter sterilize. This is Batch 1.

4. Culture the cells for another 3 days by adding 10 mL fresh L-Wnt3a conditioning medium.

5. Collect the medium and sterile filter. This is Batch 2. Discard the cells.

6. Mix Batch 1 and Batch 2 medium at a 1:1 ratio. This is the L-Wnt3a conditioned medium.

3.3 Dermal Stem Cell Culture

Day 1

1. Take out a foreskin from the transfer tube; rinse it with 70% ethanol for 1 min (see Note 2).

2. Transfer the foreskin to a sterile 100 mm culture dish, add 20 mL HBSS without Ca^{2+} and Mg^{2+}, wait for 2 min (see Note 3).

3. Open the foreskin ring with scissors and cut the foreskin into several pieces of approximately 5×5 mm^2 using a surgical scalpel blade (see Note 4).

4. Transfer the skin pieces into a 50 mL Falcon tube with 5 mL 0.48% dispase II, and incubate at 4°C overnight.

Day 2

1. Remove the Falcon tube containing the skin sample from 4°C and incubate it at 37°C for 5 min.

2. Pour the skin pieces with the dispase into a sterile 10 cm culture dish, transfer all skin pieces to a new dish using forceps.

3. Separate the epidermis from the dermis by holding the dermal part of each skin piece with one pair of forceps and gently remove the epidermal part with a surgical scalpel blade. Discard the epidermis. Repeat the procedure for each piece of skin.

4. Transfer all dermal pieces to a new dish and mince them as small as possible (see Note 5).

5. Collect the minced dermal pieces using a pipette and transfer to a 50 mL Falcon tube containing 2 mL 1 mg/mL collagenase IV. Incubate at room temperature for 24 h (see Note 6).

Day 3

1. Add 25 mL HBSS without Ca^{2+} and Mg^{2+} to the tube containing the dermis with collagenase IV. Mix well by pipetting up and down; serially filter through 100, 70, and 40 cell strainers (see Note 7).

2. Centrifuge the cell suspension at $200 \times g$ for 5 min.

3. Resuspend the cell pellet in 10 mL HESCM4 medium and seed the cells in two T25 flasks (see Note 8).

4. Put the flasks into an incubator with 5% CO$_2$ at 37°C.

5. After 48 h, aspirate 2.5 mL medium from the flask, and replace with 2.5 mL fresh HESCM4 medium. Change half the volume of the medium every 3–4 days.

3.4 Dermal Stem Cell Differentiation to Melanocytes, Neuronal Cells, Schwann Cells, Smooth Muscle Cells, Adipocytes, and Chondrocytes

1. Tissue culture-grade 4-well chamber slides (Becton Dickinson) are precoated with: 0.5 mL/well 10 ng/mL fibronectin (Advanced Biomatrix #5050; for melanocyte, chondrocyte, and adipocyte differentiation), 0.1% matrigel (BD Biosciences #354234; for neuronal and smooth muscle cell differentiation), and a mixture of 20 μg/mL laminin (BD Biosciences #354232) with 200 μg/mL poly-D-lysine (BD Biosciences #354210) (for Schwann cell differentiation) (see Note 9).

2. Collect dermal spheres and transfer into a 50 mL tube, spin down, remove the supernatant as much as possible.

3. Add 0.5 mL 1 mg/mL collagenase IV and 0.5 mL 0.25% trypsin/EDTA. Incubate at 37°C for 5 min.

4. Pipette up and down for 1 min, then add 9 mL soybean trypsin inhibitor. Spin down. Resuspend the sphere cell pellet in the various differentiation media and seed the cells in coated chamber slides (see Note 10).

5. For neuronal and Schwann cell differentiation, use medium I during the first week and switch to medium II during the second week.

6. Incubate for 2–3 weeks, replace 1/2 fresh medium twice a week. Cells are ready to fix for staining using differentiation markers.

3.5 DSCs Differentiation to Epidermal Melanocytes in Three-Dimensional Skin Reconstruct Culture

1. Coat the transwell (Organogenesis) with the collagen mixture: 0.59 mL 10× minimal essential medium (EMEM), 50 μL 200 mM L-glutamine, 0.6 mL FBS, 120 μL 7.5% sodium bicarbonate, 4.6 mL bovine collagen I and mix well. Add 1 mL of the mixture into one transwell of tissue culture trays.

2. Collect dermal spheres by centrifugation and resuspend 6,600 dermal spheres in 0.75 mL HESCM4 medium (see Note 11).

3. Trypsinize fibroblasts and collect cells by centrifugation and resuspend 0.45×10^6 cells in 0.75 mL skin reconstruct medium I.

4. Mix the following reagents in a 50 mL tube: 1.65 mL 10× MEM, 150 μL 200 mM L-glutamine, 1.85 mL FBS, 350 μL 7.5% sodium bicarbonate, 14 mL bovine collagen I, 0.75 mL dermal spheres from step 2, 0.75 mL fibroblasts suspension from step 3 and mix well. Add 3 mL to each coated transwell. Incubate for 45 min at 37°C in a 5% CO_2 tissue culture incubator. Add skin reconstruct medium I (2 mL inside and 10 mL outside of the transwell). Incubate for 4 days.

5. Harvest human keratinocytes, resuspend 3×10^6 cells in 600 μL skin reconstruct medium I.

6. Remove the skin reconstruct tray from the incubator, aspirate medium from both inside and outside of transwells.

7. Add skin reconstruct medium I (1.5 mL inside and 10 mL outside of insert). Drop 100 μL keratinocyte suspension to each inside transwell. Incubate for 2 days.

8. Remove skin reconstruct medium I from both inside and outside of transwells. Add skin reconstruct medium II (2 mL inside and 10 mL outside). Incubate for another 2 days.

9. Aspirate skin reconstruct medium II both inside and outside of transwells, add 7.5 mL skin reconstruct medium III to only the outside of the transwells (see Note 12). Change medium III every other day until day 18.

10. Harvest the skin reconstruct at day 18: Remove the transwell from the tray with forceps. Cut out the reconstruct (including

the polycarbonate filter) by tracing a circle close to the edge with a scalpel blade. Place the reconstruct in a histology cassette (Surgipath #02275-BX) between two black TBS biopsy papers (Triangle Biomedical Sciences, #BP-B) and soak the whole cassette in 10% formalin (Fisher Healthcare #245-685) for 4–6 h. Then place the cassette in 70% for paraffin embedding.

3.6   Immunostaining

For spheres:

1. Cytospin the spheres on a microscope slide using a Cytospin 2 (Shandon) (see Note 13).

2. Dry 1 h at room temperature.

3. Fix with acetone for 10 min at 4°C.

4. Wash three times with 1× PBS without Ca^{2+} and Mg^{2+}.

5. Block 30 min with 3% BSA (MP Biomedicals #810034).

6. Remove blocking solution and add primary antibodies diluted in 1× PBS without Ca^{2+} and Mg^{2+}. Incubate overnight at 4°C.

7. Remove the primary antibody solution and wash three times with 1× PBS without Ca^{2+} and Mg^{2+}.

8. Incubate slides in secondary antibodies for 45 min at room temperature in the dark.

9. Wash twice with 1× PBS without Ca^{2+} and Mg^{2+}.

10. Mount with Vectashield mounting medium for fluorescence with DAPI (Vector #H-1200) and coverslip (Fig. 1).

For monolayer differentiated cells:

1. Fix the differentiated cells in chamber slides with 4% paraformaldehyde for 20 min.

2. Wash slides three times with 1× PBS.

3. Permeabilize cells using 0.5% Triton X-100 (Sigma #T9284) for 5 min.

4. Repeat steps 4–10 for sphere staining.

5. To detect differentiated adipocytes, fix the differentiated cells with 4% paraformaldehyde, and cover them with Oil Red O solution for 16 min. After washing slides with isopropanol, stain nuclei with hematoxylin solution.

For paraffin embedded skin reconstructs:

1. Deparaffinize in xylene twice for 10 min.

2. Redehydrate in 100, 100, 95, 70, and 50% ethanol for 2 min.

3. Wash three times with 1× PBS without Ca^{2+} and Mg^{2+}.

4. Antigen retrieve with trypsin.

5. Repeat steps 4–10 for monolayer cells (Fig. 2).

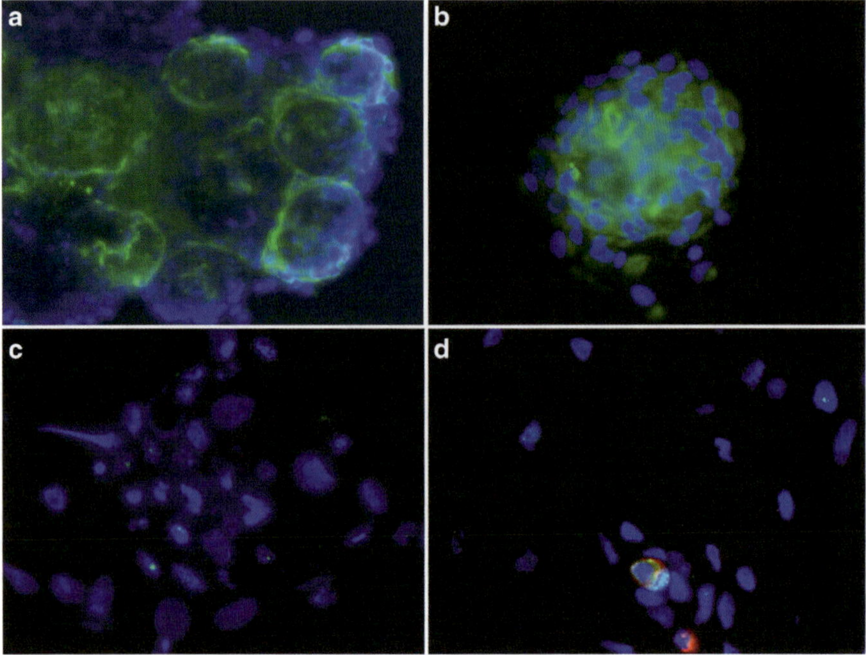

Fig. 1 Dermal spheres express stem cell markers. (**a**) Multiple dermal spheres are positive for the neural stem cell marker NGFRp75 (*green*). (**b**) A dermal sphere expressing nestin (*green*). (**c**) Embryonic stem cell marker OCT4 (*green*) expression is typically localized to nuclei. (**d**) Paraffin embedded foreskin stained with antibodies to NGFRp75 (*red*) and OCT4 (*green*). Nuclei stained with DAPI (*blue*)

4   Notes

1. Remove as much blood and organ tissue as possible to avoid contamination with other cells and to keep the pure MEF population.

2. Foreskins from the hospital are not always sterile. Use 70% ethanol to clean each foreskin to avoid contamination.

3. The purpose for this step is to completely wash out the ethanol.

4. Small size for easy enzymatic digestion to separate the epidermis from the dermis.

5. The smaller the skin is minced, the more single cells will be recovered after digestion.

6. Shake several times during this incubation for better digestion.

7. Pipet 60–80 times for better release of single cells from tissue clumps and filter with a cell strainer to remove clumps of unbroken cells and connective tissue.

8. Each T25 flask contains 3×10^6 to 4×10^6 cells in 5 mL HESCM4 medium.

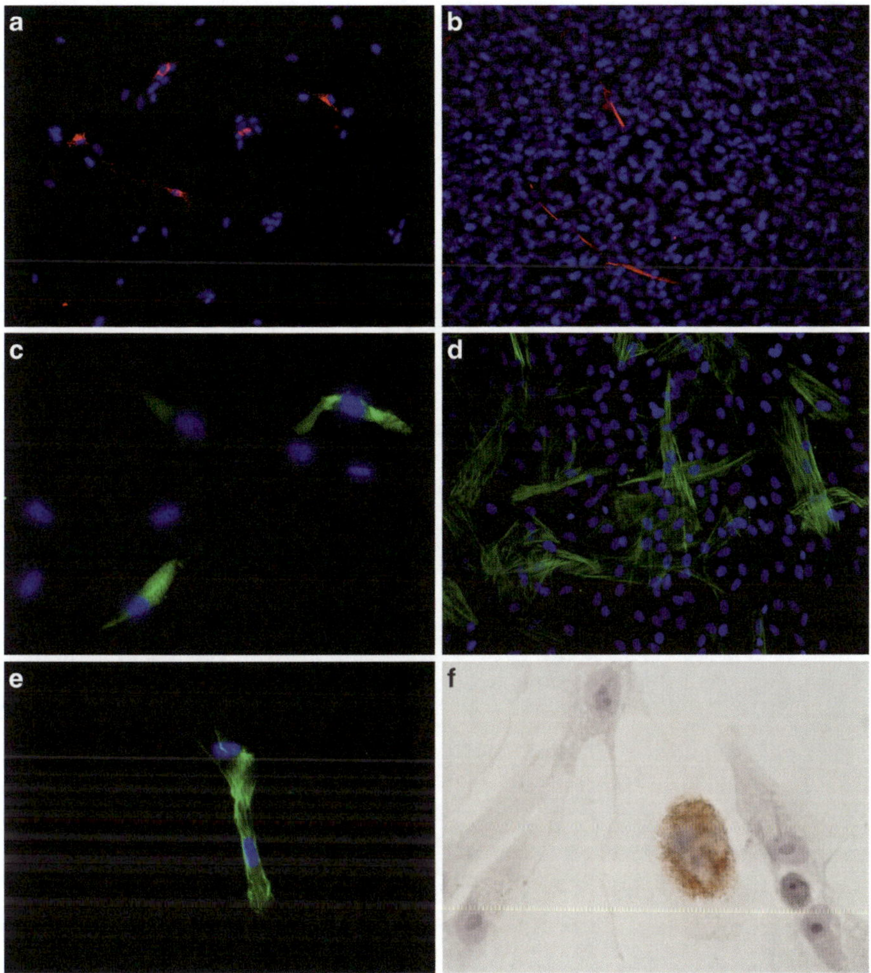

Fig. 2 DSCs can be differentiated into several cell lineages. Dermal spheres cultured in differentiation medium for 2 weeks and then stained with lineage specific antibodies. (**a**) Melanocytic differentiation is performed in melanocyte differentiation medium. Immunofluorescent staining show pigmentation marker HMB45-positive cells (*red*). (**b**) Neural differentiated cells become immunoreactive to β3 tubulin (*red*). (**c**) CNPase is used to detect differentiated Schwann cells (*green*). (**d**) Differentiated smooth muscle cells are positive for smooth muscle actin (*green*). (**e**) Differentiated chondrocytes express collagen II (*green*). (**f**) Oil-red-O staining detects a differentiated adipocyte (*red*). Nuclei stained with DAPI (*blue*, (**a**)–(**e**)) or hematoxylin (**f**)

9. Add 0.5 mL coating solution (fibronectin, or Matrigel or a mixture of laminin with poly-D-lysine), incubate at 37°C overnight before use.

10. Spheres cannot be completely dissociated. Seed single cells and spheres together in chamber slides. Aspirate coating solution from the chamber slides before seeding cells.

11. The dermal spheres are easily detached by tapping the flasks since most spheres have a slightly lower adherence to the

culture flasks. Mix well, pipet 100 µL to one well of a 96-well plate, settle for a few minutes and count using a microscope.

12. Air-lift the epidermis: Add skin reconstruct medium III only outside of the insert to expose the epidermis in air. This step induces keratinocyte differentiation to form a thick epidermis.

13. Collect spheres with medium in a 50 mL tube, wait 5 min to let the spheres settle down to the bottom of the tube, discard the top 2/3 medium which contains single dead cells. Spin down the tube with spheres and wash the sphere pellet with DPBS without Ca^{2+} and Mg^{2+}. Resuspend spheres in 1 mL DPBS without Ca^{2+} and Mg^{2+}. Drop 200 µL of the sphere suspension to each cytospin chamber with a microscope slide and spin at 800 rpm for 3 min.

References

1. Schreder A, Pierard GE, Paquet P et al (2010) Facing towards epidermal stem cells (Review). Int J Mol Med 26:171–174
2. Takahashi K, Tanabe K, Ohnuki M et al (2007) Induction of pluripotent stem cells from adult human fibroblasts by defined factors. Cell 131:861–872
3. Chang CW, Lai YS, Pawlik KM et al (2009) Polycistronic lentiviral vector for "hit and run" reprogramming of adult skin fibroblasts to induced pluripotent stem cells. Stem Cells 27:1042–1049
4. Yu J, Vodyanik MA, Smuga-Otto K et al (2007) Induced pluripotent stem cell lines derived from human somatic cells. Science 318:1917–1920
5. Yamanaka S (2009) A fresh look at iPS cells. Cell 137:13–17
6. Fernandes KJ, McKenzie IA, Mill P et al (2004) A dermal niche for multipotent adult skin-derived precursor cells. Nat Cell Biol 6:1082–1093
7. Toma JG, McKenzie IA, Bagli D et al (2005) Isolation and characterization of multipotent skin-derived precursors from human skin. Stem Cells 23:727–737
8. Stemple DL, Anderson DJ (1992) Isolation of a stem cell for neurons and glia from the mammalian neural crest. Cell 71:973–985
9. Nichols J, Zevnik B, Anastassiadis K et al (1998) Formation of pluripotent stem cells in the mammalian embryo depends on the POU transcription factor OCT4. Cell 95:379–391
10. Li L, Fukunaga-Kalabis M, Yu H (2010) Human dermal stem cells differentiate into functional epidermal melanocytes. J Cell Sci 123:853–860
11. Li L, Fukunaga-Kalabis M, Herlyn M (2011) The three-dimensional human skin reconstruct model: a tool to study normal skin and melanoma progression. J Vis Exp 54:pii: 2937. doi:10.3791/2937

Chapter 19

Isolation and Differentiation of Hair Follicle-Derived Dermal Precursors

Andrew Hagner and Jeff Biernaskie

Abstract

Several different precursor populations participate in renewal and regeneration of the mammalian skin and hair follicle. Recently, we described the existence of multipotent dermal precursors that exhibit properties of stem cells, and reside in the mesenchymal compartment of the hair follicle. When isolated and grown in vitro, these cells give rise to self-renewing, multipotent, spherical colonies of cells called Skin-derived Precursors (or "SkPs"). Here we describe methods to isolate SkPs from rodent and human skin and provide assays to determine functional differentiation of their progeny.

Key words Skin, Dermis, Stem cell, Hair follicle, Mesenchyme, Neural crest, Schwann cell, Regeneration, SkPs

1 Introduction

The mammalian epidermis and hair follicle epithelium contain several well-defined precursors that function to maintain cellular homeostasis and regeneration (1). Not surprisingly, the mammalian dermis also contains a self-renewing, multipotent precursor population and can be isolated from both rodent and human skin (2, 3). These dermal precursors, referred to as skin-derived precursors or "SkPs," originate from cells residing within the mesenchymal compartments of the hair follicle (the dermal sheath and the dermal papilla) (4). Within this hair follicle niche, their primary function is to provide the instructive signaling involved in activation/regulation of epithelial (bulge and matrix) precursors during hair follicle growth/regeneration (4). That is, when neonatal or adult SkPs are allowed to interact with epithelial keratinocytes, they initiate de novo hair follicle formation. Moreover, freshly isolated or cultured SkPs transplanted into adult skin will "home" to endogenous preexisting hair follicles and will assume inductive regulatory function within the hair follicle dermal papilla (4).

Kursad Turksen (ed.), *Skin Stem Cells: Methods and Protocols*, Methods in Molecular Biology, vol. 989,
DOI 10.1007/978-1-62703-330-5_19, © Springer Science+Business Media New York 2013

Another intriguing feature of isolated SkPs is that they exhibit a gene expression profile that is highly reminiscent of embryonic neural crest cells and when differentiated in vitro, are capable of generating a variety of neural crest derivatives (2, 3), including Schwann cells. There is enormous scientific interest in Schwann cells because of their inherent capacity to generate insulating myelin, proliferate in vitro, support growth and survival of axons, and ultimately provide improved functional recovery after nervous system injury (5–9). However, obtaining autologous nerve-derived Schwann cells requires invasive surgical biopsy and results in significant postsurgical morbidity. Therefore, SkP-derived Schwann cells represent a highly accessible and potentially valuable alternative (10, 11).

In this chapter, we provide detailed protocols for generating SkPs from rodent and human skin and subsequent assays to assess hair follicle inductive capacity and differentiation to a Schwann cell fate.

2   Materials

2.1   Isolation and Cell Culture Reagents

1. Fresh embryonic, neonatal or adult skin from mouse, rat, or human.

2. Culture medium: DMEM + Glutamax (Invitrogen) mixed 3:1 with F-12 nutrient supplement containing 1% of 10 mg/ml penicillin/streptomycin.

3. Penicillin (10,000 U/ml): streptomycin 10,000 μg/ml (Gibco).

4. Neurobasal medium (Invitrogen).

5. Hank's Balanced Salt Solution (HBSS; Ca^{2+} and Mg^+ free).

6. B27 supplement (Gibco).

7. Fungizone.

8. Dispase (2 mg/ml).

9. Trypsin diluted to 1% in HBSS and frozen at −20°C until use.

10. Trypsin EDTA (0.25%).

11. Collagenase Type XI (Sigma) diluted in sterile H_2O to a concentration of 1 mg/ml.

12. DNase I (Sigma) dissolved in sterile HBSS to a concentration of 1 mg/ml and stored as 500 μl aliquots at −20°C.

13. Heat-inactivated fetal bovine serum.

14. Basic fibroblast growth factor (bFGF).

15. Epidermal growth factor (EGF).

16. DMSO, anhydrous >99%.

17. Trypan blue.

2.2 Cell Culture Equipment	18. Fine tipped stainless steel forceps (sizes 5 and 55).
	19. 10 ml sterile pipettes.
	20. Flask 75 cm².
	21. Flask 25 cm².
	22. 15 and 50 ml conical tubes.
	23. 10 cm tissue culture-treated culture dishes.
	24. Chamber slides (4 or 8 wells) (Nunc).
	25. Cell strainer, 40 μm diameter.
	26. Cloning cylinders.
	27. Vacuum grease.
	28. #10 or #20 sterile scalpel blades.
	29. Hemocytometer.
	30. Centrifuge fitted for 15 and 50 ml conical tubes.
	31. Sterile syringe filters, 0.2 μm pore size.

---

3   Method

3.1 SkPs Medium Preparation

1. *Wash medium*: DMEM:F12 (3:1) containing 1% penicillin/streptomycin.

2. *Proliferation medium*: DMEM:F12 (3:1) containing 1% penicillin/streptomycin, 40 μg/ml fungizone, 40 ng/ml bFGF, 20 ng/ml EGF, 2% B27 supplement. For a T75 flask, we use 30–40 ml of medium. Always prepare medium fresh on the day of the experiment.

3.2 Generation of SkPs from Rodent Skin

1. Prior to dissection, fill several 10-cm plastic plates with cold sterile HBSS and place on ice.

2. Thaw aliquots of collagenase type XI. For each neonatal backskin sample, we use approximately 1 ml of collagenase (1 mg/ml).

3. Clean and sterilize all instruments by autoclave and 70% ethanol prior to dissection. Prior to and throughout the dissection procedure and dissociation, skin tissue and/or dissociated cells must be kept on ice or at 4°C.

4. Euthanize animals using an institute research ethics board-approved protocol.

5. If culturing skin from adult animals, hair on the backskin should be shaved or depilated in order to reduce the amount of hair fragments in the cultures. Wipe the skin surface with 70% ethanol to minimize bacterial/fungal contamination. Using fine forceps and scissors, dissect dorsal backskin by making incisions from shoulder to shoulder, across the rump

and down either side of the back. Remove a rectangular piece of tissue from the middle of the back.

6. Using 55 forceps gently remove all blood vessels, excess adipose, fascia, and muscle underlying the dermis. This is important to reduce contamination by other cell types in the culture. Take care not to damage follicle bulbs in the deep dermis (see Note 1).

7. To further reduce contaminating cell types, the epidermis can be removed from the dermis, the latter of which is the source of SkPs. To do this, float whole skin sections (epidermal surface face-up) on a solution of dispase (2 mg/ml) at 37°C for 30–60 min. Incubation times may need to be lengthened for older skin samples. Fine forceps (#55) can then be used to peel the epidermis off of the underlying dermis. Since epidermal cells readily adhere to the surface of the flask and do not generate spherical colonies, this step optional (see Note 2).

8. Dissected skin tissue is minced into 1–2 mm pieces using a pair of #10 or #20 scalpels and transferred into a 15-ml Falcon tube for digestion. Tissue should be submerged in collagenase type XI (2 mg/ml) at 37°C for 30–120 min. For each neonatal backskin, we use 1 ml of collagenase type XI, whereas average-sized minced adult backskin requires 3–4 ml of enzyme. Appropriate incubation times for digestion are dependent on the age of the animal from which the skin is dissected (approximate times; embryonic skin 20–30 min, neonatal skin 30–60 min, adult skin 60–120 min) and should be determined by user.

9. DNase (500 µl aliquot of DNase per 10 ml of medium) can be added to cells in collagenase prior to incubation in order to maintain a single-cell suspension. Digestion is complete when tissue becomes pale in color, and when cells can be easily liberated by mechanical impact or trituration. Tissue can be reexposed to collagenase (or trypsin) without compromising integrity of the cells.

10. While tissue is digesting, prepare wash medium.

11. Using a 10-ml disposable plastic pipette, mechanically dissociate tissue by repeatedly gently impacting tissue with the pipette tip. Do not grind tissue on wall of the tube as this will result in increased cell death. Add another 2–3 ml of medium and triturate the tissue using either a 10-ml pipette or a 1,000-µl pipette tip. Collect the supernatant, add 1 ml more wash medium, and repeat the trituration. Repeat this sequential dissociation until tissue pieces become thin and cells are no longer liberated from the tissue (see Notes 3 and 4).

12. Dilute collagenase by adding 20 ml of wash medium alone. Wash tissue extensively by repeated inversion of the tube or place on a rocker at 4°C for 5 min.

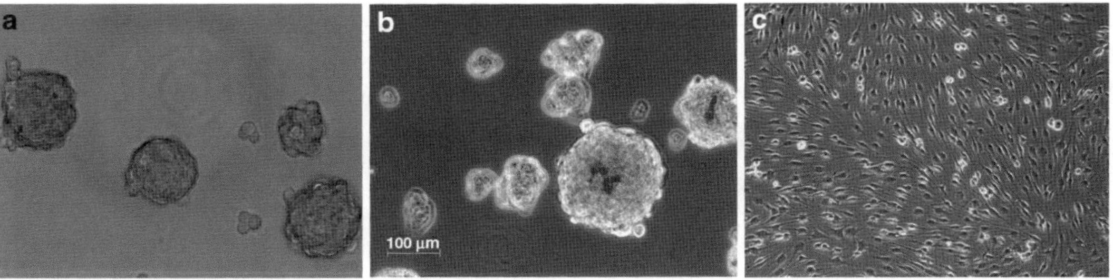

Fig. 1 In vitro growth of SkPs. SkPs can be isolated from and adult rat (**a**) and adult human (**b**) skin. Following exposure to in vitro conditions which promote glial cell survival, a subset of SkPs give rise to Schwann cells. Note the small phase-bright cell bodies, bipolar morphology, and parallel arrays typical of cultured Schwann cells (**c**)

13. Pass the dissociated cell suspension through a 40 or 70-μm cell strainer into a 15-ml conical tube. Additional wash medium may also be added to assist with filtration of a more viscous suspension.

14. Centrifuge tissue samples using a tabletop centrifuge at $150 \times g$ for 6–8 min to pellet the liberated dermal cells. Discard all supernatant, which contains the medium plus enzyme.

15. Resuspend the pellet in wash medium plus 2% B27 supplement. Resuspension volumes range from 2 to 20 ml depending upon the size of the pellet and can be adjusted to simplify quantification of cell numbers.

16. Determine the total yield of dissociated skin cells using a hemocytometer. We usually dilute 5 μl of the dissociated cell suspension in 45 μl (1:10 dilution) of trypan blue as an indicator of cell viability. Typically, one embryonic mouse backskin sample will provide approximately 10^6 cells, whereas a neonatal backskin sample will provide $3–8 \times 10^6$ cells (see Note 5).

17. Dilute the dissociated skin cells into proliferation medium to an optimum density of 5–25,000 cells/ml of medium; 30 ml total medium in a 75 cm^2 flask and 10 ml in a 25 cm^2 flask. SkPs are then grown in these flasks in a 37°C, 5% CO_2 tissue culture incubator as floating spherical colonies (see Notes 6 and 7).

18. Allow growth of spherical colonies for 7–21 days without passaging. Over this period of time, cultures must be fed every 3–5 days with the addition of 1–5 ml fresh medium containing all growth factors and supplements (bFGF, EGF, B27) at a concentration that will replenish the entirety of the culture medium. Figure 1 shows an example of an adult rat SkP cultures (a) and adult human SkP (b) following 14 days in SkPs proliferation medium (see Notes 8–11).

3.3 Generation of SkPs from Human Skin

3.3.1 Additional Reagents and Equipment Required for Human Skin

1. Liberase Blendzyme 1 (10×) (Roche): Make up stock solution by dissolving in sterile H_2O to a concentration of 9 mg/ml. Distribute into 553 μl aliquots and freeze at –20°C.

2. Cell strainer, 70 μm diameter.

3.3.2 Propagating Human SkPs

1. Human skin samples should be washed extensively in HBSS prior to dissection to remove blood cells. Carefully remove all blood vessels, fascia, and muscle underlying the deep dermis (see Note 12).

2. Prepare Liberase Blendzyme solution (0.25 mg/ml) for digestion of human skin by diluting one 5 mg in 20 ml of sterile HBSS. Keep on ice until use.

3. For human SkPs isolation, it is essential to remove the dermis from the epidermis to reduce contamination by undesired cell types. To do this, skin is cut into small, 3–5 mm^2 pieces. These skin pieces are then floated, epidermis side up, in a 10-cm plastic tissue culture dish filled with 25 ml of Blendzyme solution for 24–48 h at 4°C. Lengthening this incubation period may begin to compromise the integrity of dermal cells.

4. After 24–48 h, peel the epidermis away from the underlying dermis using fine (#55) forceps. The epidermis can be discarded (see Note 13).

5. Using two scalpel blades, mince the isolated dermal tissue into small, 1–2 mm^2 pieces. This takes approximately 30 min.

6. Place these small pieces of human dermis into a 15-ml conical tube containing 5–10 ml of fresh Blendzyme solution. DNase I (one 500 μl aliquot in 10 ml) can also be added to the suspension to reduce aggregation of cells. For most efficient digestion of the tissue we gently agitate the sample for 1–2 h at 37°C.

7. Upon completion of the digestion, add 20 ml wash medium plus 10% FBS to inactivate the Blendzyme. Repeatedly invert the tube 10–15 times. Centrifuge the tissue samples at $150 \times g$ for 6–8 min to pellet all cells and skin pieces. Discard the supernatant which contains the medium plus enzyme.

8. Add 1 ml of fresh wash medium. Using a 10-ml disposable plastic pipette, begin mechanically dissociating the tissue by gently impacting the tissue against the bottom of the tube with the tip of the pipette. Add 2–3 ml of wash medium and gently triturate the tissue further using either a 10-ml pipette or a 1,000-μl pipette tip (see Note 5). Briefly centrifuge the undissociated tissue for 20 s to pellet large pieces of skin. Remove and collect the supernatant into a 50-ml collection tube and

keep it on ice. Again, add 1 ml of wash medium to the tissue pellet and repeat the trituration step, adding the new supernatant to the collection tube.

9. Repeat this cycle until the tissue pieces become thin and cells can no longer be liberated. The supernatant should appear "milky" with suspended cells. We normally repeat these dissociation steps 4–5 times, with each dissociation step lasting 2 min.

10. Pass the dissociated cell suspension in the supernatant tube through a 70-μm cell strainer into a 50-ml conical tube. Centrifuge the filtered cell suspension in a tabletop centrifuge at $150 \times g$ for 7 min. Discard the supernatant and resuspend the cell pellet in wash medium plus 2% B27 supplement. Resuspension volumes range from 5 to 20 ml of medium depending on the size of the pellet and can be adjusted to simplify quantification of cell yield.

11. Determine the total number of skin cells using a hemocytometer. 5 μl of the dissociated cell suspension is usually diluted in 45 μl (1:10 dilution) of trypan blue as an indicator of cell viability.

12. Dilute the dissociated dermal cells into approximately 30 ml of proliferation medium for a 75 cm^2 flask and 10 ml for a 25 cm^2 flask. Conditioned medium is not typically required for primary human SkP cultures. Primary human cells are generally grown at higher cell densities of 50–100,000 cells/ml in order to enhance survival and increase yield of sphere formation. Secondary spheres can then be generated at clonal densities, according to experimental demand. Human SkPs, like rodent SkPs, are grown in these flasks in a 37°C, 5% CO$_2$ tissue culture incubator as floating spherical colonies.

13. Allow growth of spherical colonies for 7–21 days without passaging. Over this period of time, cultures must be fed every 4–5 days with the addition of 1–2 ml fresh medium containing all growth factors and supplements (bFGF, EGF, B27) at a concentration that will replenish the entirety of the culture medium (see Notes 14 and 15).

3.4 Passaging Rodent and Human SkPs

1. Remove all medium containing floating SkP spheres from the tissue culture flask and transfer it to a 10- or 50-ml conical tube, whichever is appropriate. Pellet the spheres by centrifugation at $150 \times g$ for 5 min. Remove the conditioned medium supernatant (see Note 16).

2. For murine SkPs, resuspend the cell pellet in 1 ml of wash medium and gently triturate the spheres approximately 50 times with a 1,000-μl pipette tip or until the spheres are dissociated to single cells. For rat or human SkPs, cells in the spheres are more adherent, and they require enzymatic digestion for dissociation. For these latter two cell types, resuspend the cell

pellet in 0.5 ml of collagenase type XI (1 mg/ml) for 10–20 min at 37°C, triturating intermittently. Trituration with a 100-μl tip may be required to break spheres to single cells.

3. Following successful dissociation, add 20 ml of wash medium to dilute the collagenase and mix the suspension by repeated inversion. For all types of SkPs, pellet the newly dissociated cells by centrifugation at $150 \times g$ for 6–8 min.

4. Determine the total number of SkP cells using a hemocytometer. 5 μl of the dissociated cell suspension is typically diluted in 45 μl (1:10 dilution) of trypan blue as an indicator of cell viability. Plating density can be determined according to experimental demand. For routine maintenance passaging, we will typically split one flask into two or three new flasks.

5. Resuspend the pelleted cells in 50% fresh proliferation medium containing twice the concentration of growth factors and supplements (80 ng/ml bFGF, 40 ng/ml EGF, 4% B27) mixed with 50% of filtered conditioned medium.

6. Feed passaged SkPs every 4–5 days by adding 1–2 ml of filtered conditioned medium containing all growth factors and supplements (bFGF, EGF, B27) at a concentration sufficient to replenish the medium. Figure 1a, b depicts SkPs derived from (a) adult rat skin and (b) adult human skin after 14 days in proliferation medium.

7. Repeat passaging every 7–14 days or as required. Spheres should be passaged prior to darkening of sphere cores or with increasingly large numbers of spheres in the flask, particularly if spheres are in contact with each other or have begun to aggregate. For human SkPs we typically passage once every 2 weeks, splitting cells one flask into two in order to maintain a low density of spheres (see Notes 17 and 18).

3.5 Differentiation of SkP-Derived Schwann Cells

3.5.1 Additional Materials Required for Schwann Cell Cultures

1. Forskolin dissolved to a concentration of 25 mM in DMSO, aliquoted, and frozen at –80°C (see Note 19).

2. Neuregulin-1β (R&D Systems) is dissolved in sterile PBS containing 0.1% bovine serum albumin, aliquoted at a concentration of 10 μg/ml, and stored at –80°C.

3. N2 supplement (Gibco).

4. Laminin dissolved in sterile H_2O.

5. Poly-D-lysine dissolved in sterile H_2O.

6. Chamber slides (4 or 8 wells) (Nunc).

7. 10 cm tissue culture-treated culture dishes.

8. Cloning cylinders.

9. Vacuum grease.

10. 0.25% Trypsin EDTA.

3.5.2 Schwann Cell Media Setup

Proliferation medium: DMEM/F12 (3:1) containing 1% penicillin/streptomycin, 40 µg/ml fungizone, 40 ng/ml bFGF, 20 ng/ml EGF, 2% B27 supplement, 5–10% fetal bovine serum.

SkP-derived Schwann cell differentiation medium: DMEM/F12 (3:1) containing 1% penicillin/streptomycin, 5 µm forskolin (diluted from 25 mM frozen aliquots), 50 ng/ml Heregulin-1β (diluted from frozen 50 µg/ml aliquots), 2% N2 supplement, 1–5% FBS (as determined by the experiment).

3.5.3 Schwann Cell Differentiation Procedure

1. Prepare culture plates. For successful differentiation and proliferation of SkP-derived Schwann cells, it is essential to coat tissue culture plate surfaces with laminin (0.02 mg/ml) and poly-D-lysine (0.2 mg/ml) overnight. For coating solution, add 200 µl of laminin and 100 µl of poly-D-lysine to 10 ml of sterile H_2O. For each 10 cm dish, 8 ml of coating solution should be added, whereas for four and eight chamber plastic slides, 600 µl and 300 µl of coating solution should be added to each chamber, respectively. Following incubation in a tissue culture hood for 6 h or overnight at room temperature, slides/dishes should be washed with dH_2O and allowed to dry (see Note 20).

2. Isolate SkP spheres from a culture that is ready to be passaged (see Note 15 and Step 7 of Subheading 3.4 for criteria to make this decision) by transferring the medium containing the SkPs from the flask into conical plastic tubes and then centrifuging these tubes in a tabletop centrifuge at $105 \times g$ for 3–5 min. Remove the supernatant, resuspend the pellet in 500 µl of collagenase XI, and dissociate the spheres to single cells as described in step 2 of Subheading 3.4. Following dissociation, determine cell density as described in step 12 of Subheading 3.2 (see Note 21).

3. For differentiation, cells should be grown at a density of 25–50,000 cells/ml in a 10-ml cell suspension for each 10 cm culture dish. Prepare this volume of proliferation medium. For example, to make up five 10-cm culture dishes for differentiation with a cell suspension of 8×10^6 cells/ml, make 50 ml of proliferation medium and add 313 µl of cell suspension (see Notes 22 and 23).

4. Resuspend cells in Schwann cell proliferation medium using a 1,000-µl pipette tip or 10-ml pipette and add the appropriate volume of the cell suspension to poly-D-lysine/laminin-coated chamber slides or culture dishes. For four- and eight-chamber slides, add 600 µl or 300 µl to each chamber, respectively, and for a 10-cm culture dish, use 10 ml. Culture these cells in a 37°C, 5% CO_2 incubator for 3 days.

5. After 3 days of proliferation, change the medium to Schwann cell differentiation medium. To do this, remove 50% of the

medium from the slide or dish and replace with fresh Schwann cell differentiation medium containing 2× the concentration of all growth factors and supplements. Culture the cells in the incubator, changing the medium every 2–3 days by removing 50% of the medium and replacing it with the same volume of Schwann cell differentiating medium containing 2× the concentration of supplements and growth factors.

6. SkP-derived Schwann cell colonies should appear in 2–3 weeks. These colonies may vary in number (i.e., 5–100 Schwann cells) and the majority of cells within the culture will consist of other cell types. They can also be identified morphologically as colonies of phase bright, bipolar cells (Fig. 1c). Alternatively, they can be detected immunocytochemically by their expression of marker proteins such as GFAP, S-100β, p75 neurotrophin receptor, peripheral myelin protein 0 (P0), or CNPase. If the purpose of the experiment is to isolate and expand the Schwann cells, it will be necessary to learn how to identify the colonies morphologically (see Notes 24–26).

7. To generate large numbers of purified SkP-derived Schwann cells, Schwann cell colonies are mechanically isolated using cloning cylinders as follows.

   (a) Sterilize cloning cylinders via autoclaving.

   (b) Identify colonies of SkP-derived Schwann cells morphologically under a tissue culture microscope and place a pen mark on the underside of the slide or dish to identify the location(s) of these colonies.

   (c) Remove the medium from the tissue culture dish or slides and coat one end of the cloning cylinder with sterile vacuum grease. With the aid of a low magnification inverted tissue culture microscope, place the cylinder over an individual Schwann cell colony and press it down to form a tight seal around the colony. Fill the cylinder with 200–500 μl of 0.25% trypsin-EDTA (undiluted) and incubate at 37°C for 5 min. Cells should then be easily detached from the substrate with gentle trituration.

   (d) Remove all contents from the cylinder (including the putative Schwann cells) into a conical tissue culture tube and dilute the cell suspension in 10 ml of wash medium containing 10% FBS to inactivate the enzyme. Mix the suspension by repeated inversion and then pellet cells by centrifugation in a tabletop centrifuge at 150×*g* for 6–8 min. Cells can then be resuspended in fresh Schwann cell differentiation medium (see Note 27).

8. Plate the resuspended cells in laminin/poly-D-lysine-coated chamber slides or tissue culture dishes and place them in a 37°C, 5% CO_2 incubator. These SkP-derived Schwann cells,

which grow adherently, should be fed with 80% fresh Schwann cell differentiation medium without serum every 3–4 days. We have found that serum is no longer required for proliferation and maintenance of SkP-derived Schwann cells after the initial differentiation step (see Note 28).

9. Upon reaching confluence (anywhere from 7 to 14 days, depending upon the initial density of the culture), SkP-derived Schwann cells should be passaged as follows.

    (a) Remove and discard medium, and remove cells from the substratum enzymatically by adding 2–3 ml of trypsin-EDTA (for a 10 cm dish) and incubating the dishes at 37°C for 5 min.

    (b) Rinse the cells from the surface of the plate using wash medium and collect the cell suspension in a 15 ml conical tube. Add 7 ml of fresh wash medium containing 10% FBS to deactivate the trypsin.

10. Mix the cell suspension by repeated inversion and then pellet the cells by centrifugation at $150 \times g$ for 6–8 min. The cell pellet can then be gently resuspended in new Schwann cell differentiation medium using a P1000 pipette tip. For maximum expansion and passaging, we typically split one confluent 10 cm culture dish into four or five 10-cm dishes. This would mean resuspending the cell pellet obtained from one culture dish into 50 ml fresh differentiation medium and then putting 10 ml of this suspension into each of five different 10 cm culture dishes.

3.6 Functional Assay for Hair Follicle Inductive Capacity

3.6.1 Preparation of Neonatal Epithelial Cells for Grafting

1. This procedure has been adapted from methods developed by Lichti et al. (12) and Zheng et al. (13).

2. Dorsal backskin is harvested from embryonic day 18 to postnatal day 1 (<24 h post birth) C57/Bl6 mice. C57/Bl6 mice are used because the pigment provides easy visualization of newly formed hair follicles, and so other pigmented strains can certainly be substituted (see Note 29).

3. Following removal, flatten and place the skin (epidermis side up) so that it is floating on dispase for 30–45 min at 37°C.

4. Following incubation, the dermis can then be peeled from the epidermis using fine forceps and epidermal sheets collected in cold HBSS (see Note 30). Figure 2a shows the epithelial aggregates and immature preformed follicles that remain on the underside of the epidermal sheet following removal of the dermis.

5. Transfer flat (floating) epidermal sheets to 0.25% trypsin solution (or trypsin-EDTA) and place in the incubator for 5 min; intermittently shake dish every minute or so until epithelial aggregates are mostly detached and floating. Figure 2b shows epidermal cells and epithelial aggregates following dissociation.

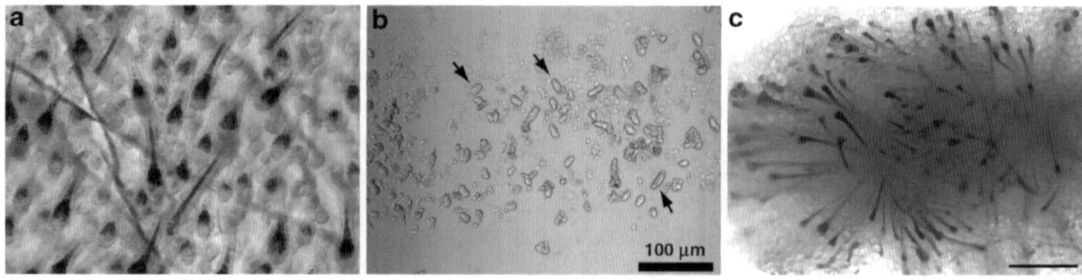

Fig. 2 Assessment of SkPs hair follicle inductive capacity. (**a**) Underside of neonatal backskin following dispase digestion and removal of the dermis. Note the epithelial aggregates as well as the immature (pigmented) hair follicles which remain on the epidermal sheets. (**b**) Liberated epithelial aggregates (*arrows*) following incubation of epidermal sheets in trypsin. (**c**) Patch of SkP-induced hair follicles in the patch assay 12 days following transplant of adult SkPs combined with neonatal epithelial aggregates. Scale bar in 2c is 200 μm

6. Add 10 ml of HBSS containing 10% FBS and DNase I (500 μl aliquot of 1 mg/ml) to the dish containing keratinocytes and epithelial aggregates to inhibit the enzyme.

7. Gently draw the underside of the remaining epidermal sheet(s) over the edge of a scalpel blade to release residual follicle buds and then collect all media containing floating cells in a 50-ml conical tube. Dilute again with 10 ml of DMEM containing 10% FBS, collect in a 50 ml conical tube. Gently triturate contents with a 10-ml Pasteur pipette to release single epithelial aggregates and place on ice.

8. Run cell suspension through a 70-μm cell strainer to remove preformed follicles and undigested epidermis.

9. Pellet the cells by centrifugation at $150 \times g$ for 5 min. If you only want epithelial aggregates and not single cells, then centrifuge for 4 min at $25 \times g$. Remove and discard the supernatant.

10. Resuspend in fresh DMEM and using a hemocytometer; calculate the total number of epithelial keratinocytes/aggregates collected (see Notes 31 and 32).

3.6.2 Preparation of SkP/Keratinocyte Suspension for Grafting

11. Collect SkPs in 50 ml conical tube and centrifuge at $150 \times g$ for 6 min.

12. Remove supernatant and resuspend SkPs in 500 μl of 1% collagenase. Incubate at 37°C for 10–15 min.

13. Triturate with a P1000 pipette tip (50×) and then with a P200 pipette tip until cells are in single-cell suspension.

14. Count cells using a hemocytometer. Typically we combine 10^5–10^6 SkPs with 1–3×10^6 epithelial keratinocytes (or 10,000 aggregates) in each graft.

15. Cell combinations are suspended in 50–100 μl of DMEM/ F12. Using a sterilized 21–27 gauge Hamilton syringe, cell

suspensions are injected subcutaneously in an adult immune-compromised mouse strain (either Nude or SCID can be used). You should observe a small "bubble" of fluid collecting under the skin where the cells have been injected.

16. Up to six grafts can be done on a single mouse. Grafts will begin to darken after 1 week indicative of a successful injection and pigmented hair follicle formation.

17. Grafts are harvested 12–14 days later and are assessed for number and phenotype of hair follicles that are generated within each graft. Figure 2c demonstrates a typical graft at 14 days post-grafting.

4 Notes

1. Dissected skin can be stored in HBSS for 24–48 h at 4°C and even shipped at this point without significantly affecting cell yield or health.

2. Experimenter should determine incubation times within their own laboratory, since enzymatic activity varies depending upon the source and batch, and this will affect digestion times. To avoid damaging the hair follicles, it may be prudent not to remove too much underlying adipose and fascia, particularly in neonatal skin where the dermis and hypodermis are intimately connected. However, this will require adding additional digestion time to the protocol.

3. The supernatant from these sequential dissociations is collected and stored on ice. The supernatant should appear "milky" with suspended cells. We normally repeat the dissociation steps 3–4 times, or until cells can no longer be liberated from tissue sample.

4. While it is important to mechanically liberate cells from the tissue, care must taken to avoid producing air bubbles during trituration and to avoid grinding the tissue with the pipette, as this will dramatically reduce cell viability.

5. We have found that in postnatal rat backskin, approximately 1–2% of cells in the skin will generate SkP spheres. As a result, the majority of cells plated in each flask will either adhere to the flask (these are mostly comprised of epidermal keratinocytes and interfollicular dermal fibroblasts) and others will die. Primary spherical colonies should be observed after 5–7 days of growth in proliferation medium. After 7–14 days, you will observe large numbers of spheres of varying size, potentially a result of the heterogeneous capacity for self-renewal between stem cells and more restricted progenitors within the SkP spheres. Quality of SkP spheres can also be verified by

immunocytochemistry for characteristic SkP markers such as nestin, versican, Sox2, PDGFRa, fibronectin, and vimentin. It is noteworthy that both rat and human SkPs show more robust growth relative to murine cells.

6. If too few cells were liberated from the skin sample, it may be the result of insufficient digestion or dissociation: (1) increase digestion time or collagenase concentration (or obtain enzyme from a new source); (2) increase trituration and homogenization of sample and ensure all supernatant during dissociation is retained; (3) since SkPs originate from hair follicle mesenchyme, be sure to remove excess fat (without damaging the hair follicles) from the underside of the dermis.

7. For lower cell densities, it is recommended to add at least 10–50% conditioned medium to promote survival and proliferation.

8. Large clusters of cells that appear within the first few days of culturing (1–3 days) are not SkPs but are instead cell aggregates and do not exhibit multipotency or self-renewal. They are an indication that either the cell density is too high and/or that the tissue was not adequately dissociated. Moreover cell densities of higher than 25,000 cells/ml may result in aggregation of cells into structures that may resemble spheres, but that are not proliferating colonies of cells.

9. The growth rate of SkPs is cell density dependent, so that the lower the cell density, the slower their growth. This can be partially circumvented with the use of conditioned medium.

10. It is extremely important to use tissue culture-treated flasks to prevent adherence of cells to the plastic. Premature adherence will typically result in a reduced yield of SkPs.

11. An absence of spheres after 10 days in proliferation media could be the result of (1) poor quality reagents (bFGF, EGF, B27), (2) an inappropriate number of cells plated, (3) poor dissociation of skin cells, or (4) cell/sphere adherence and premature differentiation of SkPs. (a) Avoid freeze/thaw of protein-based reagents and note that they have a limited shelf life, or try new stocks, a different lot, or change vendors. (b) Increase or decrease the density of cells plated, as excessively high or low densities will result in malnourished cultures and retarded proliferation. (c) Increase duration of collagenase digestion or increase trituration of tissue. (d) After 4 days in vitro, transfer all media and floating cells to a new flask to discourage cell contact between floating colonies and adherent epidermal cells.

12. Human tissue should always be treated as a biohazard and appropriate precautions taken in its handling.

13. If the epidermis does not easily peel away from the dermis, then the time of incubation in enzyme should be lengthened.

Alternatively, pieces of skin may not be small enough to allow penetration of enzyme and thus should be cut smaller.

14. Growth rates of human SkPs are significantly faster than their rodent counterparts and spherical colonies are much larger. You can expect to see the first spherical colonies at about 4–5 days following initiation of culture, and these can initially be expanded for 10–14 days.

15. After 7–14 days in proliferation medium, you will observe large, phase bright, spherical colonies. These primary spheres should then be passaged (see below), particularly if the core of the spheres has begun to darken (signifying poor health of the central cells).

16. Save the conditioned medium supernatant because it is essential for successful passaging and expansion of SkPs. This conditioned medium should be filtered through a 0.2-μm sterile filter and frozen at –20°C for storage.

17. Mass expansion of SkPs spheres after passage is enhanced if spheres are not broken completely to single cells but rather to small clusters. This is especially true at later passages (i.e., passage 5 and later) when proliferation seems to slow down.

18. If passaged cells do not form spherical colonies, they may be lacking in critical survival factors or they may have been compromised during sphere formation: (1) grow passaged cells in 100% conditioned medium containing all growth factors and supplements; (2) reduce the concentration or duration of collagenase dissociation or triturate more gently.

19. If DMSO is used as a diluent, prepare stocks so that the final concentration of DMSO in each culture does not exceed 1%.

20. This step can be done in as little as 6 h prior to the experiment or can be performed the night before differentiation is to commence.

21. While this protocol works with mouse SkPs, we generally have much more success differentiating Schwann cells from rat SkPs, consistent with the finding that primary cultures of Schwann cells are easier to generate and expand from rat than from mouse.

22. We have found that during initial stages of differentiation, addition of 1–5% FBS can enhance the survival of SkPs and ultimately their Schwann cell progeny. However, increasing levels of FBS for long periods may be detrimental because it also stimulates growth of other differentiating cell types. We suggest that following the proliferation stage, complete removal of FBS or a reduction to 1% FBS is optimal. Increasing concentration of FBS should only be considered when differentiating at low cell densities.

23. While a cell density of 25–50,000 cells/ml is optimal at this stage, lower cell densities can also be used.

24. If there is an absence of Schwann cells after 3 weeks, it may be that (1) the cell culture need to be freshly coated with laminin and poly-D-lysine; (2) there is poor survival due to a high serum concentration, or that Neuregulin-β concentration is too low; (3) cell density is too low and cells must be plated at a higher density, or else the serum concentration is too low.

25. Initially, SkP-derived Schwann cell colonies will look disorganized (shown in Fig. 1c), but with increasing numbers and density, cells will align into swirling parallel arrays, particularly following expansion of an isolated SkP-derived Schwann cell colony. It is important to note that the presence of other adherent differentiating cells within the culture will inhibit growth and proliferation of Schwann cells.

26. We have found that addition of heregulin and forskolin significantly enhances generation of SkP-derived Schwann cells relative to N2 supplement alone. Indeed, in several of our clonal sphere differentiation experiments we have observed that up to 80% of spheres generate S100β+ or glial fibrillary acidic protein (GFAP)+ cells, as opposed to <5% when differentiated in N2 supplement alone (10).

27. Plating these trypsinized cells at high density will improve cell viability and the efficiency of expansion.

28. Although the addition of factors such as neuregulin and forskolin enhances the generation of Schwann cells, our protocol for generating human SkP-derived Schwann cells has not yet been optimized.

29. Rodent pups should be no older than postnatal day 1 to minimize contaminating preformed hair follicles.

30. Dermis can be discarded or minced with scalpels and further digested in 1% collagenase at 37°C for 30 min. Dermis can then be dissociated to single cells and included as a positive control (since it will contain endogenous inductive dermal cells). As mentioned in Note 3, in neonatal skin, separating the underlying fat and fascia from the dermis may risk damaging the hair follicles, so use caution when removing fat before enzymatic digestion or extend the digestion time as necessary.

31. Each backskin from a typical neonatal C57/Bl6 mouse will yield approximately 5×10^6 to 10^7 epithelial cells after enzymatic digestion. See Fig. 2b for an example of follicle aggregate morphology for counting. If necessary, isolated keratinocytes can be collected and stored at −80°C in freezing medium (90% FBS, 10% DMSO) for future use, although fresh cells work significantly better in our hands.

32. If the number of epithelial cells collected is lower than expected, (1) increase or (2) decrease the concentration or duration of trypsin digestion and/or add DNase to reduce "stickiness" and cell aggregation.

Acknowledgments

We thank Mr. Ranjan Kumar for his assistance with images. This work was funded by a CIHR operating grant and a CIHR New Investigator Award to J.B.

References

1. Blanpain C, Fuchs E (2009) Epidermal homeostasis: a balancing act of stem cells in the skin. Nat Rev Mol Cell Biol 10(3):207–217. doi:nrm2636 [pii] 10.1038/nrm2636

2. Toma JG, Ahkavan M, Fernandes KJL, Barnabe-Heider F, Sadikot A, Kaplan DR, Miller FD (2001) Isolation of multipotent adult stem cells from the dermis of mammalian skin. Nat Cell Biol 3:778–784

3. Fernandes KJL, McKenzie IA, Mill P, Smith KM, Akhavan M, Barnabe-Heider F, Biernaskie J, Junek A, Kobayashi NR, Toma JG, Kaplan DR, Labosky PA, Rafuse V, Hui C-C, Miller FD (2004) A dermal niche for mutipotent adult skin-derived precursor cells. Nat Cell Biol 6(11):1082–1093

4. Biernaskie J, Paris M, Morozova O, Fagan BM, Marra M, Pevny L, Miller FD (2009) SKPs derive from hair follicle precursors and exhibit properties of adult dermal stem cells. Cell Stem Cell 5(6):610–623. doi:S1934-5909(09)00573-6 [pii] 10.1016/j.stem.2009.10.019

5. Pearse DD, Pereira FC, Marcillo AE, Bates ML, Berrocal YA, Filbin MT, Bunge MB (2004) cAMP and Schwann cells promote axonal growth and functional recovery after spinal cord injury. Nat Med 10(6):610–616

6. Xu XM, Chen A, Guenard V, Kleitman N, Bunge MB (1997) Bridging Schwann cell transplants promote axonal regeneration from both the rostral and caudal stumps of transected adult rat spinal cord. J Neurocytol 26(1):1–16

7. Pinzon A, Calancie B, Oudega M, Noga BR (2001) Conduction of impulses by axons regenerated in a Schwann cell graft in the transected adult rat thoracic spinal cord. J Neurosci Res 64(5):533–541

8. Paino CL, Bunge MB (1991) Induction of axon growth into Schwann cell implants grafted into lesioned adult rat spinal cord. Exp Neurol 114(2):254–257

9. Barakat DJ, Gaglani SM, Neravetla SR, Sanchez AR, Andrade CM, Pressman Y, Puzis R, Garg MS, Bunge MB, Pearse DD (2005) Survival, integration, and axon growth support of glia transplanted into the chronically contused spinal cord. Cell Transplant 14(4):225–240

10. McKenzie IA, Biernaskie J, Toma JG, Midha R, Miller FD (2006) Skin-derived precursors generate myelinating Schwann cells for the injured and dysmyelinated nervous system. J Neurosci 26(24):6651–6660

11. Biernaskie J, Sparling JS, Liu J, Shannon CP, Plemel JR, Xie Y, Miller FD, Tetzlaff W (2007) Skin-derived precursors generate myelinating Schwann cells that promote remyelination and functional recovery after contusion spinal cord injury. J Neurosci 27(36):9545–9559. doi:27/36/9545 [pii] 10.1523/JNEUROSCI.1930-07.2007

12. Lichti U, Weinberg WC, Goodman L, Ledbetter S, Dooley T, Morgan D, Yuspa SH (1993) In vivo regulation of murine hair growth: insights from grafting defined cell populations onto nude mice. J Invest Dermatol 101(1 Suppl):124S–129S

13. Zheng Y, Du X, Wang W, Boucher M, Parimoo S, Stenn K (2005) Organogenesis from dissociated cells: generation of mature cycling hair follicles from skin-derived cells. J Invest Dermatol 124(5):867–876

Chapter 20

Isolation of Mesenchymal Stem Cells from Human Dermis

Tsutomu Soma, Jiro Kishimoto, and David Fisher

Abstract

Recent studies revealed that mammalian dermis contains multipotent stem cells such as skin-derived precursors (SKPs). SKPs grow in suspension as spheres. In contrast, mesenchymal stem cells (MSCs) are adherent fibroblastic cells. Here, we describe the procedure to isolate MSCs under low-serum culture conditions. In addition, we explain the method to collect MSCs using magnetic cell sorting.

Key words Dermis, Differentiation, Magnetic cell sorting, Mesenchymal stem cells, Multipotent

1   Introduction

Epidermal and follicular bulge stem cells have been well character-ized by numerous studies over the past three decades (1, 2). In addition, melanoblasts, which are the stem cells of skin melano-cytes, have been identified in the hair follicle bulge by several stud-ies (3–5), where it was shown that their depletion is associated with age-dependent hair graying. In contrast, the existence of stem cells in the dermis is still unclear. Recent reports have described that the dermis contains multipotent stem cells such as skin-derived precur-sors (SKPs) (6–10). Furthermore, the follicular dermal papilla has been shown to be a highly enriched stem cell niche in the dermis (11, 12). Mesenchymal stem cells (MSCs) were first described as fibroblast precursor cells from bone marrow (BM-MSCs: bone marrow-derived MSCs), which differentiate into adipogenic, osteo-genic, and chondrogenic cells (13–15). Recent studies have shown that MSCs can be isolated from various tissues and organs in the human body such as adipose tissue, muscle, peripheral blood, ret-ina, central nervous system, and placenta (16). With regard to skin tissues, Jahoda et al. demonstrated that dermal sheath cells have the capability to differentiate into multiple mesenchymal lineages (17, 18). Here, we demonstrate a simple adherent culture method to isolate MSCs from human dermis using a low-serum MSC growth medium (19). Human dermal stromal cells with this

Kursad Turksen (ed.), *Skin Stem Cells: Methods and Protocols*, Methods in Molecular Biology, vol. 989,
DOI 10.1007/978-1-62703-330-5_20, © Springer Science+Business Media New York 2013

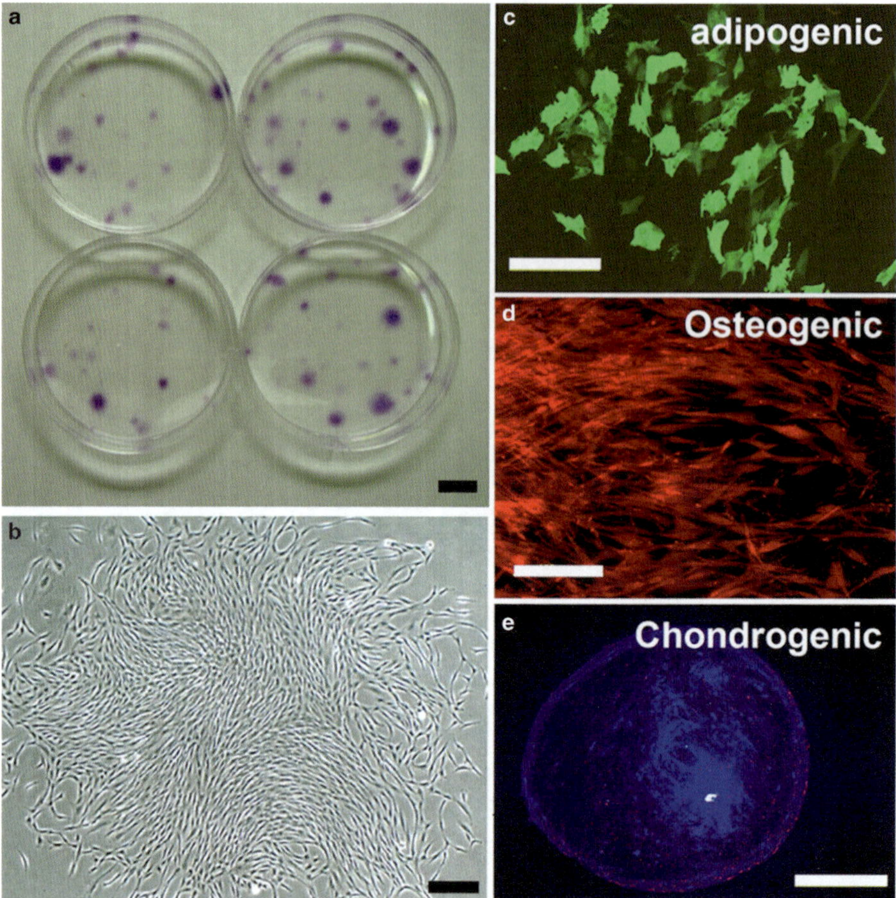

Fig. 1 Mesenchymal differentiation of stromal cells derived from human scalp skin. Stromal cells obtained from human scalp tissue were plated at 50 cells/6-cm-diameter dishes. After 14 days of cultivation, they showed colony-forming ability (**a**) and spindle-shaped fibroblastic morphology (**b**). Scale bars = 5 mm (**a**) and 500 μm (**b**). Stromal cells derived from human scalp were subjected to adipogenic, osteogenic, or chondrogenic lineage differentiation after secondary passage. Under adipogenic induction conditions, adipogenic differentiation was confirmed by anti-FABP-4 immunostaining (**c**). Two weeks after induction, an osteogenic nodule was apparent, accompanied by osteocalcin expression (**d**). In chondrogenic differentiation medium for 4 weeks, proteoglycan production was detected in the center of the cartilaginous matrices and visualized by aggrecan immunoreactivity (**e**). Scale bars: 200 μm

isolation technique exhibited high colony-forming ability and differentiation potential into mesenchymal lineages (osteogenic, chondrogenic, and adipogenic). Similar procedures have been reported for isolation of MSCs from human dermis (Figs. 1 and 2) (8, 20).

In bone marrow, BM-MSCs are a crucial component of the perivascular niche owing to their interaction with hematopoietic stem cells (HSCs) (21). However, the exact in vivo nature and localization of MSCs in other organs remains less clear. CD34 is a well-known stem cell marker for HSCs as well as other stem cells such as hair follicle stem cells, endothelial progenitor cells, and

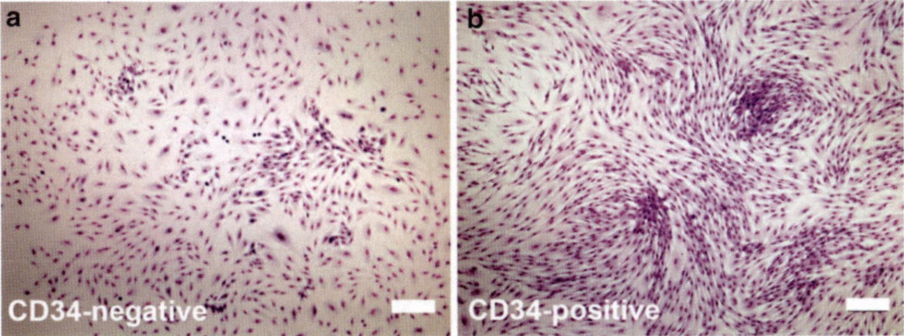

Fig. 2 Morphology of dermal mesenchymal stem cells after CD34-positive selection. CD34-positive cells collected by magnetic cell sorting showed significantly higher colony-forming ability. Compared with the CD34-negative fraction (**a**), the CD34-positive fraction contained spindle-shaped cells of uniform appearance (**b**). Scale bars: 200 μm

ADSCs (22–24). In adipose tissues, the vasculature plays a crucial role by being a progenitor niche of CD34-positive ADSCs (24). Furthermore, CD34-positive cells are well documented to participate in wound repair after skin injury (25). A recent report revealed that MSCs derived from the adult dermis were CD34-positive, similar to ADSCs (20). On the basis of these findings, we have concentrated dermal MSCs using CD34 magnetic beads (19).

2 Materials

Prepare all solutions using ultrapure water (prepared by purifying deionized water to attain a sensitivity of 18 MΩ cm at 25°C) and analytical grade reagents. Prepare and store all reagents at room temperature (unless indicated otherwise). Diligently follow all waste disposal regulations when disposing of waste materials.

2.1 Isolation of Dermal MSCs

1. 10% FBS-DMEM: Dulbecco's Modified Eagle Medium with 10% fetal bovine serum and antibiotic–antimycotic solution (Life Technologies Corporation, Grand Island, NY, USA).

2. Collagenase solution: 0.35% collagenase (Wako Pure Chemical Industries, Osaka, Japan) in 10% FBS-DMEM.

3. Low-serum MSC medium: MesenPRO RS™ medium with 2 mM L-glutamine and antibiotic–antimycotic solution (Life Technologies Corporation).

4. Giemsa stain solution (Nacalai Tesque, Kyoto, Japan): dilute tenfold with tap water before use.

5. MACS CD34-microbeads (Miltenyi Biotec, Bergisch Gladbach, Germany).

6. Separation buffer: PBS (pH7.2) with 0.5% bovine serum albumin and 2 mM EDTA.

2.2 In Vitro Mesenchymal Differentiation Assay of Dermal MSCs

1. Alpha-MEM basal medium: Minimum Essential Medium Eagle-Alpha Modification with 10% fetal bovine serum, 2 mM L-glutamine, and antibiotic–antimycotic solution (Life Technologies Corporation, Grand Island, NY, USA).

2. Type I collagen solution: acidic collagen solution for tissue culturing, pH 3.0, concentration: 0.3%, Sterile (KOKEN, Tokyo, Japan).

3. Human mesenchymal stem cell functional identification kit (see Note 1) (R&D Systems, Minneapolis, MN).

4. Adipogenic differentiation medium (see Note 2): alpha-MEM basal medium with adipogenic supplement (R&D Systems).

5. Fibronectin solution: resuspend 1 mg of human fibronectin (Becton, Dickinson and Company, Franklin Lakes, NJ, USA) in 1 ml of sterile distilled water (see Note 3).

6. Osteogenic differentiation medium (see Note 4): alpha-MEM basal medium with osteogenic supplement (R&D Systems).

7. D-MEM/F-12 basal medium: D-MEM/F-12 (Life Technologies Corporation) with ITS supplement (R&D Systems), 2 mM L-glutamine, and antibiotic–antimycotic solution (Life Technologies Corporation).

8. Chondrogenic differentiation medium (see Note 5): D-MEM/F-12 basal medium with chondrogenic supplement (R&D Systems).

9. 4% PFA in PBS: Weigh out 4 g of paraformaldehyde in the hood and add to 50 ml of water in a glass beaker with heater stirring (see Note 6). Raise the pH until the solution clears by adding a few drops of 1 N NaOH. After cooling, add 10 ml of 10× PBS and make up the volume to 100 ml with water. Filter through a 0.45 μm filter. Store at 4°C.

10. TBST: TBS containing 0.025% Triton X-100.

---

3   Methods

3.1 Isolation of Dermal Stromal Cells from Human Scalp Skin

1. Wash human skin tissue with Hanks' balanced salt solution (Life Technologies Corporation). After trimming the adipose tissue, cut into small 1–2 cm^2 pieces.

2. Put the tissue pieces into collagenase solution and place in a refrigerator overnight.

3. Incubate for 1 h at 37°C with gentle stirring in a CO_2 incubator.

4. Dissociate cells by pipetting manually with a 2 ml pipette and pass through a 40-μm cell strainer (Becton, Dickinson and Company, Franklin Lakes, NJ, USA).

5. Centrifuge at $300 \times g$ for 10 min and wash with 10% FBS-DMEM.

6. After discarding the supernatant, resuspend the cell pellet in 5 ml of low-serum MSC growth medium and then transfer to a 25 cm² untreated flask (Becton, Dickinson and Company).

7. Leave the flask for 1 h in a CO_2 incubator and then wash twice with low-serum MSC growth medium to remove non-adherent cells.

8. Incubate the flask in a CO_2 incubator to expand the dermal stromal cells with replacement of low-serum MSC growth medium every 2–3 days.

9. Before reaching full confluence, split the cells and culture further using growth medium at 1,000 cells/cm² with conventional culture vessels (see Note 7).

3.2 Determination of Colony-Forming Units of Fibroblastic Cells (CFU-F)

1. Plate the cells at 50 cells/10-cm-diameter dish to determine CFU-F.

2. After 14 days of cultivation, fix the cells with methanol for 5 min and then dry briefly.

3. Stain with Giemsa solution for 10 min at room temperature (see Note 8).

4. Wash with a large amount of water (see Note 9).

5. Count the colonies larger than 2 mm in diameter to assess CFU-F efficiency.

3.3 Magnetic Bead Cell Sorting

1. Centrifuge the cell suspension (see Note 10) at $300 \times g$ for 10 min. Aspirate the supernatant completely. Resuspend the cell pellet in 300 µl of separation buffer (see Note 11) for up to 10^8 total cells (see Note 12).

2. After adding 100 µl of FcR blocking reagent (Miltenyi Biotec, Bergisch Gladbach, Germany), add 100 µl of CD34 MicroBeads (Miltenyi Biotec).

3. Mix well and incubate for 30 min in a refrigerator (see Note 13).

4. Wash cells by adding 5 ml of buffer and centrifuge at $300 \times g$ for 10 min.

5. After aspirating the supernatant completely, resuspend the cells in 500 µl of separation buffer.

6. Place column in the magnetic field of a MiniMACS™ Separator (Miltenyi Biotec), followed by rinsing with 500 µl of separation buffer.

7. Apply cell suspension onto the column. Collect flow-through containing unlabeled cells.

8. Wash column three times with 500 µl of separation buffer (see Note 14). Collect unlabeled cells that pass through and combine with the flow-through.

9. Remove column from the separator and place it on a suitable collection tube.

10. Pipette 1 ml of separation buffer onto the column. Immediately flush out the magnetically labeled cells by firmly pushing the plunger into the column.

3.4 In Vitro Mesenchymal Differentiation Assay

1. For adipogenic differentiation, suspend cells in alpha-MEM basal medium and seed in the wells of a four-well chamber slide coated with type I collagen solution at 4×10^4 cells per well. After cultivation reaches 100% confluency (see Note 15), replace the medium with 0.5 ml of adipogenic differentiation medium (see Note 16). Culture for 14–21 days with replacement of fresh adipogenic differentiation medium (0.5 ml/well) every 3–4 days (see Note 17).

2. To differentiate into osteocytes, suspend cells in alpha-MEM basal medium and seed in the wells of a four-well chamber slide coated with fibronectin solution (see Note 18) at 8×10^3 cells per well. At 50–70% confluency, replace the medium in each well with 0.5 ml of osteogenic differentiation medium to induce osteogenesis. Culture for 14–21 days with replacement of fresh osteogenic differentiation medium (0.5 ml/well) every 3–4 days.

3. For chondrogenic differentiation, transfer 250,000 cells in the growth culture medium to a 15 ml conical tube. Centrifuge at $200 \times g$ for 5 min at room temperature. Remove the medium and resuspend the cell pellet with 0.5 ml of D-MEM/F-12 basal medium. After centrifugation at $200 \times g$ for 5 min, discard the medium by aspirating. Resuspend the cells in 0.5 ml of chondrogenic differentiation medium and centrifuge the cells at $200 \times g$ for 5 min at room temperature. Do not remove the medium. Loosen the cap of the tube to allow gas exchange and incubate at 37°C and 5% CO_2. Replace the medium with 0.5 ml of fresh chondrogenic differentiation medium every 2–3 days (see Note 19). After 21 days, mount the pellet in OCT compound (Sakura Finetek Japan, Tokyo, Japan) for cryosectioning.

3.5 Immunocytochemical Detection of Differentiation Markers

1. Wash the cells prepared on chamber slides twice with 0.5 ml of PBS (see Note 20). Hydrate the cryosections of chondrogenic cell pellets on glass slides with PBS.

2. Fix the cells with 0.5 ml of 4% PFA in PBS for 15 min at room temperature.

3. After rinsing twice with TBS, permeabilize the cells with 0.5 ml of 0.3% Triton X-100 in TBS for 20 min. Block the cells with

0.5 ml of a serum-free blocking reagent (DAKO) at room temperature for 15 min (see Note 21).

4. Incubate the cells with 300 µl of primary antibody working solution overnight at 2–8°C (see Note 22).

5. After washing three times for 15 min with TBST, incubate the cells with secondary antibody working solution (see Note 23) at 300 µl/well in the dark for 1 h at room temperature.

6. Wash three times with TBST for 10 min. Mount the cells with Vectashield Mounting Medium (Burlingame, CA, USA).

4 Notes

1. Human mesenchymal stem cell functional identification kit (catalog number SC006) includes the supplements for differentiating culture and antibodies for immunohistochemical detection of osteocytes, adipocytes, and chondrocytes.

2. Warm the adipogenic supplement vial in a 37°C water bath for 5 min, followed by vortexing until the precipitate dissolves. Add 50 µl of the adipogenic supplement to 5 ml of alpha-MEM basal medium. Store the unused supplement in a refrigerator. Adipogenic differentiation medium is also available from other suppliers such as Lonza (Basel, Switzerland) and Thermo Fisher Scientific (Waltham, MA, USA). Coat chamber slides with type I collagen solution diluted at 1:10 with HCl–H_2O (pH 3.0). Incubate at room temperature for 1 h. After aspirating, rinse three times with PBS.

3. Allow 30 min for material to dissolve (do not agitate or swirl). Transfer into small aliquots and store at –20°C (avoid multiple freeze-thaws). Use at 2.5 µg/cm^2 for coating of cultureware. To coat a four-well chamber, dilute to 10 µg/ml with PBS. Add 500 µl of diluted fibronectin to the culture surface. Incubate at room temperature for 1 h. After aspirating, rinse three times with PBS.

4. Warm the osteogenic supplement vial in a 37°C water bath for 5 min. Add 250 µl of the osteogenic supplement to 5 ml of alpha-MEM basal medium. Divide the unused supplement into 250 µl aliquots and store at –20°C (avoid repeated freeze-thaw cycles). Osteogenic differentiation medium is also available from other suppliers such as Lonza and Thermo Fisher Scientific.

5. Warm the chondrogenic supplement vial in a 37°C water bath for 5 min. Add 25 µl of the chondrogenic supplement to 2.5 ml of D-MEM/F-12 basal medium. Divide the unused supplement into 25 µl aliquots and store at –20°C (avoid repeated freeze thaw cycles). Chondrogenic differentiation medium is

also available from other suppliers such as Lonza and Thermo Fisher Scientific.

6. Heat on a hot plate with stirring in the hood (about 60°C). Make sure the solution does not boil.

7. Seed at 1,000 cells/cm² to ensure optimal growth of MSCs and use for differentiation experiments between passages two and five.

8. Use tap water but not deionized water for dilution of Giemsa stain solution.

9. Avoid prolonged washing so as not to diminish the staining.

10. It is important to obtain a single-cell suspension before magnetic labeling for optimal performance. Pass cells through a 40-μm strainer to remove cell clusters that may clog the column.

11. Work fast, keep cells cold, and use precooled solutions. This will prevent capping of antibodies on the cell surface and nonspecific cell labeling.

12. Volumes for magnetic labeling given below are for up to 10^7 total cells. When working with fewer than 10^8 cells, use the same volumes as indicated. When working with higher cell numbers, scale up all reagent volumes and total volumes accordingly.

13. The recommended incubation temperature is 2–8°C. Working on ice may require increased incubation times. Higher temperatures and/or longer incubation times may lead to nonspecific labeling.

14. Perform washing steps by adding buffer aliquots only when the column reservoir is empty.

15. Cells should be 100% confluent after overnight incubation. If they are not confluent, replace medium every 2–3 days with alpha-MEM Basal Medium until 100% confluence is reached.

16. After 5–7 days, lipid vacuoles will start appearing in the differentiated cells.

17. The adipogenic cells are fragile; medium replacement should be performed gently so as not to disrupt the lipid vacuoles.

18. Cell detachment can occur during osteogenic differentiation. Coating with fibronectin is useful to avoid cell detachment.

19. The pellet should not be attached to the tube. Use caution when replacing the medium to avoid aspirating the pellet.

20. For staining of chondrogenic differentiated cell pellets, 6 μm cryosections are treated in the same manner.

21. During the blocking, prepare primary antibody working solution by diluting primary antibodies in 3% BSA-TBST: goat

polyclonal anti-FABP-4 (1:25, R&D Systems), mouse monoclonal anti-osteocalcin (1:25, R&D Systems), and goat polyclonal anti-aggrecan (1:25, R&D Systems).

22. A negative control should be performed using 1% BSA in PBS containing 10% normal donkey serum and 0.3% Triton X-100 with no primary antibody.

23. During the washing, prepare the secondary antibody working solution by diluting the appropriate secondary antibody conjugated to Alexa Fluor® 350, 488 or 594 (1:100, Life Technologies Corporation) in 3% BSA-TBST.

References

1. Cotsarelis G (2006) Epithelial stem cells: a folliculocentric view. J Invest Dermatol 126:1459–1468

2. Blanpain C, Fuchs E (2009) Epidermal homeostasis: a balancing act of stem cells in the skin. Nat Rev Mol Cell Biol 10:207–217

3. Nishimura EK, Jordan SA, Oshima H et al (2002) Dominant role of the niche in melanocyte stem-cell fate determination. Nature 416:854–860

4. Nishimura EK, Granter SR, Fisher DE (2005) Mechanisms of hair graying: incomplete melanocyte stem cell maintenance in the niche. Science 307:720–724

5. Inomata K, Aoto T, Binh NT et al (2009) Genotoxic stress abrogates renewal of melanocyte stem cells by triggering their differentiation. Cell 137:1088–1099

6. Wislet-Gendebien S, Hans G, Leprince P et al (2005) Plasticity of cultured mesenchymal stem cells: switch from nestin-positive to excitable neuron-like phenotype. Stem Cells 23:392–402

7. Shih DT, Lee DC, Chen SC et al (2005) Isolation and characterization of neurogenic mesenchymal stem cells in human scalp tissue. Stem Cells 23:1012–1020

8. Hoogduijn MJ, Gorjup E, Genever PG (2006) Comparative characterization of hair follicle dermal stem cells and bone marrow mesenchymal stem cells. Stem Cells Dev 15:49–60

9. Lorenz K, Sicker M, Schmelzer E et al (2008) Multilineage differentiation potential of human dermal skin-derived fibroblasts. Exp Dermatol 17:925–932

10. Toma JG, Akhavan M, Fernandes KJ et al (2001) Isolation of multipotent adult stem cells from the dermis of mammalian skin. Nat Cell Biol 3:778–784

11. Wong CE, Paratore C, Dours-Zimmermann MT et al (2006) Neural crest-derived cells with stem cell features can be traced back to multiple lineages in the adult skin. J Cell Biol 175:1005–1015

12. Hunt DP, Morris PN, Sterling J et al (2008) A highly enriched niche of precursor cells with neuronal and glial potential within the hair follicle dermal papilla of adult skin. Stem Cells 26:163–172

13. Friedenstein AJ, Chailakhjan RK, Lalykina KS (1970) The development of fibroblast colonies in monolayer cultures of guinea-pig bone marrow and spleen cells. Cell Tissue Kinet 3:393–403

14. Digirolamo CM, Stokes D, Colter D et al (1999) Propagation and senescence of human marrow stromal cells in culture: a simple colony forming assay identifies samples with the greatest potential to propagate and differentiate. Br J Haematol 107:275–281

15. Colter DC, Sekiya I, Prockop DJ (2001) Identification of a subpopulation of rapidly self-renewing and multipotential adult stem cells in colonies of human marrow stromal cells. Proc Natl Acad Sci U S A 98:7841–7845

16. Salem HK, Thiemermann C (2010) Mesenchymal stromal cells: current understanding and clinical status. Stem Cells 28:585–596

17. Crisan M, Yap S, Casteilla L et al (2008) A perivascular origin for mesenchymal stem cells in multiple human organs. Cell Stem Cell 3:301–313

18. Jahoda CA, Whitehouse J, Reynolds AJ et al (2003) Hair follicle dermal cells differentiate into adipogenic and osteogenic lineages. Exp Dermatol 12:849–859

19. Yamanishi H, Fujiwara S, Soma T (2012) A perivascular localization of dermal stem cells in human scalp. Exp Dermatol 21:78–80

20. Kroeze KL, Jurgens WJ, Doulabi BZ et al (2009) Chemokine-mediated migration of skin-derived stem cells: predominant role for CCL5/RANTES. J Invest Dermatol 129:1569–1581

21. Wilson A, Oser GM, Jaworski M et al (2007) Dormant and self-renewing hematopoietic stem cells and their niches. Ann N Y Acad Sci 1106:64–75

22. Trempus CS, Morris RJ, Bortner CD et al (2003) Enrichment for living murine keratinocytes from the hair follicle bulge with the cell surface marker CD34. J Invest Dermatol 120:501–511

23. Kwamoto A, Tkebuchava T, Yamaguchi J et al (2003) Intramyocardial transplantation of autologous endothelial progenitor cells for therapeutic neovascularization of myocardial ischemia. Circulation 107:461–468

24. Tang W, Zeve D, Suh JM et al (2008) White fat progenitor cells reside in the adipose vasculature. Science 322:583–586

25. Nickoloff BJ (1991) The human progenitor cell antigen (CD34) is localized on endothelial cells, dermal dendritic cells, and perifollicular cells in formalin-fixed normal skin, and on proliferating endothelial cells and stromal spindle-shaped cells in Kaposi's sarcoma. Arch Dermatol 127:523–529

Chapter 21

Skin-Derived Mesenchymal Stem Cells: Isolation, Culture, and Characterization

M. Orciani and R. Di Primio

Abstract

The increasing interest about stem cell research is linked to the promise of developing such treatments for many life-threatening, debilitating diseases and for cell replacement therapies.

Among the various human tissues, skin represents a source characterized by great accessibility and availability with noninvasive procedures and without risks of oncogenesis after their transplantation. In this aim, the identification of suitable protocols for the isolation, characterization, and long-term storage has a pivotal role.

Key words Stem cells, Skin, Isolation, Characterization, Cryo-conservation

1   Introduction

Stem cells are undifferentiated cells that have the capacity to proliferate and to differentiate into various specialized cell types. Stem cells can be divided into two basic types, embryonic stem (ES) and adult ones. The ES cells are totipotent, able to differentiate theoretically into all cell types characteristic of the species they are taken from (1, 2).

However, their use is hindered by the ethical and technical issues that concern the ES cell field. In addition, after transplantation, in many cases they may cause teratomas due to their lost of contact inhibition mechanisms (3).

Recent studies have shown that also stem cells derived from adult human tissues seem to offer the same possibility of ES cells to respond to morphogenic signals and differentiate into any desired cell fate not only in mesenchymal lineage cells such as osteoblasts, adipocytes, and chondrocytes but also into other lineage cells, such as endothelial, neural, and hepatocytic cells (4–8) while circumventing the ethical issues.

It is now well accepted that bone marrow, peripheral blood, adipose tissue, and cord blood are some of the reservoirs of MSCs,

Kursad Turksen (ed.), *Skin Stem Cells: Methods and Protocols*, Methods in Molecular Biology, vol. 989,
DOI 10.1007/978 1 62703 330 6_21, © Springer Science+Business Media New York 2013

but cells isolated from them are often limited by availability or invasiveness of extraction or limited proliferative capacity.

To overcome these limits, new sources of MSCs have been sought (9, 10) and Bartsch et al. (11) reported the isolation of mesenchymal stem cells from human postnatal dermal tissues. Compared to bone marrow, a skin biopsy may provide an easier accessible source of stem cells, without the risk of oncogenesis (12) and with advantages for clinical use, because stem cells can be obtained from a patient's own skin biopsy, rising above the problem of immune rejection after transplantation in wound repair.

However, their recent isolation implies a lack of standardized protocols for their isolation and characterization.

In literature many studies report the isolation of MSCs from the skin, using different techniques and culture medium (13–17). In addition, most of the authors work with MSCs of animal origin, like mouse or pig. In an our previous study about MSCs isolated from human amniotic fluid (AF-MSCs) (18), we compared the growth rate, the immunophenotype, and the cryo-conservation of AF-MSCs isolated, cultured, and frozen in three different mediums: PC1, RPMI-1640, and MSCGM (Lonza Group Ltd, Basel, Switzerland). We showed that the most performing one was the MSCGM medium, which was able to enhance the proliferation; in contrast, cells frozen in PC-1 displayed an increased ability to restart culture after short-\long-term storage. We tested also skin-derived MSCs under the same conditions, obtaining the same results. Basing on these observations, we standardized an isolation protocol of MSCs from human skin using MSCGM medium; isolated cells were cultured and seeded in the same medium and then characterized by cytofluorimetric analysis using a panel of antibodies recognizing surface antigens that are specific for stem cells (19).

2 Materials

All reagents must be prepared and all procedures must be carried out using sterile technique in a laminar flow hood. All culture vessels, test tubes, pipet tip boxes, stocks of sterile Eppendorfs, etc., should be opened only in the laminar flow hood.

2.1 Cell Isolation and Culture

Assemble the following materials:

- Sterile forceps.
- Sterile curved-blade scissors.
- Sterile pipettes, individually wrapped.
- Sterile six-well plates.
- Sterile culture 25 and 75 cm² flasks.
- 100 mm sterile plastic culture dishes.

- Sterile 15 ml conical centrifuge tube.

- Cryovials.

- Mesenchymal Stem Cell Basal Medium (MSCBM™, PT-3238, Lonza Group Ltd).

- Mesenchymal Stem Cell Growth Medium (MSCGM™) SingleQuots® Kit (PT-4105, Lonza Group Ltd).

- Antibiotic-antimycotic solution (100 units/ml penicillin, 100 µg/ml streptomycin, and 0.25 µg/ml amphotericin B, Lonza Group Ltd).

- PC-1™ Liquid Base Medium (77232, Lonza Group Ltd).

- PC-1™ supplements (77232, Lonza Group Ltd).

- L-Glutamine (final concentration: 2–4 mM).

- Dimethyl sulfoxide (DMSO).

- Trypsin/EDTA solution(0.125% trypsin and 1 mM ethylene-diamine tetraacetic acid); leave an aliquot at 4°C for current use and store remaining aliquots at –20°C.

- Sterile phosphate buffered saline (PBS) without Ca^{2+} and Mg^{2+}.

2.2 Freezing Cells

- Freezing medium: PC-1 complete medium + 20% DMSO.

2.3 Cytofluorimetric Characterization

- PBS.

- Blocking solution: 1 l PBS + BSA 0.2% + 1% sodium azide. Store at 4°C.

- Mouse anti-human antibodies-FITC conjugated against HLA-ABC, HLA-DR, CD14, CD19, CD34, CD45, CD73, CD90, and CD105 (BD Biosciences).

---

3 Methods

All media preparation and other cell culture work must be performed in a laminar flow hood. Use universal precautions when handling human tissue. Dispose of contaminated materials appropriately. Cultures are maintained in a 5% CO_2 humidified incubator at 37°C.

3.1 Cell Isolation and Culture

3.1.1 Isolation

1. Decontaminate the external surfaces of the MSCBM™ bottle and MSCGM™ SingleQuots® Kit cryovials with 70% v/v ethanol or isopropanol.

2. Thaw the supplements at room temperature or quickly in a 37°C water bath (see Note 1).

3. Heat one bottle of MSCBM™ (PT-3238) in a 37°C water bath.

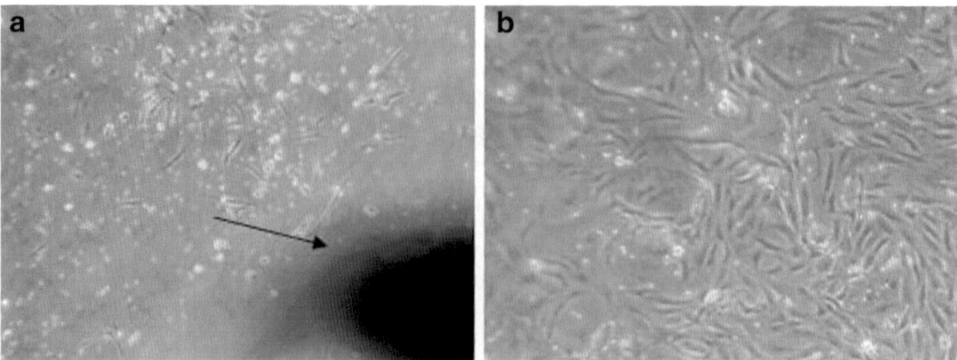

Fig. 1 Phase contrast microscopy images showing MSCs morphology at early days after the explants (*arrow*) (**a**) and after 14 days of culture (**b**) cells acquired an homogeneous, fibroblastoid morphology

4. Aseptically open the supplement. Using a pipette, prepare supplemented medium by adding the entire contents of the vials of MSCGM™ SingleQuots® Kit thawed in step 2 to one bottle of MSCBM™. Swirl gently. The supplemented medium is named MSCGM medium.

5. Store at 2–8°C in a dark location.

6. Add approximately 5 ml of PBS supplemented with antibiotic–antimycotic solution to the bottom portion of a 100-mm sterile culture dish. Keep the lid as it will be used.

7. Remove skin biopsy (range size between 10 and 15 mm) from the collection vessel with forceps. Aspirate off any accompanying media from the collection vessel and place the tissue on the dish prepared in step 6.

8. Wash tissue by agitating tissue with forceps in the supplemented PBS contained in the culture dish; place tissue epidermal side down on the overturned lid of the culture dish.

9. Crush the sample using forceps and remove fat and excess fascia using scissors and forceps.

10. Mince the remaining dermal sample into smaller strips.

11. Transfer the tissue strips in six-well plates with the prewarmed culture medium (1.5 ml/well).

3.1.2 Primary Culture and Subculture

1. Put the plates into a 5% CO_2 humidified incubator at 37°C.

2. For the first 7 days, do not aspirate culture medium but add fresh one.

3. After 7 days of culturing, numerous cells migrate from the explants; aspirate culture medium to remove non-adherent cells and residual dermal tissue and replace with fresh one (Fig. 1).

4. Change the medium twice a week.

5. For subculturing, aspirate the culture medium from the plate; wash the wells with prewarmed PBS solution.

6. Remove PBS and add trypsin/EDTA solution to each well (1 ml per each well of six-well plates). Place culture in the 37°C incubator for 2 min.

7. Monitor cells under microscope. Cells are beginning to detach when they appear rounded.

8. As soon as cells are in suspension, immediately add equal volume of MSCGM medium. Pipette over the surface of each well to resuspend all cells.

9. Transfer the cell suspension to sterile 15 ml conical tube and centrifuge gently at $300 \times g$ for 6 min.

10. Aspirate the supernatant from the cell pellets (see Note 2).

11. Resuspend pellets from step 10 in 2 ml of MSCGM medium; pipette the cell suspension up and down several times to ensure a homogeneous suspension.

12. Perform a cell count using an automated instrument (see Note 3).

13. Resuspend cells in MSCGM medium at a final density of 5×10^4 cells/ml (see Note 4).

14. Plate the cell suspension to many 25 cm^2 culture flasks as necessary and swirl the medium in the flasks to distribute the cells; transfer the flasks in the incubator.

15. Feeding is typically not necessary prior to 48 h post-seeding.

16. When cell monolayer is approximately 80% confluent (see Note 5), for subsequent subcultures repeat steps 5–13 using 75 cm^2 culture flasks.

3.2 Freezing Cells

3.2.1 Freezing Medium Preparation

1. Decontaminate the external surfaces of the PC-1 Liquid Base Medium (MSCBM™) bottle and PC-1 Supplement vial with 70% v/v ethanol or isopropanol.

2. Thaw one vial of frozen PC-1 Supplement at room temperature.

3. Heat one bottle of PC-1 Liquid Base Medium in a 37°C water bath.

4. Aseptically open the supplement. Using a pipette, prepare supplemented medium by adding the entire contents of the vials of supplement thawed in step 2 to one bottle of PC-1 Liquid Base Medium. Swirl gently. When these two components are combined, the resulting medium is named PC-1 Complete Medium (see Note 6).

5. For every cryovial, prepare 0.5 ml of freezing medium by adding DMSO at the final concentration of 20%. Keep the freezing medium on ice.

3.2.2 Freezing Protocol

1. Harvest cells as indicated in steps 5–10 of Subheading 3.1.2.

2. Resuspend cells in ice-cold PC-1 complete medium; we suggest cryopreserving the cells at $0.5–1.0 \times 10^6$ viable cells/ml. Keep cells on ice.

3. Determine the desired number of cryovials and label with date, cell type, and user's initials. Put them on ice. Each cryogenic vial will contain 1 ml of medium composed as follows: 0.5 ml of freezing medium + 0.5 ml of PC-1 complete medium in which cells have been resuspended.

4. Add ice-cold freezing medium drop by drop triturate cells until homogeneous.

5. Quickly aliquot 1 ml of cell suspension per cryogenic vial. Screw each vial closed.

6. Put vials into storage box and place box into –20°C freezer.

7. After 3 h, transfer container to –80°C freezer and store overnight.

8. Next day, put cells into appropriate rack in liquid N_2 tank.

3.3 Cytofluorimetric Characterization

1. Harvest cells as indicated in steps 5–10 of Subheading 3.1.2.

2. Wash pellets from step 2 in PBS.

3. Perform a cell count using an automated instrument (see Note 7).

4. Resuspend cells in blocking solution at a final density of 3×10^6 cells/ml (see Note 8).

5. Dispense 100 μl of the cell suspension in ten tubes suitable for flow cytometry and label each tube with the name of one primary antibody (see Note 9).

6. Add 3 μg/ml of the primary antibody into the corresponding tube.

7. Incubate for at least 30 min at 4°C.

8. Wash twice the cells by centrifugation at $400 \times g$ for 5 min and resuspend them in 500 μl of ice cold PBS. Keep the cells in the dark on ice.

9. For best results, analyze the cells on the flow cytometer as soon as possible (see Note 10).

10. Acquire data by flow cytometry (Fig. 2).

4   Notes

1. Remove these items from the water bath as soon as they are thawed and place them in the wood.

2. Be careful to not dislodge the cell pellets.

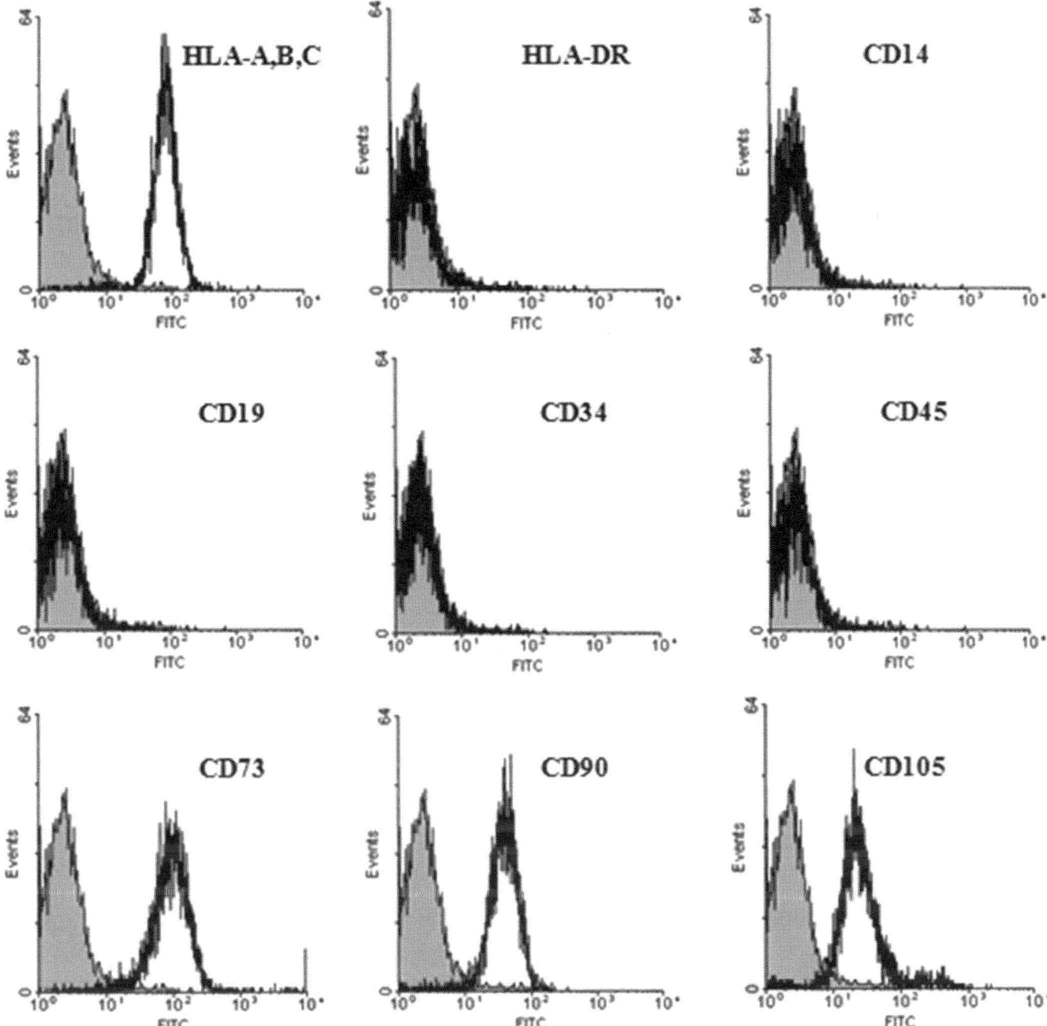

Fig. 2 Expression pattern of MSCs. FACScan analysis of the expression of HLA-ABC, HLA-DR, CD14, CD19, CD34, CD45, CD73, CD90, and CD105 antigens (*white* histograms) in MSCs derived from skin. *Solid gray* histograms refer to the negative control (IgG1 isotype control-FITC labeled)

3. May be done manually using a microscope and a counting chamber.

4. It may be necessary to determine the optimal seeding density for each specific cell sample.

5. When approaching confluency, some monolayers have a tendency to release from the vessel surface, thus making their eventual dispersal more difficult.

6. PC-1 Complete Medium is stable for 45 days at 2–8°C.

7. Cell count may be done manually using a microscope and a counting chamber.

8. Staining with ice-cold reagents/solutions in presence of sodium azide to prevent the modulation and internalization of surface antigens with consequent loss of fluorescence intensity.

9. The panel is composed by nine primary antibodies; use one tube for negative control by adding IgG1 isotype control-FITC labeled.

10. For extended storage (16 h) resuspend cells in 1% paraformaldehyde to prevent deterioration.

Acknowledgment

This work was supported by grant FIRB–RBAP10MLK7_003 from Ministero dell'Istruzione, dell'Universita e della Ricerca, Rome, Italy.

References

1. Odorico JS, Kaufman DS, Thomson JA (2001) Multilineage differentiation from human embryonic stem cell lines. Stem Cells 19:193–204

2. Schuldiner M, Eiges R, Eden A et al (2001) Induced neuronal differentiation of human embryonic stem cells. Brain Res 913:201–205

3. Wakitani S, Takaoka K, Hattori T et al (2003) Embryonic stem cells injected into the mouse knee joint form teratomas and subsequently destroy the joint. Rheumatology (Oxford) 42:162–165

4. Pittenger MF, Mackay AM, Beck SC et al (1999) Multilineage potential of adult human mesenchymal stem cells. Science 284:143–147

5. Zvaifler NJ, Marinova-Mutafchieva L, Adams G et al (2000) Mesenchymal precursor cells in the blood of normal individuals. Arthritis Res 2:477–488

6. in't Anker PS, Scherjon SA, Kleijburg-van der Keur C et al (2003) Amniotic fluid as a novel source of mesenchymal stem cells for therapeutic transplantation. Blood 102:1548–1549

7. Zuk PA, Zhu M, Ashjian P et al (2002) Human adipose tissue is a source of multipotent stem cells. Mol Biol Cell 13:4279–4295

8. Campagnoli C, Roberts IA, Kumar S et al (2001) Identification of mesenchymal stem/progenitor cells in human first-trimester fetal blood, liver, and bone marrow. Blood 98:2396–2402

9. Orciani M, Morabito C, Emanuelli M et al (2011) Neurogenic potential of mesenchymal-like stem cells from human amniotic fluid: the influence of extracellular growth factors. J Biol Regul Homeost Agents 25:15–30

10. Orciani M, Trubiani O, Vignini A et al (2009) Nitric oxide production during the osteogenic differentiation of human periodontal ligament mesenchymal stem cells. Acta Histochem 111:15–24

11. Bartsch G, Yoo JJ, De Coppi P et al (2005) Propagation, expansion, and multilineage differentiation of human somatic stem cells from dermal progenitors. Stem Cells Dev 14:337–348

12. Baker DE, Harrison NJ, Maltby E et al (2007) Adaptation to culture of human embryonic stem cells and oncogenesis in vivo. Nat Biotechnol 25:207–215

13. Jekabsons K, Riekstina U, Parfejevs V et al (2011) Culture-expanded human dermal stem cells exhibit donor to donor differences in cAMP generation. Cell Tissue Res 345:253–263

14. Orciani M, Gorbi S, Benedetti M et al (2010) Oxidative stress defense in human-skin-derived mesenchymal stem cells versus human keratinocytes: different mechanisms of protection and cell selection. Free Radic Biol Med 1:830–838

15. Orciani M, Mariggiò MA, Morabito C et al (2010) Functional characterization of calcium-signaling pathways of human skin-derived mesenchymal stem cells. Skin Pharmacol Physiol 23:124–132

16. Guan L, Yu J, Zhong L et al (2011) Biological safety of human skin-derived stem cells after long-term in vitro culture. J Tissue Eng Regen Med 5:97–103

17. Ernst N, Tiede S, Tronnier V et al (2010) An improved, standardised protocol for the

isolation, enrichment and targeted neural differentiation of Nestin+progenitors from adult human dermis. Exp Dermatol 19: 549–555

18. Orciani M, Emanuelli M, Martino C et al (2008) Potential role of culture mediums for successful isolation and neuronal differentiation

of amniotic fluid stem cells. Int J Immunopathol Pharmacol 21:595–602

19. Dominici M, Le Blanc K, Mueller I et al (2006) Minimal criteria for defining multipotent mesenchymal stromal cells. The International Society for Cellular Therapy position statement. Cytotherapy 8:315–317

Chapter 22

Isolation and Establishment of Hair Follicle Dermal Papilla Cell Cultures

Karl Gledhill, Aaron Gardner, and Colin A.B. Jahoda

Abstract

The isolation of hair follicle dermal papilla cells has become an important technique in the field of cutaneous stem cell biology. These cells can be used for a number of biological and translational purposes. They are studied to identify the cellular characteristics and molecular factors that underpin the initiation, maintenance, and modulation of hair growth; to develop new human hair replacement techniques; and as a source of cells capable of being directed down a variety of different lineages. Here, we describe the isolation of hair follicle dermal papilla cells from both human and murine sources via the microdissection techniques used in our lab.

Key words Hair follicle, Dermal papilla, Stem cells, Microdissection, End-bulb, Cell culture

1   Introduction

The hair follicle provides an easily accessible reservoir from which to isolate and culture adult stem cells. Within the hair follicle, perhaps the most widely studied population of stem cells are those of epithelial origin; however, an important population with complex progenitor and inductive properties also resides in the dermis. Hair follicle dermal papilla(e) (DP) cells are derived from embryonic dermis and exist at the base of the hair follicle, encapsulated by surrounding epidermal cells during the growing or anagen phase of the hair cycle (1). DP cells play a crucial role in the dermal–epidermal interactions that control hair production and cycling by producing and mediating the influence of systemic factors, which are known to modify this innately programmed pattern of follicle behavior (2). DP retain embryonic characteristics and these appear to be essential functional components which regulate the profound morphogenetic changes that occur during the hair growth cycle

---

Karl Gledhill and Aaron Gardner have contributed equally to this work.

Kursad Turksen (ed.), *Skin Stem Cells: Methods and Protocols*, Methods in Molecular Biology, vol. 989, DOI 10.1007/978-1-62703-330-5_22, © Springer Science+Business Media New York 2013

and determine the physical characteristics of the hair fibers produced (1). Moreover, intact papillae and cultured DP cells can induce hair growth when implanted into follicles (2, 3), can interact with skin epidermis to form new hair follicles (4) and are conducive to the activation and proliferation of epidermal cell populations.

Follicle DP cells are cultured for a number of biological and translational purposes. In the human context their multiplication and reimplantation has the potential to produce new human hair replacement techniques. They are also grown in vivo to identify those cellular characteristics and molecular factors that underpin the initiation and maintenance of hair growth, to study their hormonal physiology, and to examine their interactions with epithelial cells. Interestingly, DP cells have stem cell potential. They can be differentiated down adipogenic (5, 6), myogenic (7), neurogenic (8), and osteogenic (5, 6) lineages, have been transformed into induced pluripotent stem-cell cultures (9), and are a highly enriched source of skin-derived precursors (SKPs) (8). Moreover DP cells can be induced into a hematopoietic-like lineage and so may be useful in treating various forms of blood disorders (10).

Crucially, the retention of these inductive properties is only observed in early passage DP cells (although the timing of this loss varies between species). In this context, cultured DP cells exhibit a markedly different character to their in vivo counterparts over-time with increased proliferation, enhanced expression of α-smooth muscle actin (αSMA) (11), and the loss of veriscan and alkaline phosphatase expression (12, 13). Therefore, for many studies it is imperative that a protocol that maximizes initial cell harvest while still retaining cell viability is employed when isolating these cells for in vitro culture. There are two techniques described for isolation of dermal papilla cells; microdissection and the usage of enzymatic digestion followed by antibody-mediated sorting. As contamination of DP cultures with other mesenchymal cells, epithelial cells, matrix, or adipose tissue is undesirable, and as they exist as an aggregate population, we prefer the usage of microdissection techniques over those of enzymatic digestion and antibody-mediated sorting. Moreover, we also favor the use of minimally supplemented media for culture of these cells as we find this promotes retention of core properties. Consequently our DP cells grow slower in culture than those grown in media supplemented by additional factors or compared to inter-follicular skin fibroblasts.

2   Materials

2.1   Media

- Minimum essential medium with GlutaMAX™: Invitrogen.
- Fetal Bovine Serum: Bio Sera.
- Penicillin/Streptomycin: Invitrogen.
- Amphotericin B: Invitrogen.

2.2 Media Preparations	• 2 mM Ethylenediaminetetraacetic acid (EDTA) in phosphate buffered saline: Sigma.
2.2.1 Other Solutions	• 0.25% Trypsin/EDTA solution: Invitrogen.
2.2.2 Consumables	• Sterile plastic bacteriological petri dishes.
	• 4-well multi-dishes: Nunc.
	• T25 tissue culture vessels.
	• Sterile 1 ml syringes, without needles.
	• Sterile 25G needles.
	• Sterile plastic Pasteur pipettes.
	• 10 μl pipette tips.
2.2.3 Tools and Equipment	• Laminar flow cabinet (preferred).
	• Stereo-dissecting microscope.
	• 1× sterile, artery forceps.
	• 2× sterile, sharp watchmakers forceps.
	• 1× sterile, micro-scissors.
	• CO_2 Incubator—5% CO_2 and 37°C.
	• 20 μl pipette.

3   Methods

1. The DP resides at the base of the hair follicle within the dermis or hypodermis depending on species type and location (this should be kept in mind when sourcing samples).

2. Invert the sample to expose the base of the dermis (see Note 1); at this point it may be possible to observe the base of the hair follicle, known as the end-bulb. Using sharp forceps (see Note 2) separate the follicle from the surrounding tissue, taking care not to rip through the follicle or the epidermis.

3. Using a second pair of blunt forceps (see Note 2) gently grip the most distal part of the hair follicle, before cutting away the end-bulb with micro-scissors and immediately placing in a sterile petri dish containing dissection media (see Table 1).

4. Remove as many end-bulbs as possible from the tissue before proceeding to the next step (see Note 3). From this point on all work should be conducted under sterile conditions.

5. Using a Pasteur (dropper) pipette create drops of dissection media with a diameter of approximately 0.5 cm on a sterile surface, a petri dish lid is ideal. Transfer individual end bulbs into single drops and observe under a stereo-dissection microscope (see Note 4).

Table 1
Media preparations required for DP isolation and culture

Component	Dissection	Attachment	Expansion
Minimum essential medium	Neat	Neat	Neat
Fetal bovine serum	0%	20%	10%
Penicillin/streptomycin	100 U/100 µg/ml	100 U/100 µg/ml	100 U/100 µg/ml
Amphotericin B	0.5 µg/ml	0.5 µg/ml	0.5 µg/ml

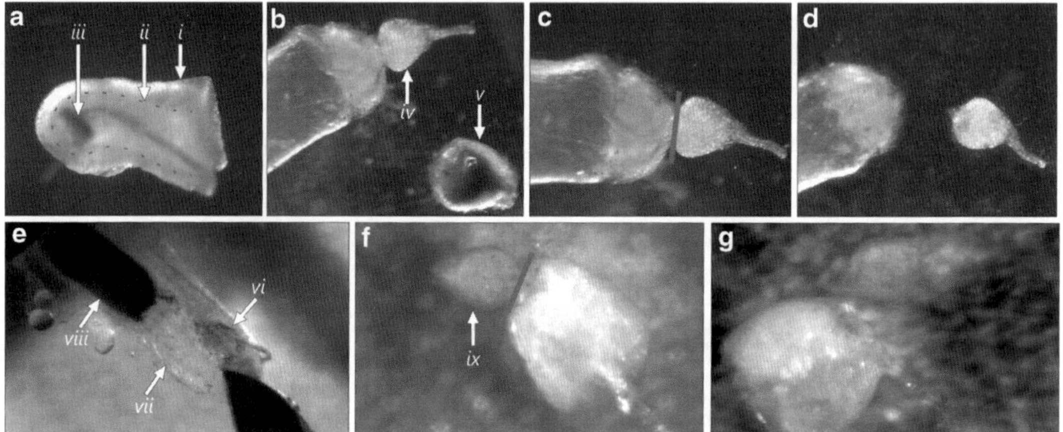

Fig. 1 Example dermal papilla isolation from rat and human end-bulbs. (**a**)–(**d**) Rat. (**a**) Intact clean end-bulb, visible within the collagen capsule (*i*) is the dermal sheath (*ii*) (delineated by *dashed line*) and the base of the hair shaft (*iii*). (**b**) Holding the end-bulb in place with one needle apply gentle pressure to the base to invert, exposing the dermal papilla (*iv*). Remaining hair matrix (*v*) may be ejected at this point. (**c**) Holding the dermal sheath with one needle a single cut should be made across the "basal stalk" as indicated by the *solid line*. (**d**) After cutting across the "basal stalk" as shown in (**c**) the intact, clean papilla can be transferred to the base of a four-well multi-dish containing attachment and growth media. (**e**)–(**g**) Human. (**e**) Intact clean end-bulb, showing the dermal papilla (*vi*) (delineated by *dashed line*) within the dermal sheath (*vii*). Residual hair matrix may still be present (*viii*) and should be ejected. (**f**) Inverted end-bulb exposing the dermal papilla (*ix*) which should be cut as indicated by the *solid line*. (**g**) Intact, clean dermal papilla

6. Using 25G needles attached to 1 ml syringes remove any residual fatty tissue stuck to the follicle end-bulb; if a hair shaft is present in the follicle, remove either by pulling the hair shaft clear with forceps or by gentle manipulation of the end-bulb.

7. At this point it may be possible to observe the DP through both the collagen capsule and dermal sheath (mouse, rat) (see Fig. 1a) or dermal sheath only (human) (see Fig. 1e). Using both needles trim excess distal follicle from just above the DP to leave a shorter end-bulb.

8. Firmly pin one side of the capsule (mouse, rat) or dermal sheath (human) at the open end in place, using the other needle gently

invert the end-bulb to expose the DP (see Note 5) (see Fig. 1b, f).

9. Remove residual contaminating tissue attached to the DP before separating from the "basal stalk" (see Fig. 1c, d, g) and transferring into a single well of a 4-well multi-dish containing attachment media (see Table 1) (400 µl per well) (see Note 6).

10. Approximately 8–10 DP should be cultured per well (see Note 7).

11. Ensure DP are situated at the base of the dish (see Note 8) and incubate at 37°C 5% CO_2 for 5–7 days without disturbing them.

12. After 7 days remove attachment media and replace with an appropriate volume of expansion media (see Table 1) from this point onwards as required (see Note 9).

13. When approximately 80% confluent, passage cells into T25 tissue culture vessels. Remove media and wash cells briefly with 2 mM EDTA solution, incubate the cells with trypsin/EDTA until the majority of cells have detached. Add an equal volume of expansion media to neutralize trypsin/EDTA activity. Pellet cells by centrifugation at $1,000 \times g$ for 5 min and resuspend in fresh expansion media (5 ml for a T25 flask).

4 Notes

1. Donor human tissue samples are separated from the source and so are easy to invert (see Fig. 2). Isolation of DP from mouse or rat whisker pads requires further preparation. Using micro-scissors make an L-shaped incision about 5 mm behind the whisker pad under the dissecting microscope (see Fig. 3a). Using the artery forceps peel back the whisker pad exposing the underside (see Fig. 3b), providing easy access to the base of the follicles (it is useful at this point to apply a few drops of dissection media to the exposed whisker follicles to prevent them from drying out. This helps with retaining maximum viability of cells). Using a pair of sharp forceps remove any connective tissue and fat between the follicles. If the correct rotational tension has been applied with the artery forceps then the follicles will stand up "to attention" (see Fig. 3c, d) aiding harvest of end-bulbs with micro-scissors.

2. A sharp pair of forceps should be used for ripping through the tissue with a blunter pair used to gently grip the distal part of the hair follicle. This helps prevent damaging the papilla with undue pressure and also limits contamination of the cleaned end-bulb.

3. As the isolation of end-bulbs occurs in a nonsterile environment, it is important to ensure that sufficient end-bulbs for the

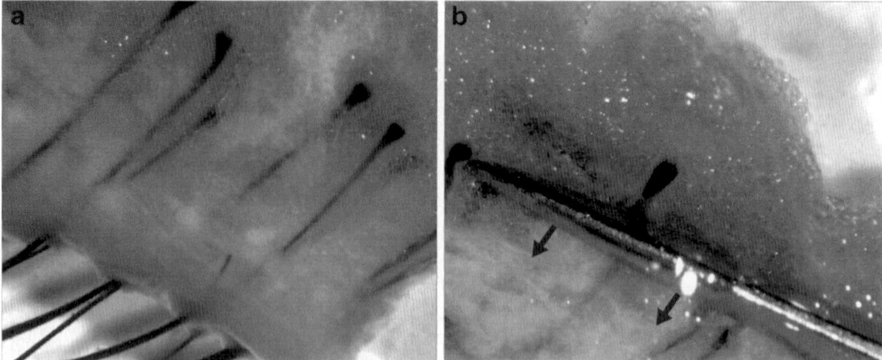

Fig. 2 Example end-bulb isolation from human facelift tissue. (**a**) Human facelift sample ready for end-bulb isolation (delineated by *dashed line*). (**b**) With the application of light pressure (using blunt forceps) to an area adjacent to an end-bulb, the end-bulb will be forced to protrude from the subcutaneous fat. This makes harvest of the end-bulb with micro-scissors easy

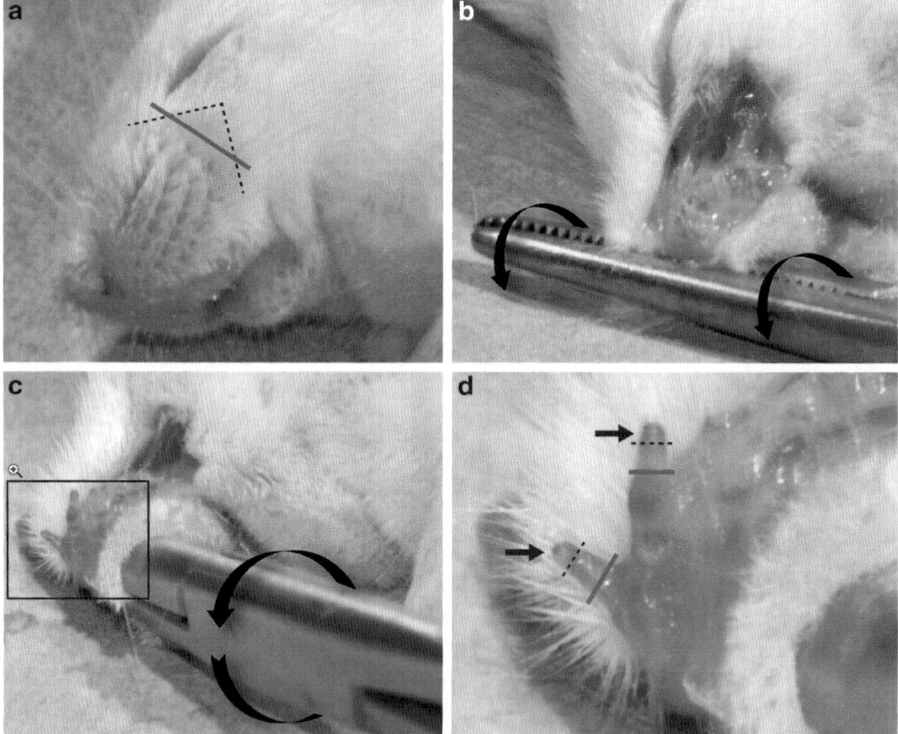

Fig. 3 Example end-bulb isolation from mouse whisker follicles. (**a**) Make two incisions along the *black dashed lines*, before gripping with artery forceps along the *solid line*. (**b**) Apply torque to pull the whisker pad away from underlying tissues. (**c**) With continued application of torque follicles will be exposed (although some removal of contaminating material may be required). (**d**) Magnified image showing appropriate location to grip distal region of follicle (*solid line*), before excising end-bulb along *black dashed lines*. The example given here depicts end-bulb isolation from the left-hand whisker pad. This is most easy for right-handed individuals. For left-handed individuals, isolation of end-bulbs is most easy from the right-hand whisker pad where an identical procedure is followed

envisioned work are taken in one sitting thus reducing the risk of contamination.

4. If drops are too small they may dry out with time as a result of evaporation, particularly if the dissection is carried out within a laminar flow hood. During the cleaning step, the drop may become murky and opaque making careful manipulation of the DP difficult; it is therefore useful to leave alternate drops "end-bulb free" so that cleaned end-bulbs can be moved to a fresh drop for DP isolation.

5. An alternative approach to releasing the DP involves making an incision on the outer hair follicle capsule at a level adjacent to the DP. Gentle pressure can then be applied to the hair follicle at a level just above the end-bulb on the lower part of the hair shaft. The DP is squeezed out of the follicle through the incision as a result. The "basal stalk" which attaches the DP to the connective tissue may then be severed using a dissection needle.

6. In our lab we use a number of different methods to transfer the isolated DP from the drop to the well of a multi-dish. One method involves transferring the DP in 5 µl of isolation media from the drop using a pipette. The dissection media containing the DP is then expelled into the attachment media contained within the well of the multi-dish. Care must be taken to ensure that the DP does not stick to the inside of the pipette tip. In order to ensure that this does not occur, the pipette tip is equilibrated in the dissection media before the DP is taken up into the pipette. Moreover, it is important to ensure that no bubbles are produced during expulsion of the DP from the pipette. The DP may stick to any bubbles produced and this will inhibit attachment of the DP to the base of the multi-dish.

7. The number of DP required per well will vary depending on their size. The larger rat DP contain more cells and therefore may only require six DP for successful culture. Human or mouse DP are considerably smaller and so contain fewer cells thus may require a larger number to be cultured. It must be noted that not all the isolated DP will necessarily adhere to the base of the well and therefore produce colonies of cells. Therefore, it is wise to add more DP to each well than is required. On occasions the DP can be scored with a needle across the base to encourage attachment.

8. The differentiation of cultured DP cells has been demonstrated by several groups. We have previously demonstrated that through the usage of specific media it is possible to drive DP cells towards adipogenic and osteogenic lineages (5), with a similar method being employed by other groups to generate neuronal like SKPs (8) and myogenic cells (7). We have also demonstrated DP cells can be induced down erythroid and

myeloid lineages, and perhaps more importantly reconstitute the hematopietic system of irradiated mice (10).

9. Enzymatic dissociation of a sample followed by fluorescent-activated cell sorting (FACS) into distinct populations can be used as an alternative to microdissection (and may be preferred for isolating very small DP from pelage hairs). This process is perhaps most elegantly described in a paper by Rendl et al. (14). In this method using cell type-specific transgenic expression of red and green fluorescent proteins in combination with antibody-mediated sorting with CD34, 45, and 117, Rendl et al. were able to isolate relatively pure populations of DP and other surrounding cell types.

Acknowledgments

The authors would like to acknowledge the photographical assistance provided by Rebecca P. Hill and Simon C. Cork in producing Figs. 1 (e–g) and 3 (respectively) and proof reading by Laura R. J. Ions. We are very grateful to DEBRA and the MRC for grant support.

References

1. Oliver RF, Jahoda CA (1988) Dermal-epidermal interactions. Clin Dermatol 6(4):74–82

2. Driskell RR, Clavel C, Rendl M, Watt FM (2011) Hair follicle dermal papilla cells at a glance. J Cell Sci 124(8):1179–1182

3. Oliver RF (1967) The experimental induction of whisker growth in the hooded rat by implantation of dermal papillae. J Embryol Exp Morphol 18(1):43–51

4. Oliver RF (1970) The induction of hair follicle formation in the adult hooded rat by vibrissa dermal papillae. J Embryol Exp Morphol 23(1):219–236

5. Richardson GD, Arnott EC, Whitehouse CJ, Lawrence CM, Hole N, Jahoda CAB (2005) Cultured cells from the adult human hair follicle dermis can be directed toward adipogenic and osteogenic differentiation. J Invest Dermatol 124(5):1090–1091

6. Jahoda CAB, Whitehouse J, Reynolds AJ, Hole N (2003) Hair follicle dermal cells differentiate into adipogenic and osteogenic lineages. Exp Dermatol 12(6):849–859

7. Rufaut NW, Goldthorpe NT, Wildermoth JE, Wallace OAM (2006) Myogenic differentiation of dermal papilla cells from bovine skin. J Cell Physiol 209(3):959–966

8. Hunt DP, Morris PN, Sterling J, Anderson JA, Joannides A, Jahoda C, Compston A, Chandran S (2008) A highly enriched niche of precursor cells with neuronal and glial potential within the hair follicle dermal papilla of adult skin. Stem Cells 26(1):163–172

9. Higgins CA, Itoh M, Inoue K, Richardson GD, Jahoda CAB, Christiano AM (2012) Reprogramming of human hair follicle dermal papilla cells into induced pluripotent stem cells. J Invest Dermatol 132(6):1725–1727

10. Lako M, Armstrong L, Cairns PM, Harris S, Hole N, Jahoda CA (2002) Hair follicle dermal cells repopulate the mouse haematopoietic system. J Cell Sci 115(20):3967–3974

11. Jahoda CA, Reynolds AJ, Chaponnier C, Forester JC, Gabbiani G (1991) Smooth muscle alpha-actin is a marker for hair follicle dermis in vivo and in vitro. J Cell Sci 99(3):627–636

12. Kishimoto J, Ehama R, Wu L, Jiang S, Jiang N, Burgeson RE (1999) Selective activation of the versican promoter by epithelial- mesenchymal interactions during hair follicle development. Proc Natl Acad Sci U S A 96(13):7336–7341

13. McElwee KJ, Kissling S, Wenzel E, Huth A, Hoffmann R (2003) Cultured peribulbar dermal sheath cells can induce hair follicle development and contribute to the dermal sheath and dermal papilla. J Invest Dermatol 121(6):1267–1275

14. Rendl M, Lewis L, Fuchs E (2005) Molecular dissection of mesenchymal-epithelial interactions in the hair follicle. PLoS Biol 3(11):e331

Chapter 23

Enrichment of Oral Mucosa and Skin Keratinocyte Progenitor/Stem Cells

Kenji Izumi, Cynthia L. Marcelo, and Stephen E. Feinberg

Abstract

The isolation of human oral mucosa/skin keratinocytes progenitor/stem cells is clinically important to regenerate epithelial tissues for the treatment of oral mucosa/skin defects. Researchers have attempted to isolate a keratinocyte progenitor/stem cell population using cell markers, rapid adherence to collagen type IV, and other methods. In this regard, one of the specific characteristics of keratinocyte progenitor/stem cells is that these cells have a smaller diameter than differentiated cells. This chapter describes methods used in our laboratory to set up primary human oral mucosa and skin keratinocytes in a chemically defined culture system devoid of animal derived products. We utilized the cells in a FDA-approved human clinical trial that involved the intraoral grafting of an ex vivo produced oral mucosa equivalent to increase keratinized tissue around teeth. We also provide two protocols on how to sort keratinocytes using physical criterion, cell size, using a cell sorter and a serial filtration system.

Key words Keratinocyte, Oral mucosa, Skin, Cell size, Cell sorting, Progenitor/stem cells

1 Introduction

In the oral cavity, reconstructions of oral mucosa defects require suitable graft materials to encourage wound healing as well as the functions of oral cavity. The lack of available oral mucosa tissue has limited the ability of surgeons to reconstruct the oral cavity. The use of skin grafts in the oral cavity has several drawbacks. For instances, skin has adnexal structures such as hair follicles and sweat glands and shows different keratinization pattern. Recent clinical applications has involved the use of in vitro combinations of cultured human cells with supporting scaffolds and bioactive signals produces tissue-engineered constructs, human cell/tissue- based products. Our research team successfully developed a tissue-engineered oral mucosa equivalent (ex vivo produced oral mucosa equivalent: EVPOME) suitable for intraoral grafting procedure. The EVPOME is constructed in a defined medium without the use

Kursad Turksen (ed.), *Skin Stem Cells: Methods and Protocols*, Methods in Molecular Biology, vol. 989, DOI:10.1007/978-1-62703-330-5_23, © Springer Science+Business Media New York 2013

of bovine serum, pituitary extract or use of a xenogeneic feeder layer (1–3). Currently, an unsorted human adult oral mucosa keratinocyte population is used to fabricate the EVPOME. The success with manufacturing EVPOME provides compelling empiric evidence that a subpopulation exists within a cell population that represents the progenitor/stem (P/S) cell compartment. In addition to the intraoral applications there are other "mucosal" areas of the body that are also amenable to our mucosal keratinocyte technology in the areas of urology, oculoplastic and gynecologic surgery. Previously, investigators have produced cell-based devices consisting of oral keratinocytes, utilized to treat patients, that have shown good results for extraoral applications (4–7). These clinical achievements in regenerative medicine using oral mucosa keratinocytes are attributed to not only the presence of a keratinocytes P/S cell subpopulation but also the hyperproliferative characteristic of cells, an easily accessible/obtainable tissue and an "autologous" graft fabrication. Therefore, the P/S cell subpopulation needs to be isolated in a simple and cost effective manner to make this technology more clinically viable. A more practical approach to isolate and expand primary oral keratinocytes is using only physical and pharmacological means under chemically defined conditions consistent with FDA regulatory guidelines. This approach will enhance their ability to fabricate a superior EVPOME graft for extraoral grafting and can lead to greater advances in cell replacement therapy. Several studies showed that cells possessing significantly higher clonogenicity have a smaller diameter than larger cells, suggesting the subpopulation of keratinocyte progenitor/stem cells is capable of maintaining their size small (8, 9). Mammalian target of rapamycin (mTOR) regulates several anabolic and catabolic processes such as cell growth (an increase of the size of cells), cell proliferation (an increase of the number of cells), cell cycle, and autophagy. mTOR forms two distinct multi-protein complexes referred to as mTORC1 and mTORC2 (10). We previously showed the activation of mTORC1 specifically increases cell growth and differentiation in suprabasal layer cells of oral mucosa as well as cultured oral keratinocytes (11). The immunosuppressant rapamycin binds and inhibits mTORC1. Thus, the use of rapamycin to downregulate mTORC1 signaling pathway may keep the cells in a more primitive state of differentiation that would mandate a smaller size (11).

In this chapter, we describe our techniques to isolate oral mucosa keratinocytes enriched for a P/S cell subpopulation that can be subsequently, pharmacologically, manipulated with rapamycin to alter their replication and differentiation (11–13). This technique has been utilized to further investigate biological characteristics of oral mucosa keratinocytes as well as to generate new cell based tissue-engineered devices to improve patient care for a translational research.

2 Materials

2.1 Primary Keratinocyte Culture	#10 Scalpel, Forceps (w/teeth).

#10 Scalpel, Forceps (w/teeth).

Tissue culture dishes (35, 60, 100, 150 mm).

Serological pipettes (5, 10, 25 ml).

Tissue culture flasks (T-25, T-75).

Centrifuge tubes (15, 50 ml).

Micropipette tips (yellow).

100 µm strainer.

Pasteur pipette.

DPBS w/o Ca^{2+}, Mg^{2+}.

0.25% Trypsin solution.

Trypsin inhibitor solution (DTI; Life Technologies).

Trypan Blue.

Antibiotic-Antimycotic (Life Technologies).

EpiLife keratinocyte culture medium (w/$CaCl_2$ solution; Life Technologies).

EDGS (Life Technologies).

Hemocytometer.
Inverted microscope.

2.2 Serial Keratinocyte Culture

0.05% Trypsin/EDTA solution.

DPBS w/o Ca^{2+}, Mg^{2+}.

Serological pipettes (5, 10, 25 ml).

Tissue culture flasks (T-75, T-150).

Centrifuge tubes (15, 50 ml).

Micropipettes tips (yellow).

Trypsin inhibitor solution (DTI; Life Technologies).

Trypan Blue.

Gentamicin.

Fungizone.

EpiLife keratinocyte culture medium (w/$CaCl_2$ solution; Life Technologies).

EDGS (Life Technologies).

Hemocytometer.

Inverted microscope.

2.3 Cell Sorting

Syringe.

0.22 µm syringe filter.

12 × 75 mm round bottom tube.

HBSS.

DPBS w/o Ca^{2+}, Mg^{2+}.

Bovine serum albumin.

Sodium azide.

Filter system (250 ml).

Autoclaved 240 μm mesh filter.

Propidium Iodide.

RNase.

Citric acid.

Micropipette tips (yellow).

Size standard beads (8, 15, 21, 30, 40 μm).

2.4 Gravity Assisted Cell Sorting

Sterile Forceps (w/o teeth) and Scissors.

Nylon net filter (11, 20, 30, 40 μm) (Millipore).

70% Ethanol.

DPBS w/o Ca^{2+}, Mg^{2+}.

Staple, stapler.

3   Methods

Obtain an approval from the institutional Internal Review Board to procure human tissue samples. Give sufficient information to healthy patients regarding the tissue procurement, and have the consent form signed by all individuals. Oral mucosa sample(s) will be harvested without causing any morbidity by physicians.

3.1 Primary Oral Keratinocyte Culture

Day 1: Oral mucosa sample washing

1. Prepare 0.04% trypsin solution by mixing 0.25% Trypsin solution with DPBS. For primary skin keratinocyte culture, please refer Note 1.

2. Obtain the tissue sample from the physician in a 15 ml (or 50 ml depending on the size of the mucosa sample) centrifuge tube containing 5 ml (or 15 ml) of EpiLife basal culture medium with calcium chloride, and keep it in the refrigerator (at 4°C).

3. Pour the tissue sample as well as the entire medium from the 15 ml centrifuge tube into the 60 mm (or 100 mm) tissue culture dish. For primary skin keratinocyte culture, please refer Note 1.

4. Add 3 ml (or 8 ml) of EpiLife basal culture medium with calcium chloride to three 35 mm (or 60 mm) tissue culture dishes.

5. Hold the sample with the forceps with teeth and scrape the sample with scalpel to remove any blood residue on the surface of submucosal tissue as well as the epithelial surface. Transfer

the sample into the first 35 mm tissue culture dish, using the same forceps.

6. Repeat the previous step for the remaining two 35 mm tissue culture dishes.

7. Transfer the sample to the third 35 mm tissue culture dish and soak until ready to start soaking the sample in the 0.04% trypsin solution.

8. Dispense 0.04% trypsin solution into another 35 mm (or 60 mm) tissue culture dish, and add Antibiotic-Antimycotic at the final concentration of 1.5%. For primary skin keratinocyte culture, please refer Note 1.

9. Using the same forceps used, transfer the tissue sample to the tissue culture dish with trypsin solution, and place the tissue culture dish lid.

10. Leave the tissue culture dish in the laminar flow hood for 16 ± 1 h. The soaking time can be adjusted by the step 7, depending on the time to start the following procedure.

Day 2: Oral keratinocyte plating

(1) Isolating oral keratinocyte

1. Add 10 ml of trypsin inhibitor solution to the 60 mm tissue culture dish, and 8 ml of EpiLife complete culture medium (containing EDGS, calcium chloride and preferred antibiotics and antifungal reagents) to the flipped lid of the same dish.

2. Transfer the trypsinized tissue sample into the tissue culture dish.

3. Hold the submucosal layer with sterilized forceps with teeth and gently scrape off the epithelial layer with a #10 scalpel. The basal layer cells should be scraped off and released into the trypsin inhibitor solution.

4. Transfer the scraped tissue sample into the flipped lid of the same tissue culture dish. Repeat the previous step.

5. Place a sterile 100 μm strainer over the opening of the 50 ml centrifuge tube.

6. Using a 10 ml pipette, dispense the cell suspension into the strainer.

7. Using the same pipette, rinse the tissue culture dish surface with 4 ml of EpiLife complete culture medium, and pipette the rinse solution into the filter.

8. Repeat the previous step with another 4 ml of EpiLife complete culture medium.

9. Discard the submucosal layer in the overnight soaking plate into biohazard trash, or subsequently harvest oral mucosa fibroblasts by the appropriate method.

(2) Flask plating

10. Tap the strainer on the edge of the 50 ml conical tube.

11. After shaking the tube, aspirate a 15 μl cell suspension, mix with 15 μl trypan blue, and load 10 μl of the suspended cells into each of the two chambers of the hemocytometer. Count viable cells under the inverted microscope and calculate the total number of cells harvested.

12. Secure cap on the 50 ml centrifuge tube, and set the centrifuge and spin cells (800 rpm, 3 min).

13. Remove the supernatant above the cell pellet with a Pasteur pipette and be sure not to disturb the cell pellet.

14. Resuspend the pellet with 5 ml of EpiLife complete culture medium by gentle pipetting.

15. Dispense the 5 ml of cell suspension in which up to 5.0×10^6 cells are contained into one T-25 tissue culture flask. When the number of cells harvested is ranging from 5.0×10^6 to 15.0×10^6 cells, dispense the 15 ml of cell suspension in one T-75 flask. For primary skin keratinocyte culture, please refer Notes 2 and 3.

16. Feed the initial seeding cells (referred to as P_0 cells) every other day.

17. Use the identical procedures with a few minor modifications (Notes 1, 2, 3) for establishing primary skin keratinocytes culture.

3.2 Serial Oral Keratinocyte Culture (Subculture, Passage)

Split a 70% confluent monolayer of oral keratinocytes ($P_0, P_1, P_2,...$) into a larger tissue culture flask.

3.2.1 Removal (Harvesting) of Adherent Oral Keratinocytes

1. Prepare 0.025% trypsin/EDTA solution by diluting 0.05% Trypsin solution with the same volume of DPBS.

2. Using a 10 ml pipette, aspirate the spent medium from the tissue culture flask and transfer to a sterile 50 ml conical tube.

3. Using a 5 ml pipette, add 2 ml of 0.025% trypsin/EDTA solution for a T-25 flask or 4 ml of trypsin/EDTA for a T-75 flask, or 8 ml of trypsin/EDTA for a T-150 flask.

4. Secure cap of flask and place in incubator for 8–12 min.

5. After 8 min, take a look at the flask under the inverted microscope to examine if cells are detaching from the flask surface. Gently tap the flask bottom to detach cells. View flask under inverted microscope to determine if most cells are detached. Return flask to the incubator for an additional few minutes if cells are still attached. View flask under the inverted microscope to determine that cells are now free floating.

6. Using a 5 ml pipette, add trypsin inhibitor solution into the flask with the same amount of 0.025% trypsin/EDTA solution.

7. Using the 10 ml pipette, aspirate the cellular suspension from the flask and dispense into another 50 ml centrifuge tube.

8. Rinse the flask with the spent medium kept at the step 2 and add the rinse medium to the 50 ml centrifuge tube.

9. After shaking the tube, aspirate a 15 µl cell suspension, mix with 15 µl trypan blue, and load 10 µl of the suspended cells into each of the two chambers of the hemocytometer. Count viable cells under the inverted microscope and calculate the total number of cells harvested.

3.2.2 Replating Cells into New Flasks

10. Secure the cap of a 50 ml centrifuge tube, place in centrifuge and balance contents.

11. Centrifuge cells at $200 \times g$ for 3–5 min at room temperature with the brake on.

12. Remove the supernatant above the cell pellet with a Pasteur pipette and be sure not to disturb the cell pellet.

13. Based on the cell count, determine the number and type of flasks required for seeding. For passage 1 (P_1) and after, standard cell seeding density is $0.7–1.3 \times 10^4$ cells/cm^2.

    For instances, T-25 flask $= 0.25 \times 10^6$ cells/flask, T-75 flask $= 0.75 \times 10^6$ cells/flask.

14. Resuspend the pellet with the required amount of EpiLife complete culture medium by gentle pipetting, and dispense the cell suspension into the desired types and numbers of flasks.

15. Secure caps and gently rock the flask to evenly distribute the cellular suspension. Be careful not to allow the suspension to flow into the flask neck.

16. Feed cells every other day.

17. Use the identical procedures for serial skin keratinocytes culture.

3.3 Cell Sorting by Cell Size (Fluorescence Activated Cell Sorting; FACS)

Using an appropriate cell sorter, sort cultured oral keratinocytes (P_0, P_1, P_2,...) by a physical means (cell size = forward scatter), and estimate the range of cell size sorted. Oral keratinocytes progenitor/stem cell population will be enriched in the small-sized cell population.

1. Prepare sterile 1% BSA solution with 0.1% sodium azide in HBSS, referred to as a staining buffer, and sterile PI solution (20 mg citric acid, 1 mg PI, 1 mg RNase in 20 ml DPBS; filtered).

2. Continue from the step 12 of Subheading 3.2.2. Based on the total number of cells harvested, resuspend the cell pellet with the staining buffer at the cell density of 10×10^6 cells/ml.

3. Place an autoclaved 240 μm mesh over the opening of the 50 ml centrifuge tube, and filter the staining buffer cell suspension through the mesh.

4. Prepare 12×75 mm round bottom tubes.

5. For unstained sample, add 30 μl (0.3×10^6 cells) of cell suspension into one tube, and then add 250 μl of DPBS.

6. For PI stained sample, add 10 μl (0.1×10^6 cells) cell suspension into another tube, followed by adding 200 μl of DPBS, and then add 10 μl of PI solution.

7. Centrifuge the rest of cell suspension ($200 \times g$, 2 min), aspirate supernatant, and then resuspend the pellet with DPBS as sort sample(s) at the final cell density of a 5×10^6 cells/ml.

8. Keep all tubes on ice until cell sorting. For sort sample, aliquot cell suspension into multiple tubes when 3 ml and more sorting sample is prepared. Add PI solution into the sort sample tube at the concentration of a 25 μl/ml at least 15 min prior to cell sorting.

9. Prepare three different size beads in three individual tubes with adding a few drops into a 500 μl of DPBS.

10. Run unstained sample followed by the PI stained sample.

11. Create a sorting gate on the left side (small-sized, 20–25% of total viable cell population) in the dot plots (based on forward scatter), but do not include debris. Check the forward scatter of both sides of the gate.

12. Sort cells.

13. Run three different size beads and check the forward scatter.

14. Using Excel software, plot the forward scatter and the size of beads (μm) on chart, and develop the regression line. From the linear equation, the range of sorted cell size can be estimated. Continue subsequent experiments.

15. When sort cells by "absolute" cell size, start steps 13 and 14 first, then complete steps 11 and 12 after step 10.

3.4 Cell Sorting by Cell Size (Gravity Assisted Cell Sorting; GACS)

1. Prepare a funnel-shaped nylon net filter (30 and 20 μm mesh size filters) by folding the round-shaped filter twice and then stapled at the edge of the overlapped portion. Soak several filters in 95% Ethanol overnight for disinfection. Prior to use, thoroughly rinse filters in sterile DPBS. In addition, prepare two 50 ml centrifuge tubes (A and B) with 15–20 ml of EpiLife complete culture medium.

2. Continue from the step 12 of Subheading 3.2.2. Resuspend the cell pellet with EpiLife complete culture medium at the cell density of $0.6–1.0 \times 10^6$ cells/ml by gentle pipetting.

3. Place the funnel-shaped 30 μm filter over the opening of the 50 ml centrifuge tube (C). Using a 10 ml pipette, aspirate the cell suspension.

4. Hold the pipette tip about 5 cm above the filter, and at a rate of two drops per second, drop the cell suspension into the funnel, where they passed through into a 50 ml centrifuge tube (C).

5. Replace the filter every 5–6 ml or the cell suspension did not go though the filter.

6. When replaced, using sterile forceps, close the funnel-shaped filter and pick up its rim. Cut off the middle part of the filter over the 50 ml centrifuge tubes (A) prepared at step 1, and drop the lower half of the filter into the EpiLife complete culture medium.

7. Using the forceps, pick up the edge of the filter to unfold the nylon net filter. Rinse the filter vigorously (up and down) in the EpiLife complete medium for wash off cells into the medium.

8. Repeat steps 3–7. Wash off cells in the same tube (A). Cells released into the 50 ml centrifuge tube (A) are referred to as 30 μm filter trapped cells.

9. Place the funnel-shaped 20 μm filter over the opening of the 50 ml centrifuge tube (D). Using a 10 ml pipette, aspirate the cell suspension (C) which had been filtered through the 30 μm filter.

10. Hold the pipette tip about 5 cm above the filter, and at a rate of two drops per second, drop the cell suspension into the funnel, where they passed through into a 50 ml centrifuge tube (D).

11. Replace the filter every 5–6 ml or the cell suspension did not go though the filter.

12. When replaced, using sterile forceps, close the funnel-shaped filter and pick up its rim. Cut off the middle part of the filter over the 50 ml centrifuge tubes (B) prepared at step 1, and drop the lower half of the filter into the EpiLife complete culture medium.

13. Using the forceps, pick up the edge of the filter to unfold the nylon net filter. Rinse the filter vigorously (up and down) in the EpiLife complete medium for wash off cells into the medium.

14. Repeat steps 9–13. Cells released into the 50 ml centrifuge tube (B) are referred to as 20 μm filter trapped cells, and cells went through both 30 and 20 μm filters in tube (D) are referred to as non-trapped cells.

The order of cell size distribution is (A) 30 μm filter trapped cells, (B) 20 μm filter trapped cells and (D) non-trapped cells; (larger, medium, smaller). Notes 5, 6, 7.

4 Notes

1. For primary culture of skin (interfollicular epidermis) keratinocytes, use a higher concentration (0.125%) of 0.25% trypsin solution by diluting with DPBS at 1:1, and larger size tissue culture dishes, such as 100 or 150 mm in diameter, may be required depending on the tissue size obtained. In contrast, antibiotic and antifungal agents are not necessarily to add into the 0.125% trypsin solution as far as the skin surface was thoroughly scrubbed with Betadine.

2. When the number of skin keratinocytes obtained is ranging from 7.0×10^6 to 20.0×10^6 cells, resuspend the cell pellet with 15 ml EpiLife culture medium and plate it into one T-75 flask.

3. On primary culture, the plating efficiency of skin keratinocytes is significantly lower compared with oral keratinocytes. If preferred to increase the plating efficiency, add 0.3 ml of FBS into the 15 ml of the cell suspension (2% FBS) when plating in a T-75 flask although FBS is not favorable to use in our chemically defined culture system. FBS is not necessary to be added after the first medium change.

4. FACS, a specialized form of flow cytometry, has become a widespread and vital resource in the medical and biological sciences. The major purpose is to separate and retrieve populations of interest from a heterogeneous mixture of cells, based on their fluorescent light-scattering properties. Thus, it is an extremely powerful and reliable tool that allows the individual measurement of physical and chemical characteristics of cells/particles. There are, however, some limitations in using this apparatus. Since this technology is basically designed for cell counting and cell sorting, excessive physical force, to avoid clogging and cell clumps, can damage cells. If the cell sorter needs to run for long durations, this may pose quality issues in FDA regulatory framework because such long runs may cause the decrease of cell viability for basic scientific experiments as well as for fabricating cell-based device. There is a potential for contamination, which affects their ability to be used in human clinical trials. In addition, it is expensive to run and time consuming even a high-speed sorter.

5. In contrast to FACS, GACS is less time-consuming, and less labor-intensive and more cost-effective, indicating GACS is more cell-friendly and more predictable for use in clinical trials under FDA guidelines. However, GACS also has shortcomings. First, nonviable cells cannot be eliminated during cell-sorting. Secondly, the distribution of cell size in the small-sized cell population is inconsistent every GACS and cannot be determined.

Lastly, cross-contamination is unavoidable. Especially, a lot of small-sized cells come into the larger cell population.

6. In general, because skin keratinocytes are smaller than oral mucosa keratinocytes, we recommend to use 20 and 11 μm mesh size filters for GACS.

7. According to our previous studies, smaller-sized cell populations sorted by FACS and GACS showed a higher clonogenicity, survived longer days in culture and had an ability to regenerate an epithelial tissue even a less seeding density.

Acknowledgments

We thank Drs. Blake Roessler and Yasushi Fujimori for their support. This study was supported by National Institutes of Dental and Craniofacial Research at the National Institute Health, Grant number DE 13417, to S.E.F.

References

1. Izumi K, Neiva RF, Feinberg SE (2011) Intraoral grafting of tissue-engineered human oral mucosa. Oral Craniofac Tissue Eng 1:103–111

2. Hotta T, Yokoo S, Terashi H et al (2007) Clinical and histopathological analysis of healing process of intraoral reconstruction with ex vivo produced oral mucosa equivalent. Kobe J Med Sci 53:1 11

3. Izumi K, Feinberg SE, Iida A et al (2003) Intraoral grafting of an ex vivo produced oral mucosa equivalent: a preliminary report. Int J Oral Maxillofac Surg 32:188–197

4. Bhargava S, Patterson JM, Inman RD et al (2008) Tissue-engineered buccal mucosa urethroplasty—clinical outcomes. Eur Urol 53:1263–1269

5. Ohki T, Yamato M, Murakami D et al (2006) Treatment of oesophageal ulcerations using endoscopic transplantation of tissue-engineered autologous oral mucosal epithelial cell sheets in a canine model. Gut 55:1704–1710

6. Nishida K, Yamato M, Hayashida Y et al (2004) Corneal reconstruction with tissue-engineered cell sheets composed of autologous oral mucosal epithelium. N Engl J Med 351:1187–1196

7. Tsai CY, Ueda M, Hata K et al (1997) Clinical results of cultured epithelial cell grafting in the oral and maxillofacial region. J Craniomaxillofac Surg 25:4–8

8. Barrandon Y, Green II (1985) Cell size as a determinant of the clone forming ability of human keratinocytes. Proc Natl Acad Sci U S A 82:5390 5394

9. De Paiva CS, Pflugfelder SC, Li DQ (2006) Cell size correlates with phenotype and proliferative capacity in human corneal epithelial cells. Stem Cells 24:368–375

10. Suzuki T, Inoki K (2011) Spatial regulation of the mTORC1 system in amino acids sensing pathway. Acta Biochim Biophys Sin 43:671–679

11. Izumi K, Inoki K, Fujimori Y et al (2009) Pharmacological retention of oral mucosa progenitor/stem cells. J Dent Res 88:1113–1118

12. Izumi K, Tobita T, Feinberg SE (2007) Isolation of human oral keratinocyte progenitor/stem cells. J Dent Res 86:341–346

13. Fujimori Y, Izumi K, Feinberg SE et al (2009) Isolation of small-sized human epidermal progenitor/stem cells by Gravity Assisted Cell Sorting (GACS). J Dermatol Sci 56:181–187

Chapter 24

Skin Stem Cells

Keita Inoue and Kotaro Yoshimura

Abstract

The skin stem cells are located at two distinct locations, the basement of epidermis and the hair follicle bulge. The bulge stem cell is considered to be of higher hierarchy in terms of the stemness than the epidermal stem cell. Recently, hair follicle bulge cells can be successfully isolated using fluorescence-activated cell sorting (FACS) by labeling of a specific combination of surface antigens. In this chapter, we describe a standard method by which the bulge subset in the human follicle can be isolated for subsequent functional analyses.

Key words Skin, Hair follicle, Flow cytometry, CD200, K15

1 Introduction

The skin stem cells are located at two distinct locations, the basement of epidermis and also the hair follicle bulge near the insertion of the arrector pili muscle. The bulge stem cell is thought to be of the higher hierarchy in terms of the stemness than the epidermal stem cells, because the bulge stem cells can give rise to both the skin and hair epithelia, while the epidermal stem cells differentiate only into the epidermis in the physiological situation(1–3).

In the modern era of tissue engineering and regenerative medicine, there is a growing need for isolating stem cell population from living tissues so as to utilize them in further functional analyses, manipulations, and in vivo transplantation. Although the discovery of skin stem cell niche, the bulge, in the mammalian hair follicle has made a great impact in the field of skin biology, we have long been devoid of the practical technique of isolating them. Surgical dissection of the hair follicle under a microscope was classically the only way to obtain the bulge cells. More recently, using a genetically modified mice having gradient of reporter fluorescence expression, a minor subset of hair follicle cells were isolated using fluorescence-activated cell sorting (FACS) for various functional

Kursad Turksen (ed.), *Skin Stem Cells: Methods and Protocols*, Methods in Molecular Biology, vol. 989,
DOI 10.1007/978-1-62703-330-5_24, © Springer Science+Business Media New York 2013

studies including microarray analysis (4). In addition, human hair follicle bulge cells were successfully isolated using FACS and magnetic-activated cell sorting (MACS) by labeling of a specific combination of antibodies against surface antigens (5).

We will introduce here a practical method for isolating functional bulge components of a human hair follicle (6). The method is simply composed of two steps: (1) microdissection and enzymatic digestion of hair follicles, followed by cell isolation by FACS and (2) functional analyses (colony-forming assay). This protocol enables the isolation of viable stem cell populations from hair follicle bulge and allows subsequent functional analyses which are difficult with other methods.

2   Materials

2.1 Enzymatic Digestion of Skin and Hair Follicle

1. Stereoscopic microscope.

2. Autoclaved spring microscissors and forceps (#5) (Fine Science Tools).

3. Sterile gauze.

4. Sterile culture plates (3.5, 6 and 10 cm diameter).

5. Dulbecco's modified Eagle's medium (DMEM), high glucose.

6. Fetal bovine serum (FBS).

7. Dispase I (Sanko Junyaku, Japan). Dissolve in high-glucose DMEM supplemented with 10% FBS and use at the final concentration of 1,000 protease unit (PU)/ml. Use freshly made solution for every experiment. An alternative product is available as "Dispase Grade I" from Roche. Its 1 U equals to approximately 416 PU of Sanko-Junyaku Dispase I.

8. Dulbecco's phosphate-buffered saline without calcium and magnesium (PBS).

9. 100× penicillin (5,000 U/ml)–streptomycin (5,000 μg/ml) mixture (Penstrep).

10. 10× trypsin(0.5%) + EDTA(0.2%).

11. 40-μm-pore nylon mesh cell strainer.

2.2 Flow Cytometry and Cell Sorting

Equipment: FACS machine (e.g., BD FACS Aria).
Software: FACS software compatible to the FACS machine (e.g., BD FACS Diva).
Reagents:

1. Use the following antibodies as the typical positive and negative markers for the bulge cell population. Positive control: R-phycoerythrine (PE) conjugated anti-CD200 antibody (clone MRC OX-104, 1:100). Negative control: Allophycocyanin (APC) conjugated anti-CD34 antibody (clone MRC 8G12, 1:100).

2. Dead cell staining: 7-Aminoactinomycin D, unconjugated (7-AAD, available as a ready-to-use product from BD pharmingen). Store at 4°C and protect from light.

3. Prepare sterile 0.2% bovine serum albumin (BSA) in PBS as "wash buffer" and 0.5%BSA in PBS as "sample buffer." Sterile 7.5% BSA solution in PBS is commercially available. If you prepare the wash/sample buffers from powders of BSA, pass the dissolved solution through 0.22 μm filter for sterilization prior to use. The solution can be stored at 4°C up to 2 weeks.

4. PBS (calcium and phenol red free).

5. Sample tube (5-ml, 40-μm-pore filter top tubes, BD Falcon).

6. Collection tube (15-ml collection tubes, BD Falcon).

2.3 Cell Culture and Colony Assay

1. Defined keratinocyte serum-free medium (DK-SFM) (Gibco).

2. Laminin-coated 6-well multiwell plate (BioCoat, BD).

3. 4% paraformaldehyde (PFA) in PBS.

4. Rhodanile blue: 2% rhodamine B (Sigma) and 2% nile blue (Sigma) diluted in ethanol.

---

3   Methods

3.1 Enzymatic Digestion of Hair Follicle

Carry out all the procedures under a laminar flow hood whenever the sample is exposed to the open air.

1. Obtain a human hairy scalp sample (approximately 2×2 cm) from surgery (e.g., facelift surgery). Store the scalp sample wrapped in a wet gauze with PBS at 4°C until the next step (see Note 1 below).

2. Cut hair shaft short (approximately 2 mm) and rinse the scalp in 10 ml of PBS containing $2 \times$ Penstrep (100 U/ml penicillin and 100 μg/ml streptomycin) in a 50-ml conical tube for 5 min. Repeat rinsing once. Transfer the scalp to a new 10-cm culture plate containing 5 ml of PBS without Penstrep.

3. Remove the subcutaneous fat using spring microscissors and forceps under a stereoscopic microscope (Fig. 1). Cut the scalp into strips of about 3-mm width (see Note 2). Transfer the scalp strips into a new 6-cm culture plate containing 8 ml of Dispase solution (Dispase I 1,000 PU/ml in high-glucose DMEM supplemented with 10% FBS). Put the lid on and seal the plate by a Parafilm strip.

4. Incubate overnight at 4°C.

5. Wash and rinse the scalp strips in 10 ml of fresh PBS twice. Transfer the strips into a new 10-cm culture plate containing 5 ml of PBS.

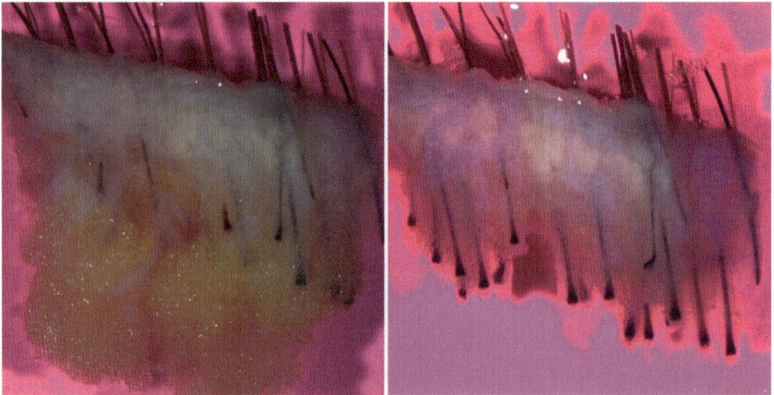

Fig. 1 Human scalp skin strip from facelift surgery. Before (*left*) and after (*right*) the removal of the subcutaneous fat tissue

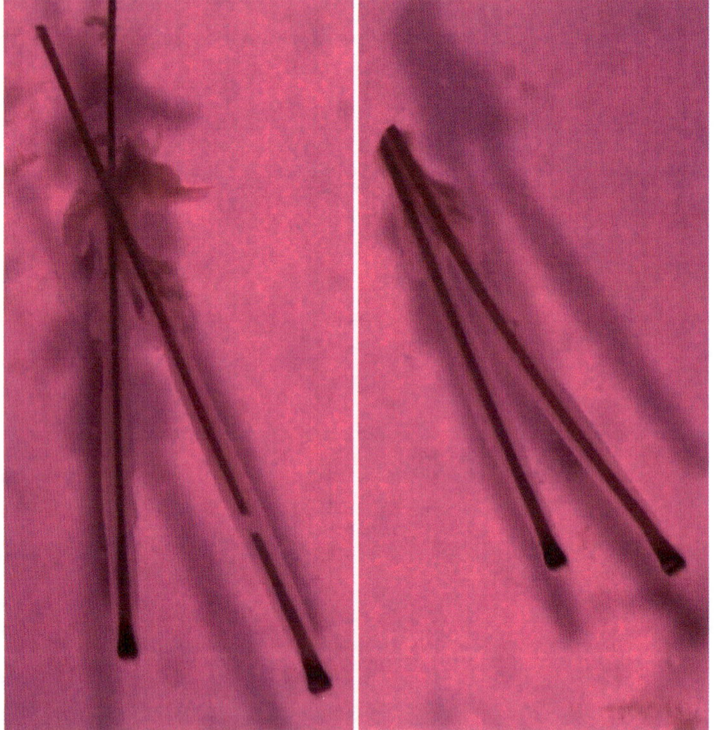

Fig. 2 Plucked hair follicles after dispase digestion. Follicular epithelial components of pilosebaceous unit, including follicular epithelium, sebaceous grand, and epidermis, are kept attached (*left*). Prepared follicles to be digested by trypsin (*right*); the epidermis and infundibulum distal to the sebaceous grand duct have been cut away

6. Separate the epidermis/follicles from the dermis using #5 forceps under a dissection microscope (see Note 3). Then cut the follicles at the level of the infundibulum to separate them from the epidermis (Fig. 2). Keep the epidermis as well, if

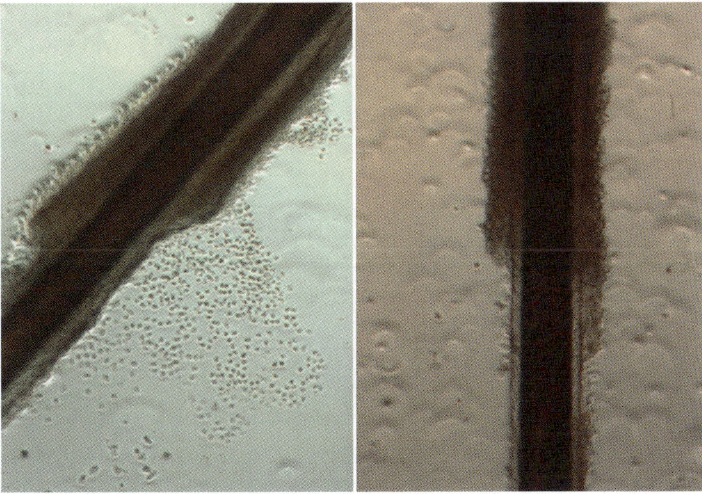

Fig. 3 Trypsin digestion of hair follicles. Cells start to detach from hair follicle just after applying trypsin (*left*) and at the end of digestion (30 min, *right*)

required. Transfer the follicles (and epidermis) into a new 3.5-cm plate containing 1 ml of trypsin (0.05%)/EDTA (0.02%) mixture in PBS.

7. Incubate at 37°C for 30 min. Briefly pipette follicles, using a 1 ml pipetter, in the enzyme solution every 5 min during the digestion (Fig. 3) (see Note 4).

8. Transfer all of the follicles and solution into a 50-ml conical tube containing 10 ml of PBS with 10% FBS to stop digestion. Pipette ten times using a 10-ml pipette.

9. Place a 40-μm-pore nylon mesh cell strainer on top of a new 50-ml conical tube. Pass the resultant solution through the strainer (see Note 5).

10. Centrifuge the filtrated cell suspension at $170 \times g$ for 5 min.

11. Discard supernatant and suspend cells in an appropriate volume of cooled sample buffer (PBS with 0.5% BSA) (see Note 6). Keep them on ice.

12. Cells are counted with an automated cell counter or a manual counting. For the manual cell counting, take 20 μl of the sample solution to stain the cells by adding equal volume of Trypan blue. Count the number of viable cells by excluding blue-stained dead cells using a hemocytometer under a microscope.

3.2 Flow Cytometry/ Cell Sorting

3.2.1 Labeling of Cells with Fluorescent-Conjugated Antibodies

1. Adjust the cell concentration to 1×10^6 cells/ml in sample buffer.

2. Label 5-ml sample tubes with a 40-μm-pore filter top for each antibody, including "unstained control," "7-AAD," "isotype IgG control," "single staining (CD200 or CD34)," and "multiple staining (CD200 and CD34 for sorting)" (see Note 7).

3. Dispense 100 μl (1×10^5 cells) of the cell suspension into each sample tube for analysis. When sorting is performed, dispense up to 1,000 μl (1×10^6 cells) of cells per tube. Multiply the sorting tubes, if required. Keep them on ice throughout the following steps.

4. Add antibody to each tube (see Note 8). Incubate on ice for 30 min in the dark.

5. Add 4 ml of cooled wash buffer (PBS with 0.2% BSA). Then, centrifuge at $170 \times g$ for 5 min.

6. Aspirate the supernatant (see Note 9).

7. Resuspend pellets in 300 μl of sample buffer and keep the cells on ice in the dark until analysis is carried out. When cell sorting is performed, suspend pellets in 3 ml of sample buffer. Make sure every tube has more than 300 μl of solution, because less volume may cause bubbling in the sheath line.

3.2.2 Cell Sorting

1. Set up the fluidics with a 100 μm nozzle. Do not use a nozzle with smaller diameter (i.e., 70 μm) since it may cause to damage cells by shear stress.

2. Turn on the FACS Aria and laser exciter, then computer, and FACS Diva software. Prime the sample sheath and fluidics. Calibrate the cell sorter in "low pressure" setting. Do not use higher pressure setting since it may damage cells by shear stress.

3. Add 1 μl of 7-AAD to each sample, except the "unstained control," and incubate 15 min on ice in the dark. Proceed to the following steps within 30 min (see Note 10). Pass cells through a 40-μm filter top just before analysis or sorting.

4. Run the "unstained control" first, followed by the "7-AAD" sample. Adjust detector voltages. Determine gating on FSC and SSC parameters to exclude cellular debris (small) and large differentiated keratinocytes. Use the ratio area/width on FSC for doublet exclusion. Determine the gating on 7-AAD parameter to exclude 7-AAD-positive dead cells.

5. Run the "isotype IgG control" samples and determine the appropriate gates for each fluorochrome. Less than 0.1% of the cell fraction can be regarded as negative.

6. Run the "single stain" samples and determine the appropriate compensation values.

7. Run the "multi-stain" samples and define the gate to sort the cell population of interest based on the standard FSC/SSC profile, as well as each fluorochrome (e.g., CD200+CD34− human bulge cells) (Fig. 4).

8. Set up 15-ml collection tubes, each containing 8 ml of DK-SFM. It is important that the sorted cells reach directly the

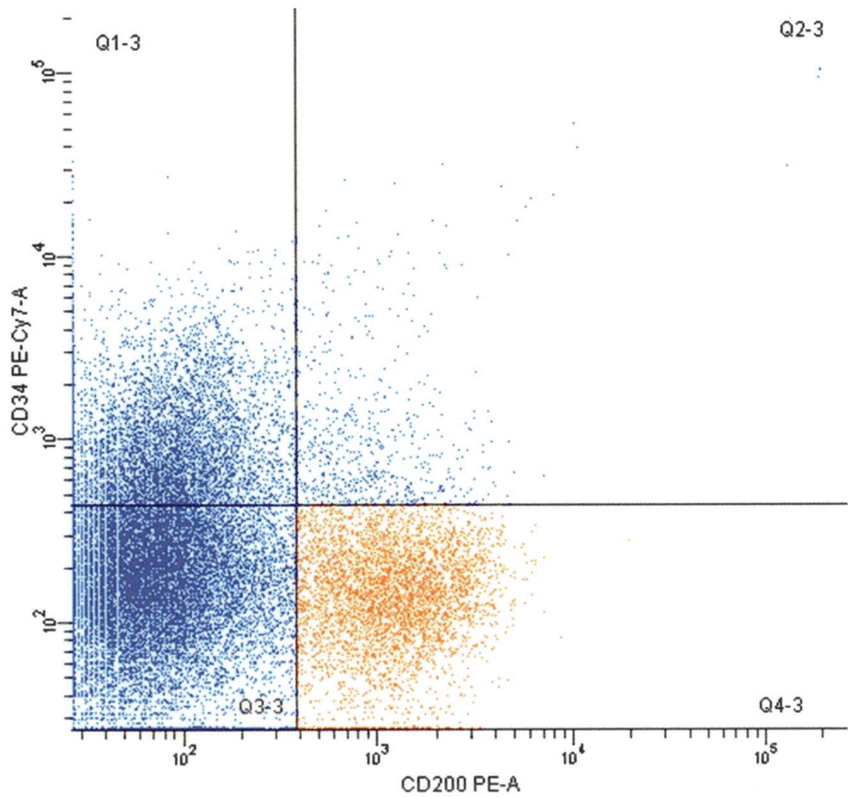

Fig. 4 Typical FACS dot plot of digested hair follicle cells. The bulge cells are enriched in the CD200+CD34– fraction (*right lower quadrant*, colored *orange*)

surface of the medium to prevent cells from being lost on the wall of the collection tube. Cells should be sorted at a low flow rate (i.e., below 2,000 cells/s). Record the sort cell number. Keep tubes on ice until plating for culture.

3.3 Cell Culture and Colony-Forming Assay

3.3.1 Culture of Sorted Keratinocytes

1. Centrifuge the sorted cells at $170 \times g$ for 5 min. Suspend the cell pellet in 1 ml of fresh DK-SFM.

2. For colony-forming assay, plate cells at three different concentrations; 10^3, 3×10^3, and 10^4 cells, in laminin-coated 6-well plates containing 2 ml of serum-free media (see Note 11). For serial subculture, plate 5×10^4 cells. Culture at 37°C in 5% CO_2 for 1–2 weeks. Change the culture medium every second day (see Note 12).

3. For serial subculture, trypsinize the primary culture and seed them on new plates. First, suck the media, wash cells twice using warm (37°C) 4 ml PBS, add 0.4 ml of warm (37°C) 1× trypsin (0.05%) and EDTA (0.02%), and incubate at 37°C for exactly 5 min. Do not digest cells for a longer time.

Then, neutralize the trypsin by adding 4 ml of 10% FBS-supplemented DK-SFM. Collect the cells with whole solution into 15-ml centrifuge tube and spin down the cells at $170 \times g$ for 5 min. Suck the supernatant by a Pasteur pipette and resuspend the pellet with 1 ml of DK-SFM. Count the cells as above. Adjust cell number to the concentration of 5×10^4 in 2 ml DK-SFM per well for subculture.

4. For other purposes than serial subculture, the cell densities may vary depending on the downstream experiments (e.g., colony forming assay, cell cycle analysis, 3D culture for in vitro skin equivalent, in vivo transplantation, hair follicle reconstitution assay, and RNA/Protein collection).

3.3.2 Rhodanile Blue Staining of Colonies

1. Rinse the culture plate with 4 ml PBS for each well and fix cells with 2 ml of 4% PFA for 5 min at room temperature. Add 4 ml PBS gently and suck. Repeat twice.

2. Stain the cells by adding 2 ml of rhadanile blue solution. Incubate at room temperature for 5 min.

3. Immerse the plate under running tap water for 2–3 min to wash out the dye. Although the colony would hardly come off, keep the plate away from direct flow of the water. Air-dry with the plate upside down for a day.

4. Scan the wells and measure colony number and area using an image processing software (e.g., Image-J or Adobe Photoshop).

4   Notes

1. The scalp skin sample should include at least 50 follicles; otherwise FACS isolation and culture may be difficult due to shortage of cells. Start the isolation within a few hours after the sample collection. Wring the gauze after wetting it by PBS, wrap the sample, and store in a closed small container at 4°C until processing. Do not leave the sample in water (PBS or saline).

2. Take great care to keep hair follicles intact when dissecting the scalp. The thickness and width of the strip is a critical factor for successful dispase digestion.

3. The epidermis and follicle are supposed to separate from the dermis without any resistance if the enzyme digestion is sufficient. If not, in most cases, the scalp is too thick and needs more trimming before processing.

4. The basal keratinocytes of hair follicles are first digested by trypsinization and they are the most important cells. Suprabasal are not important and it is hard to dissociate them completely. Do not try to digest them completely, since a prolonged incubation will affect the viability of basal cells.

5. Gently contact the tip of the pipette on the strainer mesh to give the cell suspension a pressure to pass through.

6. The typical number of cells obtained is around $5-8 \times 10^3$ from one follicle. For the following FACS experiment, you should prepare a solution of 10^6 cells/ml. If the initial sample contains 100 hair follicles, isolated cells can be resuspended in 500 µl of sample buffer.

7. For example, if you want to analyze and sort cells using PE-conjugated anti-CD200 and APC-conjugated anti-CD34, prepare sample tubes for "unstained," "PE-conjugated isotype IgG," "APC-conjugated isotypeIgG," "PE-CD200," "APC-CD34," and "PE-CD200 plus APC-CD34."

8. The working dilution of the antibody should be optimized by researchers. Usually it is 1:100 to 1:50.

9. Do not try to aspirate all of the supernatant; otherwise you may suck and lose the cells. You can leave in the tube some amount of supernatant.

10. Prolonged incubation times will result in excess staining with 7-AAD. Usually the proportion of dead cells is between 15 and 40% of the total cells. If more dead cells are observed, it is possible that any step of the isolation procedure has damaged the cells or that 7-AAD staining has been excessive.

11. Keratinocytes, especially of differentiated population, does not grow and form colonies if plated too sparse in serum-free media. To assess the colony-forming ability, it is desirable to prepare plates with different cell concentrations at this point.

12. The culture duration may vary depending on the initial cell concentration. Usually it is 1–2 weeks to glow to 70% confluent. Passage cells before they get fully confluent to prevent unwanted differentiation of keratinocytes. For example, CD200+CD34– population can be passaged every 1 week up to 6 passages.

References

1. Cotsarelis G, Sun T-T, Lavker RM (1990) Label-retaining cells reside in the bulge of pilosebaceous unit: implications for follicular stem cells, hair cycle, and skin carcinogenesis. Cell 61:1329–1337

2. Oshima H, Rochat A, Kedzia C et al (2001) Morphogenesis and renewal of hair follicles from adult multipotent stem cells. Cell 104:233–245

3. Fuchs E (2009) The tortoise and the hair: slow-cycling cells in the stem cell race. Cell 137:811–819

4. Rendl M, Lewis L, Fuchs E (2005) Molecular dissection of mesenchymal-epithelial interactions in the hair follicle. PLoS Biol 3:e331

5. Ohyama M, Terunuma A, Tock CL et al (2006) Characterization and isolation of stem cell–enriched human hair follicle bulge cells. J Clin Invest 116:249–260

6. Inoue K, Aoi N, Sato T, Yamauchi Y, Suga H, Eto H, Kato H, Araki J, Yoshimura K (2009) Differential expression of stem-cell-associated markers in human hair follicle epithelial cells. Lab Invest 89:844–856

INDEX

Kursad Turksen (ed.), *Skin Stem Cells: Methods and Protocols*, Methods in Molecular Biology, vol. 989,
DOI 10.1007/978-1-62703-330-5, © Springer Science+Business Media New York 2013

Printed by Printforce, the Netherlands